Mixture of Human and Machine Intelligence in Digital Manufacturing

Mixture of Human and Machine Intelligence in Digital Manufacturing

Guest Editors

Yang Lu
Ming Zhang
Ziwei Wang

MDPI

Basel • Beijing • Wuhan • Barcelona • Belgrade • Novi Sad • Cluj • Manchester

Guest Editors

Yang Lu
Department of Computer Science
Loughborough University
Loughborough
UK

Ming Zhang
Digital and Intelligent Technology Department
Beijing Bohua Xinzhi Technology Co., Ltd.
Beijing
China

Ziwei Wang
School of Engineering
Lancaster University
Lancaster
UK

Editorial Office
MDPI AG
Grosspeteranlage 5
4052 Basel, Switzerland

This is a reprint of the Special Issue, published open access by the journal *Designs* (ISSN 2411-9660), freely accessible at: https://www.mdpi.com/journal/designs/special_issues/R78POOKP14.

For citation purposes, cite each article independently as indicated on the article page online and as indicated below:

Lastname, A.A.; Lastname, B.B. Article Title. *Journal Name* **Year**, *Volume Number*, Page Range.

ISBN 978-3-7258-5063-1 (Hbk)
ISBN 978-3-7258-5064-8 (PDF)
https://doi.org/10.3390/books978-3-7258-5064-8

Contents

About the Editors

Yang Lu

Yang Lu is a Senior Lecturer in Computer Science at Loughborough University. Her research interests include human-in-the-loop system design, digital interventions, cybersecurity, artificial intelligence, and decision making. She has led and contributed to interdisciplinary projects funded by UKRI, EPSRC, STFC, and the UK Space Agency, applying methods such as privacy-preserving technologies, machine learning, and user-centred design to domains including digital health, energy sustainability, and food systems. Her work focuses on creating ethical, trustworthy, and impactful AI-enabled solutions through collaboration with government, industry, and academic partners.

Ming Zhang

Ming Zhang is the Chief Technology Officer at Beijing Bohua Xinzhi Technology Co., Ltd. He specializes in industrial artificial intelligence (AI), prognostics and health management (PHM), and predictive maintenance, with a particular focus on developing large language models and AI agents for industrial applications. He has led the development of AI-driven diagnostic and prognostic systems, industrial digital twin platforms, and intelligent operation and maintenance solutions. During his time at Aston University, he served as the technical lead for the EU H2020 project RECLAIM (Remanufacturing and Refurbishment of Large Industrial Equipment) and the UKRI project DIATOMIC (Digital Innovation Transformative Change). Dr. Zhang has published 47 academic papers, garnering over 1,000 citations and an H-index of 13. He currently serves as an Associate Editor of the IEEE Transactions on Instrumentation and Measurement and as a guest editor for several SCI-indexed journals. He holds eight granted Chinese invention patents and has contributed a book chapter published by Springer.

Ziwei Wang

Ziwei Wang is a Lecturer in Robotics at Lancaster University and Co-Leads in Fundamental and Additive Manufacturing of the Lancaster Intelligent, Robotic and Autonomous Systems Centre. He uses AI-based telerobotics and control technologies to enhance remote manipulation in extreme environments, which have been validated in laser welding, brazing and clinical trials. In the last 3 years, he has attracted funding for over 1M£ from Innovate UK, The Royal Society and the European Commission. Recent awards include New Generation Star Project IROS 2024, a Young Author Award at the IFAC Workshop on Control Applications of Optimization in 2018, the Best Paper Award of the IEEE International Conference on Automation and Computing in 2022, and the Best Poster Award of the IROS Workshop 2022. He has served as an Associate Editor of the IET Journal of Engineering, Evolving Systems, Biomimetic Intelligence and Robotics (BIROB) and Guest Editor of IEEE Transactions on Industrial Cyber-Physical Systems.

Preface

The Mixture of Human and Machine Intelligence in Digital Manufacturing Reprint presents a collection of innovative research and practical insights into how human expertise and computational intelligence can be combined to transform modern manufacturing. The scope of this Reprint spans human–machine interface design, human-in-the-loop systems, intelligent automation, decision support, and socio-technical integration, with contributions from both academic researchers and industrial practitioners.

The motivation for compiling this Reprint lies in the rapid evolution of Industry 4.0 technologies, where the collaboration between humans and autonomous systems is no longer optional but essential for achieving resilient, adaptive, and sustainable manufacturing processes. By examining advances in perception, control, arbitration, and sensory data analytics, the Reprint aims to highlight pathways for embedding human factors into digital manufacturing in ways that enhance efficiency, safety, and innovation.

This Reprint is intended for engineers, researchers, system designers, and decision-makers who are engaged in developing, deploying, or studying digital manufacturing systems. It offers a rich source of knowledge for those seeking to understand both the technical and human dimensions of intelligent manufacturing, and how these can be harmonised to create the next generation of productive, agile, and human-centred manufacturing systems.

Yang Lu, Ming Zhang, and Ziwei Wang
Guest Editors

designs

MDPI

Review

A Review of Game Theory Models to Support Production Planning, Scheduling, Cloud Manufacturing and Sustainable Production Systems

Paolo Renna

School of Engineering, University of Basilicata, 85100 Potenza, Italy; paolo.renna@unibas.it

Abstract: Cyber-physical systems, cloud computing, the Internet of Things, and big data play significant roles in shaping digital and automated landscape manufacturing. However, to fully realize the potential of these technologies and achieve tangible benefits, such as reduced manufacturing lead times, improved product quality, and enhanced organizational performance, new decision support models need development. Game theory offers a promising approach to address multi-objective problems and streamline decision-making processes, thereby reducing computational time. This paper aims to provide a comprehensive and up-to-date systematic review of the literature on the application of game theory models in various areas of digital manufacturing, including production and capacity planning, scheduling, sustainable production systems, and cloud manufacturing. This review identifies key research themes that have been explored and examines the main research gaps that exist within these domains. Furthermore, this paper outlines potential future research directions to inspire both researchers and practitioners to further explore and develop game theory models that can effectively support the digital transformation of manufacturing systems.

Keywords: game theory; manufacturing systems; cooperation; decision making; network

1. Introduction

The digital transformation, driven by the principles of Industry 4.0, offers unprecedented opportunities for industrial companies to revolutionize efficiency and customer satisfaction [1–3]. This evolution hinges on interoperable physical and cyber systems, decentralization, and real-time data analytics. These advancements empower companies to establish geographically dispersed multi-factory supply chains, enhancing flexibility while reducing labor and logistics costs [4]. Companies achieve this through the strategic distribution of production capacity or by forging collaborative multi-entity supply chains. Industrial applications of multi-site production planning and scheduling abound, spanning semiconductor manufacturing [5], automotive [6], pharmaceutical [7], and TFT-LCD [8]. However, this paradigm shift introduces complexity into planning and scheduling models, requiring solutions that can be found quickly and efficiently. Metaheuristic approaches have shown promise [4], and game theory, in particular, offers a powerful analytical tool that is well-suited to address interactions among multiple decision makers engaged in multi-objective optimization. Game theory is a branch of mathematics that studies strategic decision making in situations where multiple agents interact with each other. It provides a framework for analyzing and predicting the behavior of players in competitive and cooperative scenarios. Game theory is a powerful tool for analyzing strategic decision making in various contexts. By understanding its key elements, types, applications, and benefits, we can gain valuable insights into complex interactions and design effective strategies for different scenarios. Game-theory-based approaches have been successfully applied to solve various complex engineering problems, including power systems, collaborative product design, and production planning, enhancing solving efficiency [9]. A notable advantage of game theory is that it solves distributed algorithms in less time and with less computations compared to heuristic-based approaches [10].

Driven by the significance of game theory for production planning and scheduling in cooperation among multi-factory and cyber systems, there is a growing interest in leveraging its principles to optimize manufacturing processes. As manufacturing systems become increasingly complex and interconnected, game theory offers a powerful framework to model strategic interactions and decision making among various entities involved in production activities. This includes manufacturers, suppliers, distributors, and even autonomous cyber systems operating in smart factories. By applying game theory, researchers and practitioners aim to enhance efficiency, resource utilization, and overall performance in modern manufacturing environments characterized by interconnectedness and interdependence. This research proposes an overview of the recent applications of game theory to answer the following questions: What are the current research trends on the use of game theory for production planning and scheduling? What, then, are the existing research gaps and what are the potential contributions for future research? The structure of this paper is organized as follows: Section 2 outlines the review methodology. Section 3 describes the literature review following the areas of production planning problems, scheduling, sustainable production systems and smart and cloud manufacturing. Section 4 discusses the key findings of the literature review and identifies the limits. Section 5 concludes with a summary and a discussion of future research needs.

2. Research Methodology

The research methodology follows the guidelines proposed by Durach et al. [11]: (1) defining the research question, (2) determining the required characteristics of primary studies, (3) retrieving a sample of potentially relevant literature, (4) selecting the pertinent literature, (5) synthesizing the literature, and (6) reporting the results.

2.1. Question Formulation and Keywords

The first stage of this research involves defining the research questions. The main question focuses on understanding the current state of game theory models and their relevance in addressing the evolving needs of Industry 4.0. The research will then delve deeper into exploring how game theory models can facilitate various aspects, such as production and capacity planning, scheduling, sustainable production systems, and cloud manufacturing. Production planning is the process of organizing and coordinating resources to ensure efficient manufacturing of products. It involves determining what to make, how much to make, and when to make it to meet customer demand while minimizing costs. Production scheduling is the process of sequencing and timing production activities to optimize efficiency and meet customer demand. It involves determining the start and end dates for each task, as well as the resources needed to complete each task. Sustainable production systems refer to manufacturing processes and practices that are designed and operated in a manner that minimizes negative environmental impacts, conserves resources, promotes social equity, and ensures long-term economic viability. Cloud manufacturing is the use of cloud computing technologies to deliver manufacturing services on demand. It is a new paradigm that enables manufacturers to access and use manufacturing resources, such as software, hardware, and data, from a cloud-based platform. Then, it is necessary to define the keywords involved in this study. The main keywords are the following: "Game Theory" and "distributed production planning", "industry 4.0", "cloud manufacturing", "scheduling", "sustainable manufacturing", "energy saving", "energy reduction", "cooperative game", "non-cooperative game"," industry 4.0", and "smart manufacturing".

2.2. Inclusion/Exclusion Criteria

To focus the study and limit the literature search, a set of criteria was developed to identify the most relevant articles. These criteria are presented in Table 1.

Table 1. Inclusion and exclusion criteria used to select papers.

	Criteria	Justification
Inclusion	Papers published between 2013 and the first half of 2023	Most recent papers that focus on Industry 4.0
	Publications in peer-reviewed journals and conference papers	Peer-reviewed journals assures the quality of the research discussed
Exclusion	Production planning models not related to Industry 4.0	The purpose of this research is approaches for Industry 4.0 context.
	Studies in a language other than English	This assures that this research can be read by more researchers.

2.3. Database for Relevant Literature

A comprehensive search of three prominent academic databases—Google Scholar, Web of Science, and Scopus—was conducted to identify relevant articles. The initial search utilized predefined keywords to generate a preliminary set of articles. To refine this set further, a strict inclusion/exclusion criteria filter was applied. Additionally, only articles published in conferences indexed by Scopus or Web of Science were included to ensure a high level of quality.

2.4. Selecting the Pertinent Literature

To broaden the article selection process, the references of the initial set of evaluated articles were examined. This allowed for the identification of key authors who have made significant contributions to the main themes of this review. By analyzing their work, a secondary search was conducted, which enriched and diversified the scope of the reviewed literature.

2.5. Synthesizing the Literature

Following the identification of relevant articles, the next phase involved literature synthesis. Aligned with the research questions established earlier, the articles were categorized based on the specific issues they addressed and the research methodologies employed. The primary areas of focus included production and capacity planning, scheduling, sustainable production systems, and smart and cloud manufacturing.

2.6. Reporting the Results

The following sections analyze the selected literature, focusing on production and capacity planning, scheduling, sustainable production systems, and smart and cloud manufacturing. This analysis aims to identify current trends, research gaps within each area, and explore promising future research directions. The review encompasses 35 papers, selected from an initial pool of over 100 articles. The majority of the articles were sourced from the Scopus index, with only two articles originating from conference publications. The selection process also involved examining references, with an emphasis on the number of citations that each article received. Notably, the reference analysis did not identify specific authors as key contributors to this field.

3. Literature Review

The literature review is structured based on the classification outlined in the paper, which categorizes the research into distinct groups: production planning problems, scheduling, sustainable production systems, and cloud manufacturing. This approach allows for a systematic examination of the existing body of literature, enabling a comprehensive analysis of each specific area, facilitating the identification of key research themes, trends, and gaps within these domains.

3.1. Production and Capacity Planning

This section addresses the challenges of production and capacity planning in distributed geographic networks. These networks enable enterprises to pool capacity, services, and technology, enhancing efficiency, responsiveness, and competitiveness. By leveraging this network, enterprises can effectively respond to unforeseen events such as demand fluctuations, machine breakdowns, rush orders, and supplier delays.

Argoneto and Renna [12] proposed a model to support capacity sharing for a set of independent firms that were geographically distributed, combining their resources and predicting demand to improve production and cost efficiency. The partners of the network are independent and share partial information. A multi-agent architecture has been developed to support cooperation activities in this context. The coordination model uses the Gale–Shapley algorithm to find a stable matching among plants in the network, including the information that each partner decides to share using the preferences function.

Krenczyk and Olender [13] studied the problem of production planning in a virtual manufacturing network with geographically distributed manufacturers. The objectives were to minimize cycle time and production costs. The proposed approach uses a multi-agent system that solves the problem with a non-cooperative game. The selection of alternative routes for a set of production orders is modeled as a non-cooperative game f-player non-zero-sum game with complete information. The model proposed is a framework of a potential application but any numerical test is provided.

Yin et al. [14] proposed a non-cooperative model to allocate production to multi-suppliers from one manufacturer. The model considers quality and demand variations. The proposed approach is a non-cooperative game based on the Stackelberg equilibrium, where the manufacturer is regarded as a leader and the suppliers as followers. As argued by the authors, the model needs to be studied on a larger scale to evaluate its application in real industrial cases.

Olender and Krenczyk [15] proposed the use of a game theory approach to support the production planning problem in a virtual manufacturing network. The objectives were the minimization of production and transport costs using a non-cooperative game. A very limited numerical case was discussed.

Hafezalkotob et al. [16] addressed the approach of coalitions of production plants for cooperative production planning problems. They proposed several methods of cooperative game theory, including the Shapley value. The numerical results highlight how cooperation ensures the satisfaction of production plants, reducing total costs.

Bigdeli et al. [17] proposed a game theory model to support a production planning problem with fuzzy variables. Duality theory in the single-objective and weighted sum methods in multi-objective games is proposed to obtain the payoffs of the players.

Renna [18] studied capacity and resource allocation in flexible production networks. The objective is to obtain a trade-off between the costs and flexibility of the network to satisfy the customer demand. A dynamic allocation of the flexibility is proposed based on the game theory approach using the Gale–Shapley algorithm. The proposed model allows the performance of the network to improve compared to the long-chain approaches proposed in the literature.

Nishizaki et al. [19] addressed two-stage stochastic linear production planning with partial cooperation, involving resource pooling, technology transfer, and product transshipment. Manufacturers determine production levels individually in the first stage, and then collaborate to produce products using pooled resources in the second stage. Additional profits from cooperative game theory are distributed among all manufacturers. We developed a method to maximize total profits by finding a Nash equilibrium point valuated by numerical examples.

Table 2 summarizes the primary contributions of recent research on production and capacity planning, outlining key characteristics such as the addressed problem (capacity or production planning), the development of multi-agent System (MAS) architecture to support activities, the level of cooperation (non-cooperative or coalition), and the specific algorithm proposed (Gale–Shapley or Shapley Value). Notably, recent studies show that while one study utilized a coalition approach, the majority focused on non-cooperative models. Additionally, capacity planning has been addressed to a lesser extent compared to production planning.

Table 2. Production and capacity planning works.

	Capacity	Production Planning	MAS	Non-Cooperative	Coalition	Gale–Shapley	Shapley Value	Nash
Argoneto and Renna [12]	X		X			X		
Krenczyk and Olender [13]		X	X	X				
Yin et al. [14]		X		X				
Olender and Krenczyk [15]		X		X				
Hafezalkotob et al. [16]		X			X		X	
Bigdeli et al. [17]		X		X				
Renna [18]	X					X		
Nishizaki et al. [19]		X						

3.2. Scheduling

In today's competitive environment, scheduling models need to be highly responsive to real-time events, leveraging the vast amount of data available from Industry 4.0 technologies. However, the abundance of information also brings about increased computational complexity, necessitating more efficient scheduling algorithms to capitalize on this opportunity. Game theory emerges as a promising model to address scheduling challenges with greater efficiency.

Sun et al. [20] studied the flexible job-shop scheduling problem subject to machine breakdown, considering the objectives of robustness and stability. To optimize these two objectives, they modeled the problem as a non-cooperative game and the Nash equilibrium was derived to optimize the two objectives.

Chandrasekaran et al. [21] studied the n-job, m-machine job-shop scheduling problem using a game theory model to find the optimal makespan, mean flow time, and mean tardiness values. The approach proposed is a simplified heuristic derived from game theory tested in a reduced scheduling problem.

Han et al. [22] studied the flow-shop scheduling problem with component altering times, which is a particular problem for sequence-dependent setup times. They developed six rules for the machine assignment of jobs and proposed a Nash equilibrium model to manage these rules. The numerical results show how the game theory model performs better than a model using a genetic algorithm.

Renna [23] proposed a reconfigurable machine scheduling method based on a Gale–Shapley model. The Gale–Shapley model forms a coupled of overloaded and underloaded machines to allocate the modules for the reconfigurable machines. The numerical results of the simulation model show how the game theory model improves all performance measures with a restricted number of machine reconfigurations.

Wang et al. [24] proposed a multi-agent architecture to support real-time scheduling in flexible job-shop systems. A bargaining game model based on the Nash equilibrium was developed to support coordination among the agents of the architecture.

Nie et al. [25] modeled the flexible job-shop scheduling problem as a game theory model where the manufacturer wants to minimize the makespan of all jobs, and the job wants to minimize its tardiness. The game is solved by searching the Nash equilibrium, supported by a genetic algorithm.

Renna et al. [26] studied the dual resource scheduling problem in job-shop manufacturing systems. They proposed a Gale–Shapley model to support worker assignment for dual resource-constrained job-shop problems. The simulation experiments highlight how the Gale–Shapley model leads to better results, particularly when the workers have different efficiency levels.

Atay et al. [27] studied open-shop scheduling problems to minimize the total completion times. They proposed a cooperative TU game and allocate the affected jobs for each alliance to minimize the makespan.

Han et al. [28] modeled the flow-shop scheduling problem with multiple batches as a game model. The method proposed is based on multi-player cooperation and a static game with complete information. The proposed method allows the waiting time to be reduced and improves other performance measures of the flow line.

Wei et al. [29] addressed the multi-objective dynamic flexible job-shop scheduling problem when unforeseen events such as machine breakdown occur. They developed a model that approximates the Nash equilibrium solution to balance Pareto optimality and fairness between the two objectives of production efficiency and stability. The numerical results of several problem sizes are compared to three meta-heuristics proposed in the literature.

Table 3 summarizes the key findings from recent research on scheduling problems in manufacturing systems. The table highlights various characteristics, including the type of manufacturing system studied (flow line, job-shop, and reconfigurable), the game theory approaches employed (cooperative, Nash equilibrium, and Gale–Shapley), the use of multi-agent system (MAS) architecture, and the integration with genetic algorithms.

Table 3. Studies on scheduling.

	Flow Line	Job Shop	RMS	Open Shop	Cooperative Model	Nash Eq.	Gale–Shapley	MAS	Genetic Algorithm
Sun et al. [20]		X				X			
Chandrasekaran et al. [21]		X			X				
Han et al. [22]	X					X			
Renna [23]			X				X		
Wang et al. [24]		X				X		X	
Nie et al. [25]		X				X			X
Renna et al. [26]		X					X		
Atay et al. [27]				X	X				
Han et al. [28]	X				X				
Wei et al. [29]		X			X				

Notably, recent research reveals that only one study has explored reconfigurable manufacturing systems and open-shop scenarios. Additionally, the integration of game theory with other optimization techniques, such as genetic algorithms, has only been proposed in a single article.

3.3. *Sustainable Production Systems*

In recent years, the growing relevance of climate change, coupled with rising energy costs, has prompted manufacturing systems managers to prioritize energy efficiency and the utilization of renewable energy sources. These factors underscore the increasing significance of sustainable production systems.

Zhang et al. [30] proposed a dynamic game model based on the Nash equilibrium to improve production efficiency further and reduce processing costs, including total energy consumption for flexible job-shop problems. The numerical test highlighted reducing makespan, the total workload of machines, and the total energy consumption compared to genetic algorithm solutions.

Renna [31] developed a model to allocate the power to machines using the Gale–Shapley algorithm. The model exchanges the power from the underloaded to overloaded machines. The simulation results show how the model can improve the performance of a manufacturing system under a constraint power limit.

Wang et al. [32] studied the real-time scheduling problem in a job shop with the application of Internet of Things technology to improve production efficiency and reduce energy consumption. An infinitely repeated game optimization approach is developed, and the numerical results show that game theory can improve results compared to other dynamic scheduling methods.

Schwung et al. [33] presented a multi-agent architecture for a decentralized control design of modular production units. The interactions among the agents are supported by a game theory approach. The numerical tests show promising results for improvements in production efficiency in terms of energy consumption as well as throughput times.

Wang et al. [34] proposed a scheduling model for a flexible job shop in real-time. An evolutionary game-based solver method was proposed to support the scheduling model improving energy efficiency.

Sun et al. [35] proposed a digital twin framework to support process planning and scheduling in job-shop systems. Then, a dynamic game theory was adopted to improve production efficiency and reduce energy consumption. The model considered two sub-games, the process planning sub-game and scheduling sub-game, integrated with the Nash equilibrium solution.

Zhao et al. [36] proposed an optimization method for shared energy storage in micro-grids using negotiation game theory. This establishes a cooperative interaction mechanism between Microgrid Cluster Operator (MGCO) and Shared Energy Storage Operator (SESO), leading to an optimization framework for microgrid clusters. The dynamic leasing of shared energy storage is considered, resulting in a negotiation game-based capacity configuration model for MGCO and SESO, demonstrating a cost reduction for MGCO and revenue increase for SESO.

Table 4 summarizes the key findings from recent research on applying game theory to sustainable production systems. While the majority of studies focus on job-shop systems, only one addresses peak power constraints. Significantly, no studies explored using game theory to optimize the adoption and integration of renewable energy sources within manufacturing systems.

Table 4. Studies on sustainable production systems.

	Job Shop	**Nash Equilibrium**	**Gale–Shapley**	**MAS**	**Cooperative Model**	**Energy**	**Peak Power**
Zhang et al. [30]	X	X				X	
Renna [31]	X		X	X			X
Wang et al. [32]	X				X	X	
Schwung et al. [33]	X			X	X	X	
Wang et al. [34]	X				X	X	
Sun et al. [35]	X	X				X	
Zhao et al. [36]					X	X	

3.4. Cloud Manufacturing

Cloud manufacturing represents an emerging paradigm where distributed resources are encapsulated into cloud services and centrally managed. This network of shared resources enables customers to access on-demand services supporting the entire product lifecycle. The efficiency of cloud manufacturing is heavily dependent on coordination models.

Su et al. [37] studied the problem of manufacturing resource allocation, in which the manufacturing service demander and cloud manufacturing service platform operator are considered gamers. They proposed a non-cooperative game approach to support the problem of resource allocation.

Liu et al. [38] proposed a model of resource and service sharing in cloud manufacturing sharing based on the Gale–Shapley algorithm. The results of the proposed model highlighted that there are always enterprises of the network that perform worse.

Carlucci et al. [39] proposed a coordination model based on a minority game to allocate resources/services among partners of a cloud manufacturing system. The proposed model was tested in a simulation environment compared to a model with complete information among the partners.

Xiaoning et al. [40] investigated three resource-sharing strategies: independently, as an alliance, and by cooperating with a cloud platform operator. The interactions between the operator and suppliers were modeled as a two-stage Stackelberg game that contains a simultaneous sub-game. They found the highest system profit when the suppliers cooperate with the operator.

Xiao et al. [41] proposed a cloud manufacturing multi-task scheduling model based on game theory from a customer perspective. The model is derived from the Nash equilibrium game. The simulation results highlight how the proposed model leads to better results compared to basic biogeography-based optimization algorithms, genetic algorithms, and particle swarm optimization.

Wang et al. [42] studied decentralized decision making in the management of manufacturing service allocation in cloud manufacturing systems. The model is based on an evolutionary game approach able to converge to equilibrium.

Zhang et al. [43] considered a cloud manufacturing system where each manufacturer provides manufacturing resources; when the cloud manufacturing received an order, it coordinated manufacturing resources to satisfy order requirements. To solve the scheduling problem of cloud manufacturing, they proposed a genetic algorithm with the use of the Nash equilibrium for a non-cooperative game model.

Liu et al. [44] studied the application of a cloud manufacturing approach for 3D printing services. They proposed a non-cooperative game model for a 3D printing service scheduling problem. The non-cooperative game is based on Nash equilibrium points supported by a genetic algorithm.

Liu et al. [45] proposed a game-theory-based collaborative scheduling approach for cloud manufacturing (CMfg), addressing dynamics and uncertainties. It optimizes manufacturing and logistic resources efficiently, considering fuzzy uncertain task migration. The model achieves the Nash equilibrium through a decision tree optimization algorithm, enhancing transportation efficiency. Simulation results validate its effectiveness and performance in dynamic CMfg environments.

Koochaksaraei et al. [46] presented a novel approach for cloud service providers (CSPs) to efficiently allocate resources through a barter-based auction market, using evolutionary game theory. CSPs estimate and bid their resources without monetary exchange, fostering cooperation and reducing SLA violations. The simulation results demonstrate improved social welfare and fewer contracts.

Zhang et al. [47] proposed a real-time strategy for a flexible job-shop scheduling problem-based on game theory. The solution and optimization strategy for process tasks using the Nash equilibrium was designed and developed to implement the dynamic optimization model. A case study is presented to demonstrate the efficiency of the proposed strategy and method.

An emerging issue concerns using the circular economy to improve the sustainability of different enterprise sectors, such as manufacturing systems [48], the apparel industry [49], and civil engineering [50].

Tushar [51] provided a recent overview of the literature on cyber–physical systems supported by different game theory models. They argued that multi-agent and game theory are adapted to support cyber–physical systems.

Table 5 summarizes the key findings from recent research on applying game theory to cloud manufacturing systems. Notably, the Nash equilibrium is the dominant approach, with fewer studies exploring alternative methods such as the Gale–Shapley model or minority game models. This suggests potential avenues for future research.

Table 5. Studies on cloud manufacturing systems.

	Resources	Service	Nash Eq.	Gale–Shapley	Cooperative Model	Non Cooperative Model	Minority Game
Su et al. [37]	X					X	
Liu et al. [38]	X	X		X			
Carlucci et al. [39]	X	X					X
Xiaoning et al. [40]	X				X		
Xiao et al. [41]	X		X				
Zhang et al. [45]	X		X				
Wang et al. [42]		X	X				
Zhang et al. [43]	X		X			X	
Liu et al. [44]	X		X				
Liu et al. [45]		X	X		X		
Koochaksaraei et al. [46]		X				X	

4. Main Findings

This section synthesizes the findings from the reviewed literature. Based on this analysis, we will identify research gaps and discuss potential avenues for future research. Table 6 summarizes the distribution of the reviewed papers across different areas in recent years (2014–2023) and in terms of percentages of the total reviewed papers.

Table 6. Summary of the literature review.

	2014–2016	2017–2019	2020–2022	2023	No.	%
Production and capacity planning	4	1	2	1	8	22.86%
Scheduling	3	3	3	1	10	28.57%
Sustainable production systems	-	2	3	2	7	20.00%
Cloud manufacturing	1	4	3	2	10	28.57%
Total	8	10	11	2	35	

The analysis of the reviewed literature reveals that research on production and capacity planning peaked between 2014 and 2016, with a decline in subsequent years. Conversely, the application of game theory approaches to scheduling problems has remained consistent throughout the study period. Interestingly, sustainable production systems and cloud manufacturing have emerged as growing areas of interest, with cloud manufacturing attracting the most recent research efforts. Figure 1 visually depicts this trend, highlighting the surge in research focused on sustainable manufacturing systems and cloud manufacturing in recent years.

4.1. *Key Outcomes for Production and Capacity Planning*

A critical gap identified in recent production and capacity planning research is the lack of models that facilitate coalition formation among enterprises. Such models could enhance efficiency, responsiveness, and resource sharing, ultimately leading to improved customer satisfaction. Small and Medium Enterprises (SMEs) could leverage these coalitions to compete more effectively on the global stage [52]. However, cooperative game theory models, which are well suited for analyzing coalition formation, have not been extensively explored in this area.

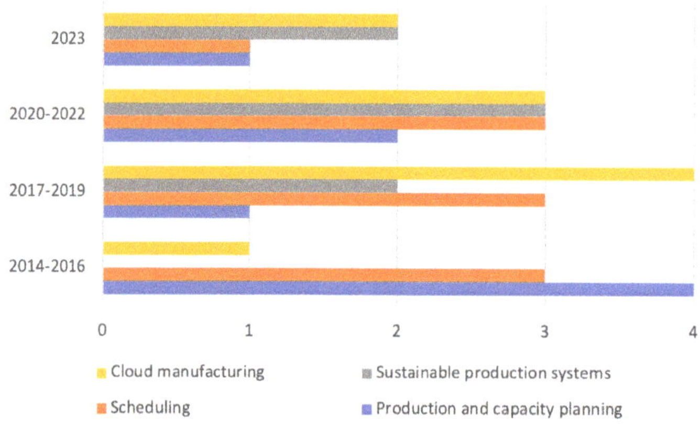

Figure 1. Trend of the literature review.

Another promising avenue for applying cooperative game theory lies in the realm of strategic alliances, often referred to as 'co-opetition' [53]. These alliances unite competing enterprises with complementary capabilities to pursue shared objectives such as expanding market share, improving efficiency, and fostering innovative solutions. By integrating cooperative game theory into the context of capacity and production planning, researchers can develop models that optimize coalition formation and resource sharing among enterprises. This approach holds immense potential for SMEs, allowing them to gain a competitive edge and make meaningful contributions to the global manufacturing landscape.

4.2. Key Outcomes on Scheduling

Recent research on scheduling for manufacturing systems has primarily focused on job shops, with reconfigurable and open-shop systems receiving less attention. However, these under-explored systems offer significant potential for leveraging game theory models to improve decision making and operational efficiency. In particular, reconfigurable manufacturing systems boast a modular and adaptable design, allowing them to swiftly adapt to changing production needs. Game theory models can play a pivotal role in allocating and reconfiguring machines and equipment. By optimizing resource utilization within a limited availability, decision makers can strike a balance between resource allocation and production objectives.

Unlike job-shop systems, open-shop systems accommodate a broader spectrum of process types and job priorities. Game theory models step in to provide efficient solutions while minimizing computational time. By dynamically considering job arrivals and process requirements, game theory guides decision makers in scheduling tasks. The goals? Minimizing idle time, reducing production lead times, and boosting system responsiveness.

From the point of view of integration challenges, there is a critical gap in the existing literature—an underexplored fusion of game theory with other methodologies.

Two promising avenues are as follows: fuzzy logic is renowned for handling imprecise and uncertain information and can effectively complement game theory. Integrating these approaches takes into account the inherent variability and complexities of real-world manufacturing environments. Genetic algorithms are algorithms that search for optimal solutions through evolutionary processes. When combined with game theory, they explore a broader array of scheduling alternatives, ultimately identifying the most efficient paths. In summary, bridging game theory with fuzzy logic and genetic algorithms can pave the way for more robust and adaptable scheduling models, ultimately enhancing decision-making efficiency in manufacturing systems.

Reconfigurable manufacturing systems, with their modular and adaptable designs, can leverage game theory models to optimize resource allocation and the reconfiguration of machines and equipment. This is particularly valuable in scenarios with limited resource availability. By applying game theory, decision makers can make informed choices that balance resource allocation with production objectives, ultimately maximizing overall system efficiency. Open-shop systems, characterized by a wider range of process types and job priorities, present another promising area for applying game theory. In these systems, game theory models can efficiently identify solutions while minimizing computational time. Game theory's strength lies in its ability to account for the dynamic nature of job arrivals and process requirements. This allows it to guide decision makers in real-time task scheduling, minimizing idle time, reducing production lead times, and ultimately enhancing system responsiveness. However, a critical gap exists in the current research—the limited exploration of how game theory can be combined with other methodologies. Fuzzy logic, known for its ability to handle imprecise information, can complement game theory in real-world manufacturing environments where data may be uncertain. Similarly, genetic algorithms, which utilize evolutionary processes to find optimal solutions, can be integrated with game theory to explore a wider range of scheduling options and identify the most efficient paths. By combining these approaches, researchers can develop more robust and adaptable decision models for scheduling problems.

Fuzzy logic, with its ability to handle imprecise and uncertain information, can be integrated with game theory to account for the inherent variability and complexities of real-world manufacturing environments. This integration can improve the robustness and adaptability of scheduling models to changing conditions and disruptions. Genetic algorithms, with their ability to search for optimal solutions through evolutionary processes, can be combined with game theory to explore a wider range of scheduling alternatives and identify the most efficient solutions. This integration can enhance the computational efficiency of scheduling models and lead to more robust and effective scheduling strategies. By fully leveraging the power of game theory and integrating it with complementary methodologies, researchers can address the challenges of scheduling in reconfigurable and open-shop systems, unlocking new avenues for enhancing manufacturing efficiency and productivity.

4.3. Key Outcomes on Sustainable Production Systems

Sustainability has emerged as a key driver in manufacturing, prompting research into sustainable production systems that minimize environmental impact while maintaining economic viability. Recent studies have focused on job-shop manufacturing systems, primarily considering total energy consumption. However, a crucial aspect—the costs associated with peak power constraints—has received less attention. Additionally, the utilization of renewable energy sources has not been extensively explored in recent research. The allocation of demand energy among various renewable energy sources presents a promising research direction that can be effectively addressed using game theory models. By incorporating game theory's principles of strategic decision making, researchers can develop models to optimize the utilization of renewable energy sources while ensuring the overall energy demand is met. This approach holds immense potential for minimizing reliance on conventional energy sources and reducing the environmental footprint of manufacturing operations. Game theory models can capture interdependencies among different renewable energy sources, considering their fluctuating availability, variability in generation patterns, and associated costs. By analyzing these interactions, researchers can identify optimal strategies for scheduling the use of renewable energy sources, ensuring that demand is met while maximizing the utilization of these sustainable resources. Furthermore, incorporating game theory into the optimization of renewable energy utilization can facilitate collaboration among different stakeholders, such as energy providers, manufacturers, and consumers.

By modeling their decision-making processes and aligning their interests, game theory can promote the efficient coordination of renewable energy resources, leading to a more sustainable and resilient energy landscape. By embracing game theory as a tool for optimizing renewable energy utilization in sustainable production systems, researchers can pave the way for a future where manufacturing operations operate in harmony with the environment, reducing their carbon footprint and promoting the transition towards a sustainable future.

4.4. Key Outcomes on Cloud Manufacturing

Cloud manufacturing, a transformative paradigm, has emerged in manufacturing, enabling distributed and collaborative production capabilities. While recent research has primarily investigated the allocation of manufacturing resources and services within cloud manufacturing systems, two crucial aspects demand further investigation: cooperative models and design stage support. Cooperative models are essential for facilitating cooperation among the diverse stakeholders within cloud manufacturing ecosystems, including cloud providers, manufacturers, and customers. Game theory offers a powerful framework for modeling these interactions. By analyzing strategic decision making, game theory can aid in the design of mechanisms that promote efficient cooperation. These mechanisms can optimize resource allocation, maximize utilization of cloud manufacturing capabilities, and align the interests of all parties involved, ultimately contributing to a more sustainable and equitable cloud manufacturing landscape.

Design stage support is critical for reducing the gap between customer requirements and the actual production process in cloud manufacturing. Game theory can be leveraged to develop collaborative design models that facilitate the active participation of customers, designers, and manufacturers. These models enable a participatory design process, empowering customers to express their specific needs and preferences while designers and manufacturers can offer expert guidance, feasibility assessments, and technical expertise.

By incorporating cooperative models and design stage support into cloud manufacturing systems, researchers can pave the way for a more agile, responsive, and customer-centric manufacturing paradigm. This approach empowers customers to actively participate in the design and production of their desired products and services. In turn, manufacturers gain valuable insights, improve customer satisfaction, and potentially unlock new avenues for innovation and competitive advantage.

5. Conclusions

The adoption of new technologies and digitization generate vast amounts of real-time data, which necessitates the adoption of novel organizational and cooperative models. Game theory models can expedite cooperation among independent partners, reducing computational time compared to alternative methodologies.

This review surveys recent research (2014–2023) utilizing game theory models in production and capacity planning, scheduling, sustainable production systems, and cloud manufacturing. The literature review examines the application of game theory models in these areas, identifying key research gaps and proposing future directions.

Coalition models are an important area for study, offering support across various topics. Cooperative game models facilitate collaboration among multi-site or manufacturing resources within the same enterprise, while non-cooperative game models with incomplete information aid independent enterprises in temporary collaborations. Integrating game theory models with other methodologies, such as genetic algorithms, fuzzy logic, and Monte Carlo simulations, can enhance multi-objective solutions and reduce computational complexity for real-time data.

The management and integration of renewable energy sources are a critical area that can benefit from game theory models. Decision support models based on game theory can allocate energy demand among various sources, including solar, wind, and storage options, aiming to minimize total energy consumption, address peak power constraints, and maximize renewable sources.

Future Research Paths

The main key points for future research paths can be summarized as follows.

Future research in manufacturing systems will focus on open-shop and reconfigurable manufacturing systems, which stand to gain significant benefits from game theory models. Advancing these open research areas is essential for expanding knowledge on the use of game theory models and supporting industry and practitioners in transitioning towards decision support systems for new organizational paradigms.

Existing studies on capacity and production planning have neglected the application of cooperative game theory in facilitating coalition formation and resource sharing among enterprises, especially SMEs. Cooperative game theory offers valuable insights for designing and managing strategic alliances to optimize mutual benefits and minimize potential conflicts. It provides a structured framework for analyzing and negotiating resource allocations, fostering collaboration, and enhancing overall operational efficiency.

Research on scheduling has predominantly centered on job shops, overlooking reconfigurable and open-shop systems. Game theory models offer effective solutions for optimizing resource utilization, minimizing idle time, reducing lead times, and enhancing responsiveness in these systems. Integrating game theory with complementary methodologies like fuzzy logic and genetic algorithms can further improve scheduling model efficacy.

While recent research on sustainable production systems has emphasized job-shop systems and total energy consumption, it has often neglected costs related to peak power constraints and the integration of renewable energy sources. Game theory models offer opportunities to optimize the utilization of renewable energy sources, ensuring demand fulfillment while minimizing reliance on conventional energy sources. Moreover, game theory can foster collaboration among stakeholders to efficiently coordinate renewable energy resources.

While research on cloud manufacturing has predominantly centered around resource allocation, there has been limited exploration of cooperative models and design stage support. Cooperative game theory presents an opportunity to develop efficient mechanisms for resource allocation and engage customers in the design and production processes of their desired products and services. Integrating cooperative models and design stage support can enhance the flexibility, responsiveness, and customer centricity of cloud manufacturing.

Funding: This research received no external funding.

Institutional Review Board Statement: Not applicable.

Informed Consent Statement: Not applicable.

Data Availability Statement: Not applicable.

Conflicts of Interest: The author declares no conflicts of interest.

References

1. Gilchrist, A. Introducing Industry 4.0. In *Industry 4.0*; Apress: Berkeley, CA, USA, 2016; pp. 195–215; ISBN 978-1-4842-2046-7. [CrossRef]
2. Koleva, N. Industry 4.0′s opportunities and challenges for production engineering and management Innovations. *DEStech Trans. Eng. Technol. Res.* **2018**, *6*, 17–18.
3. Lennon Olsen, T.; Tomlin, B. Industry 4.0: Opportunities and challenges for operations management. *Manuf. Serv. Oper. Manag.* **2019**, *22*, 113–122. [CrossRef]
4. Lohmer, J.; Lasch, R. Production planning and scheduling in multi-factory production networks: A systematic literature review. *Int. J. Prod. Res.* **2021**, *59*, 2028–2054. [CrossRef]
5. Wang, P.-S.; Yang, T.; Yu, L.-C. Lean-Pull Strategy for Order Scheduling Problem in a Multi-Site Semiconductor Crystal Ingot-Pulling Manufacturing Company. *Comput. Ind. Eng.* **2018**, *125*, 545–562. [CrossRef]
6. Gnoni, M.G.; Iavagnilio, R.; Mossa, G.; Mummolo, G.; Di Leva, A. Production Planning of a Multi-Site Manufacturing System by Hybrid Modelling: A Case Study from the Automotive Industry. *Int. J. Prod. Econ.* **2003**, *85*, 251–262. [CrossRef]
7. De Matta, R.; Miller, T. Production and Inter-Facility Transportation Scheduling for a Process Industry. *Eur. J. Oper. Res.* **2004**, *158*, 72–88. [CrossRef]

8. Chen, W.-L.; Huang, C.-Y.; Lai, Y.-C. Multi-Tier and Multi-Site Collaborative Production: Illustrated by a Case Example of TFT-LCD Manufacturing. *Comput. Ind. Eng.* **2009**, *57*, 61–72. [CrossRef]
9. Li, W.D.; Gao, L.; Li, X.Y.; Guo, Y. Game theory-based cooperation of process planning and scheduling. In Proceedings of the 12th International Conference on Computer Supported Cooperative Work in Design, Xi'an, China, 16–18 April 2008; pp. 841–845.
10. Chen, X.; Jiao, L.; Li, W.; Fu, X. Efficient multi-user computation offloading for mobile-edge cloud computing. *IEEE/ACM Trans. Netw.* **2016**, *24*, 2795–2808. [CrossRef]
11. Durach, C.F.; Kembro, J.; Wieland, A. A new paradigm for systematic literature reviews in supply chain management. *J. Supply Chain. Manag.* **2017**, *53*, 67–85. [CrossRef]
12. Argoneto, P.; Renna, P. Capacity sharing in a network of enterprises using the Gale–Shapley model. *Int. J. Adv. Manuf. Technol.* **2013**, *69*, 1907–1916. [CrossRef]
13. Krenczyk, D.; Olender, M. Simulation Aided Production Planning and Scheduling Using Game Theory Approach. *Appl. Mech. Mater.* **2015**, *809–810*, 1450–1455. [CrossRef]
14. Yin, S.; Nishi, T.; Zhang, G. A game theoretic model for coordination of single manufacturer and multiple suppliers with quality variations under uncertain demands. *Int. J. Syst. Sci. Oper. Logist.* **2016**, *3*, 79–91. [CrossRef]
15. Olender, M.; Krenczyk, D. Practical application of game theory based production flow planning method in virtual manufacturing networks. *IOP Conf. Ser. Mater. Sci. Eng.* **2016**, *145*, 022031. [CrossRef]
16. Hafezalkotob, A.; Chaharbaghi, S.; Lakeh, T.M. Cooperative aggregate production planning: A game theory approach. *J. Ind. Eng. Int.* **2019**, *15* (Suppl. S1), 19–37. [CrossRef]
17. Bigdeli, H.; Tayyebi, J.; Hassanpour, H. Production Planning Games in Uncertain Environment. *New Math. Nat. Comput.* **2022**, *19*, 757–771. [CrossRef]
18. Renna, P. Capacity and resource allocation in flexible production networks by a game theory model. *Int. J. Adv. Manuf. Technol.* **2022**, *120*, 4835–4848. [CrossRef]
19. Nishizaki, I.; Hayashida, T.; Sekizaki, S.; Furumi, K. A two-stage linear production planning model with partial cooperation under stochastic demands. *Ann. Oper. Res.* **2023**, *320*, 293–324. [CrossRef]
20. Sun, D.-H.; He, W.; Zheng, L.-J.; Liao, X.-Y. Scheduling flexible job shop problem subject to machine breakdown with game theory. *Int. J. Prod. Res.* **2014**, *52*, 3858–3876. [CrossRef]
21. Chandrasekaran, M.; Lakshmipathy, D.; Sriramya, P. GT heuristic for solving multi objective job shop scheduling problems. *ARPN J. Eng. Appl. Sci.* **2015**, *10*, 5472–5477.
22. Han, Z.; Zhu, Y.; Ma, X.; Chen, Z. Multiple rules with game theoretic analysis for flexible flow shop scheduling problem with component altering times. *Int. J. Model. Identif. Control* **2016**, *26*, 1–18. [CrossRef]
23. Renna, P. Decision-making method of reconfigurable manufacturing systems' reconfiguration by a Gale-Shapley model. *J. Manuf. Syst.* **2017**, *45*, 149–158. [CrossRef]
24. Wang, J.; Zhang, Y.; Liu, Y.; Wu, N. Multiagent and Bargaining-Game-Based Real-Time Scheduling for Internet of Things-Enabled Flexible Job Shop. *IEEE Internet Things J.* **2019**, *6*, 2518–2531. [CrossRef]
25. Nie, L.; Wang, X.; Pan, F. A game-theory approach based on genetic algorithm for flexible job shop scheduling problem. *J. Phys. Conf. Ser.* **2019**, *1187*, 032095. [CrossRef]
26. Renna, P.; Thürer, M.; Stevenson, M. A game theory model based on Gale-Shapley for dual-resource constrained (DRC) flexible job shop scheduling. *Int. J. Ind. Eng. Comput.* **2020**, *11*, 173–184. [CrossRef]
27. Atay, A.; Calleja, P.; Soteras, S. Open shop scheduling games. *Eur. J. Oper. Res.* **2021**, *295*, 12–21. [CrossRef]
28. Han, Z.; Bian, X.; Ding, Z.; Sun, D. Optimisation of group batch scheduling in flexible flow shop based on multi-player cooperative game. *Int. J. Model. Identif. Control* **2022**, *40*, 114–126. [CrossRef]
29. Wei, L.; He, J.; Guo, Z.; Hu, Z. A multi-objective migrating birds optimization algorithm based on game theory for dynamic flexible job shop scheduling problem. *Expert Syst. Appl.* **2023**, *227*, 120268. [CrossRef]
30. Zhang, Y.; Wang, J.; Liu, Y. Game theory based real-time multi-objective flexible job shop scheduling considering environmental impact. *J. Clean. Prod.* **2017**, *167*, 665–679. [CrossRef]
31. Renna, P. Peak Electricity Demand Control of Manufacturing Systems by Gale-Shapley Algorithm with Discussion on Open Innovation Engineering. *J. Open Innov. Technol. Mark. Complex.* **2020**, *6*, 29. [CrossRef]
32. Wang, J.; Yang, J.; Zhang, Y.; Ren, S.; Liu, Y. Infinitely repeated game based real-time scheduling for low-carbon flexible job shop considering multi-time periods. *J. Clean. Prod.* **2020**, *247*, 119093. [CrossRef]
33. Schwung, D.; Reimann, J.N.; Schwung, A.; Ding, S.X. Smart Manufacturing Systems: A Game Theory based Approach. In *Intelligent Systems: Theory, Research and Innovation in Applications*; Studies in Computational Intelligence; Jardim-Goncalves, R., Sgurev, V., Jotsov, V., Kacprzyk, J., Eds.; Springer: Cham, Switzerland, 2020; Volume 864. [CrossRef]
34. Wang, J.; Liu, Y.; Ren, S.; Wang, C.; Wang, W. Evolutionary game based real-time scheduling for energy-efficient distributed and flexible job shop. *J. Clean. Prod.* **2021**, *293*, 126093. [CrossRef]
35. Sun, M.; Cai, Z.; Yang, C.; Zhang, H. Digital twin for energy-efficient integrated process planning and scheduling. *Int. J. Adv. Manuf. Technol.* **2023**, *127*, 3819–3837. [CrossRef]
36. Zhao, Q.; Liu, G.; Wang, Z.; Yuan, H.; Ma, H. Capacity Optimization Configuration of Multi-Microgrid Shared Energy Storage Based on Negotiation Game. In Proceedings of the 2023 IEEE Sustainable Power and Energy Conference (iSPEC), Chongqing, China, 28–30 November 2023.

37. Su, K.; Xu, W.; Li, J. Manufacturing resource allocation method based on non-cooperative game in cloud manufacturing. *Comput. Integr. Manuf. Syst.* **2015**, *21*, 2228–2239.
38. Liu, Y.; Zhang, L.; Tao, F.; Wang, L. Resource service sharing in cloud manufacturing based on the Gale–Shapley algorithm: Advantages and challenge. *Int. J. Comput. Integr. Manuf.* **2017**, *30*, 420–432. [CrossRef]
39. Carlucci, D.; Renna, P.; Materi, S.; Schiuma, G. Intelligent decision-making model based on minority game for resource allocation in cloud manufacturing. *Manag. Decis.* **2020**, *58*, 2305–2325. [CrossRef]
40. Cao, X.; Bo, H.; Liu, Y.; Liu, X. Effects of different resource-sharing strategies in cloud manufacturing: A Stackelberg game-based approach. *Int. J. Prod. Res.* **2023**, *61*, 520–540. [CrossRef]
41. Xiao, J.; Zhang, W.; Zhang, S.; Zhuang, X. Game theory—Based multi-task scheduling in cloud manufacturing using an extended biogeography-based optimization algorithm. *Concurr. Eng.* **2019**, *27*, 314–330. [CrossRef]
42. Wang, T.; Li, C.; Yuan, Y.; Liu, J.; Adeleke, I.B. An evolutionary game approach for manufacturing service allocation management in cloud manufacturing. *Comput. Ind. Eng.* **2019**, *133*, 231–240. [CrossRef]
43. Zhang, W.; Xiao, J.; Zhang, S.; Lin, J.; Feng, R. A utility-aware multi-task scheduling method in cloud manufacturing using extended NSGA-II embedded with game theory. *Int. J. Comput. Integr. Manuf.* **2021**, *34*, 175–194. [CrossRef]
44. Liu, S.; Zhang, L.; Zhang, W.; Shen, W. Game theory based multi-task scheduling of decentralized 3D printing services in cloud manufacturing. *Neurocomputing* **2021**, *446*, 74–85. [CrossRef]
45. Liu, S.; Li, L.; Zhang, L.; Shen, W. Game-Based Collaborative Scheduling with Fuzzy Uncertain Migration in Cloud Manufacturing. *IEEE Trans. Autom. Sci. Eng.* **2023**. [CrossRef]
46. Ghasemian Koochaksaraei, M.H.; Toroghi Haghighat, A.; Rezvani, M.H. An efficient cloud resource exchange model based on the double auction and evolutionary game theory. *Clust. Comput.* **2023**. [CrossRef]
47. Zhang, Y.; Wang, J.; Liu, S.; Qian, C. Game theory based real-time shop floor scheduling strategy and method for cloud manufacturing. *Int. J. Intell. Syst.* **2017**, *32*, 437–463. [CrossRef]
48. Basbam, N.; Taleizadeh, A. A hybrid circular economy—Game theoretical approach in a dual-channel green supply chain considering sales effort, delivery time, and hybrid manufacturing. *J. Clean. Prod.* **2020**, *250*, 119521.
49. Rogers, L.; Wang, Y. Understanding the Impact of Game Theory on Circular Economy within the Apparel Industry. In *Advanced Manufacturing and Automation X*; IWAMA 2020, Lecture Notes in Electrical Engineering; Wang, Y., Martinsen, K., Yu, T., Wang, K., Eds.; Springer: Singapore, 2020; Volume 737. [CrossRef]
50. Alcantar, P.; Hunt, D.; Rogers, C. The complementary use of game theory for the circular economy: A review of waste management decision making methods in civil engineering. *Waste Manag.* **2020**, *102*, 598–612. [CrossRef] [PubMed]
51. Tushar, W.; Yuen, C.; Saha, T.K.; Nizami, S.; Alam, M.R.; Smith, D.B.; Poor, H.V. A Survey of Cyber-Physical Systems from a Game-Theoretic Perspective. *IEEE Access* **2023**, *11*, 9799–9834. [CrossRef]
52. Renna, P. Negotiation policies and coalition tools in e-marketplace environment. *Comput. Ind. Eng.* **2010**, *59*, 619–629. [CrossRef]
53. Brandenburger, A.; Nalebuff, B. *Co-Opetition: A Revolution Mindset That Combines Competition and Cooperation*; Harvard Business Press: Cambridge, MA, USA, 1996.

designs

MDPI

Article

From Meaning to Expression: A Dual Approach to Coupling

Lukas Van Campenhout *, Ward Vancoppenolle and Ivo Dewit

Product Development, Faculty of Design Sciences, University of Antwerp, Prinsstraat 13, 2000 Antwerp, Belgium
* Correspondence: lukas.vancampenhout@uantwerpen.be

Abstract: Coupling is a key concept in the field of embodied interaction with digital products and systems, describing how digital phenomena relate to the physical world. In this paper, we present a Research through Design process in which the concept of coupling is explored and deepened. The use case that we employed to conduct our research is an industrial workplace proposed by Audi Brussels and Kuka. Our aim was to enrich this workplace with projection, or Spatial Augmented Reality, while focusing on operator interaction. We went through three successive design iterations, each of which resulted in a demonstrator. We present each of the three demonstrators, focusing on how they propelled our understanding of coupling. We establish a framework in which coupling between different events, be they physical or digital, emerges on four different aspects: time, location, direction, and expression. We bring the first three aspects together under one heading—coupling of meaning—and relate it to ease of use and pragmatic usability. We uncover the characteristics of the fourth aspect—coupling of expression—and link it to the psychological wellbeing of the operator in the workplace. We conclude this paper by highlighting its contribution to the embodied interaction research agenda.

Keywords: design for interaction; embodied interaction; coupling; workplace design; spatial augmented reality; research through design

1. Introduction

This paper is about coupling and how it drives design for interaction. The concept of coupling is a recurring theme in the knowledge domain of embodied interaction. In previous work [1], we formulated a definition of coupling based on the MCRpd interaction model [2] and the Interaction Frogger Framework [3]. In our definition, coupling is the relationship between different events that make up a user-product interaction routine. Events are representations of digital phenomena in the real world. These representations can have a physical or a digital character. For example, a display hangs on the wall and shows an image of a landscape. The display is controlled via a physical button, which is mounted next to it. When the user pushes the button, another landscape appears on the display. This interaction routine contains two events: the pushing of the button and the changing of the images on the display. It is clear that the two events are related, since pushing the button causes the display to change its scenery. The successive pictures on the display feel digital because they are visible but intangible, can easily be replaced, and can disappear instantly. They form a digital event. The user's pushing of the button is a physical, tactile interaction. The button is pushed, and the user feels this movement. The physical button is persistent. Unlike the on-screen images, it will not suddenly vanish or change its appearance. The movement of the button forms a physical event. The relationship between both events is referred to as coupling. In this paper, we aim to further define our understanding of coupling, and its different aspects, against the background of embodied interaction. We complete this task by presenting a Research through Design (RtD) project [4] that we conducted in the area of industrial workplaces and Spatial Augmented Reality.

1.1. Embodied Interaction

Embodied interaction is a perspective on human interaction with digital phenomena and has roots in phenomenology [5,6] and ecological psychology [7]. It advocates a deep integration of digital phenomena into the physical world through giving them material form, i.e., embodying them. In this way, these digital phenomena become tangible [8] and graspable [9]. In a large part of literature, the rationale for embodied interaction starts from the idea that people are familiar with the physical world [10], i.e., they have highly developed knowledge and skills for interacting in and with it. The purpose of embodied interaction is, thus, to leverage this familiarity in their interaction with digital products and systems. As such, this interaction becomes more natural and intuitive than interacting with today's generation of PCs or tablet computers [11].

In this paper, we apply a particular perspective from the embodied interaction research agenda. This perspective deals with strong specific products [12], i.e., digital products that are specifically designed to fulfil one single function. Buxton [13] claims that the design of single-purpose products has more merit than the "one-size-fits-all" approach of the personal computer. The reason for this merit is specialization [10]. When a product is dedicated to a specific task, its user interface can be designed around that task and its design can reflect a physical commitment to it [14]. For the design process of these products, this approach means that both physical and digital components can be developed simultaneously, i.e., as one whole. Such a design process works radically against the common design practice of today's consumer electronics, where the physical and the digital are designed separately from each other and sequentially. In this process, a generic hardware structure is created (a PC, a smartphone or a tablet; only after that are software applications are designed and added.

1.2. Spatial Augmented Reality in Workplace Design

Augmented Reality (AR) [15] is driving the vision of industrial workplaces for maintenance [16] and manufacturing [17–19]. The aim of AR in these workplaces is to increase the accessibility of work instructions for the operator or service engineer through providing an augmented layer with real-time, task-relevant information within their field of view [20]. In our Research through Design project, we focused on a measuring workplace with Spatial Augmented Reality (SAR) [21]. SAR uses one or more projectors to project digital images directly onto a physical object, environment or, in this context, workplace.

The workplaces in industry today that employ SAR mainly consider the design of the augmented layer as something that happens only after the design of the workplace itself is completed, and only after the workplace is realized. We value the versatility and flexibility that this design methodology offers; however, we believe that the resulting operator interaction is far from natural and intuitive. We started our Research through Design project with a speculative view of the design of such workplaces [22,23]. We wanted to apply embodied interaction in the design process and, in particular, the perspective of strong specific products described above. In other words, we wanted to conduct an in-context exploration [24] of the strong specific workplace: a workplace with a limited set of functions, the physical and digital form of which are designed in conjunction with each other. Our initial research question was as follows: what coupling possibilities emerge when a strong specific workplace is enriched with SAR? Our research goal was to seamlessly integrate projected images in the physical workplace, and design an interaction with these images that feels natural, rather than detached or stuck on [25]. Our use case was provided by Audi Brussels and Kuka. They proposed a workplace for a specific measuring task, which was augmented with projection.

1.3. Objectives and Structure of This Paper

In this paper, we present three demonstrators that we built during our RtD process, and discuss the insights that the demonstrators gave us regarding the concept of coupling in spatially augmented workplaces. We define coupling as a relationship that groups and connects different events that occur in the interaction between the operator and different components of the workplace. We advocate that the design of this relationship is a key part of the overall design process. The number of coupled events can be more than two, and whether these events are physical or digital, or product or user related, is not important. What is important, however, is the added value for the operator that the coupling brings to the interaction flow. This added value has a dual nature, as we propose a taxonomy of different couplings consisting of two main groups: couplings of meaning, related to ease of use; and couplings of expression, related to emotional well-being.

This paper is divided into eight sections. In the second section, we put our research into perspective through positioning it between several areas of knowledge about coupling. In the next three sections, we present three demonstrators of spatially augmented cobot workplaces, each representing a stage of our RtD process. For each demonstrator, we present its functionality and interaction routine, and discuss the different coupling themes each demonstrator reveals. In the discussion, we look back on our work, and articulate our key findings. In our conclusion, we come full circle by showing what our initial research question and the subsequent work add to the current notion of coupling and the design of spatially augmented workplaces. Finally, we outline directions for future research.

2. Coupling in Perspective

In this section, we discuss three frameworks on coupling, including our own.

2.1. The MCRpd Interaction Model

In 1997, tangible interaction, which was an early precursor to embodied interaction, was established [8]. In order to articulate ideas about representation and control of digital phenomena in an interactive system, the MCRpd (Model-Control-Representation, physical and digital, Figure 1) interaction model is proposed [2]. The MCRpd makes the following distinction. On one hand, there is the underlying digital information, i.e., the bits and bytes, referred to as model in the diagram. On the other hand, there are different representations of this model in the real world. These representations can take two forms: physical (REP-P) and digital (REP-R) representations. Physical representations are artifacts that embody digital information. Digital representations are computationally mediated entities without physical form (e.g., video projection and audio). Both physical and digital representations are said to be perceptually coupled. This coupling is indicated via the horizontal line in the diagram.

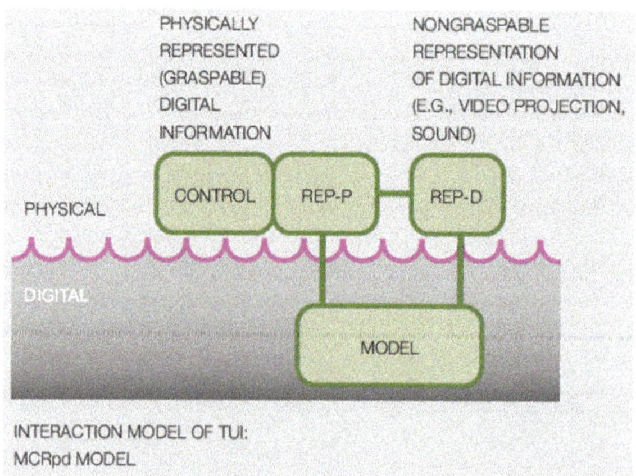

Figure 1. MCRpd Interaction Model. Reprinted with permission from [2], Tangible Media Group, MIT Media Lab, 2000.

From this model, we have inherited the idea that digital phenomena are represented in the physical world through two kinds of representations. These representations are those that feel physical, and those that feel digital. We called them physical events and digital events. The coupling between them—the green horizontal line on the diagram—is the coupling that we investigate in our research.

2.2. The Interaction Frogger Framework

Wensveen et al. [3] translated the concept of coupling into a framework for product designers. The interaction frogger framework was developed to make people's interaction with digital products natural and intuitive. The reference for natural interaction is the interaction with simple mechanical products (e.g., cutting paper with a pair of scissors). The framework focuses on the coupling between user action and product function. In an interaction routine, the two factors constantly alternate. A user action results in a product function, and vice versa. The coupling is the mutual relationship between the two factors. It describes how user action and product function are related to each other, and to what extent the user perceives them as similar. The framework suggests that when user action and product function are united on six aspects, they are experienced as naturally coupled. In the context of this paper, we retain three of these six aspects. These three aspects are:

- Time: User action and product function coincide in time;
- Location: User action and product function occur in the same location;
- Direction: User action and product function have the same direction of movement.

2.3. The Aesthetics of Coupling

In our previous work, we deepened the concept of coupling by considering it as a source of aesthetic experience [1]. As described earlier in this paper, we distinguish physical and digital events. Whether these events are initiated by the user or the product, as emphasized by the Interaction Frogger Framework, is of secondary importance to us. We argued that the aesthetics of coupling uniquely play between two events: physical and digital events. The intrinsic difference in the nature of both events makes up the essence of the aesthetics. Physical events are persistent and static, while digital events are temporal and dynamic. The designer brings both events very close to one another, so that they are perceived by the user as one coherent, harmonious user experience. At the same time, the user is very aware of the different natures of the two events, which clearly makes a complete unification impossible. This inherent paradox, which is the tension field between being apart and together at the same time, causes feelings of surprise and alienation that constitute the aesthetics of coupling. The first author, together with Floor Van Schayik, designed a night lamp that illustrated this concept (Figure 2). A video of this night lamp can be found in Supplementary Material: Video S1-NightLamp. When the user pulls up the sphere at the top of the lamp, the light inside the lamp moves with it and jumps into the sphere at the end of the movement. As such, the reading lamp becomes a night lamp. In this example, the aesthetics of coupling lie in the contrast between the physicality of the sphere and the intangibility of the light inside the lamp.

(**a**) (**b**) (**c**)

Figure 2. (**a**) Lamp in reading mode. (**b**) user pulls the sphere upwards. (**c**) Lamp in night mode.

3. The Audi Demonstrator

Demonstrator video can be seen in Supplementary Material: Video S2-AudiDemonstrator.

3.1. Background

We explored SAR in a human–cobot workplace [26]. A cobot, or collaborative robot, is a relatively small industrial robot designed for direct human–robot interaction. The context for our research project was provided by Audi Brussels (https://www.audibrussels.be, accessed on 17 May 2023) and Kuka (https://www.kuka.com, accessed on 17 May 2023). In the production line for the body panels of the Audi A1 in Brussels, Audi needed a workplace that could measure and inspect the adhesion quality of the fold glue joints for each panel (doors, hood and trunk). In this workplace, a single operator works together with a cobot: the Kuka LBR iiwa. This cobot tracks the contour of the body panel using contact force feedback, and checks the adhesion quality of the fold glue joints through means of an ultrasonic sensor.

We conducted a few preliminary ideations on the operator's interaction with the workplace and the cobot. We adopted the stance that interaction with projected content in workplaces is not going to replace today's interaction styles, but will coexist with and complement them [21]. This assumption resulted in a preliminary concept for the Audi workplace, defined in 3D CAD (Figure 3). The concept consisted of a circular table and a control unit, containing a display-button setup. The Audi A1 body panels are placed manually on the table, one at a time. Above them, a Kuka LBR iiwa cobot hangs upside down from a horizontal bridge, with its work envelope covering the entire table surface. On either side of the cobot, two projectors are mounted on the bridge, augmenting the table and body panels with projected images. The idea is that the operator steers and controls the cobot with the control unit. The cobot moves over the body panel while touching its contour and ultrasonically senses the adhesion quality of the glue joints. Relying on several SAR studies [27–29], we decided that the visual feedback on this measurement should be projected in real time on the body panel itself, rather than appearing on a separate, isolated display.

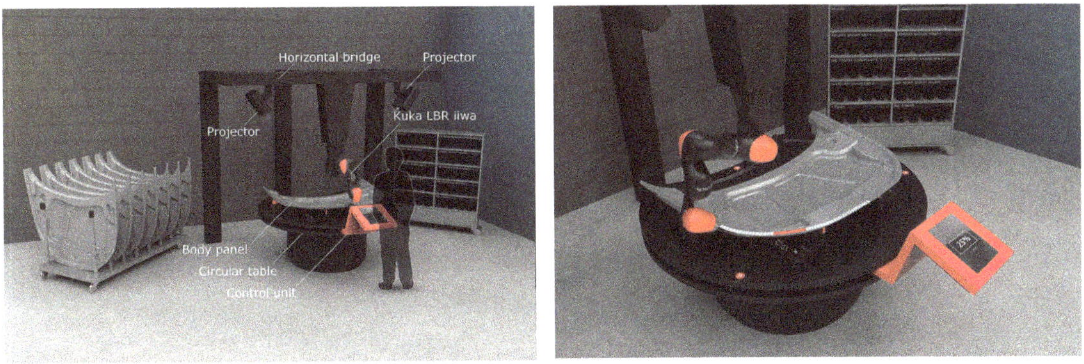

(a) (b)

Figure 3. (**a**) Preliminary concept for Audi workplace. (**b**) Projection on table and body panel.

3.2. Description of the Demonstrator

We designed and built a 1/6 scale model of the Audi workplace (Figure 4). The purpose of this scale model was to inform and inspire people from the Flemish manufacturing industry. With this model, demonstrations are given by one operator to an audience of around ten people. The setup is made of sheet metal and SLS (Selective Laser Sintering) parts. It features a horizontal surface, on which the circular tabletop is mounted, and a vertical wall containing a mini beamer. Adjacent to the circular tabletop, a rectangular surface represents the control unit. The beamer simulates the control unit's multi-touch display and projects images onto the table surface and the objects that are placed on it: an adapted replica of the Audi A1 hood and two positioning supports. Finally, the setup includes an adapted scale model of the Kuka LBR iiwa, which is used by the operator as a puppet in a theatre.

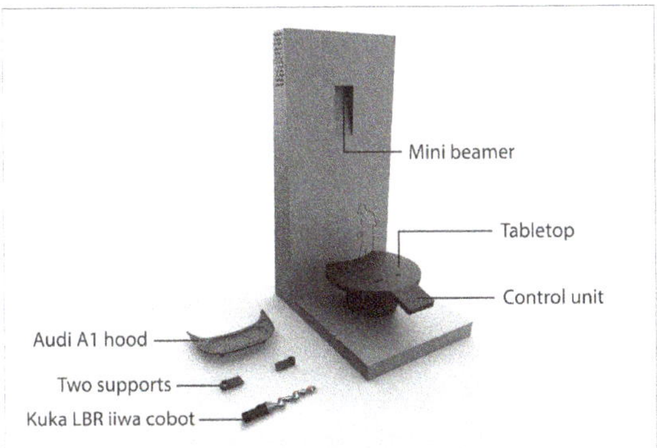

Figure 4. A 1/6 scale model of Audi workplace.

3.3. Description of the Interaction

The interaction starts at the control unit. The operator stands in front of it and performs menu navigation actions on the multi-touch display. He/she selects the type of car, in this case a five-door Audi A1, and the body panel, in this case the hood (Figure 5).

Figure 5. Operator selects type of car and body panel.

When the hood is selected, the hood icon moves upwards, leaves the display, and slides onto the augmented tabletop, where it grows into a full-scale positioning contour (Figure 6). The operator moves along with this motion, leaves the control unit, and stands at the tabletop. From there on, he/she is guided through projected work instructions. The contours of both supports are shown in red (Figure 7a). The operator positions both supports on the table (Figure 7b). When the system detects the correct placement of each support, the red pulsating projection turns white and adapts to the shape of the supports (Figure 7c). Next, the outline of the hood is projected in red (Figure 8a). The operator places the hood on both supports, which help to ensure a correct positioning. Next, the operator presses a start button on the control unit, and the cobot begins to track the hood contour and measures the adhesion quality of the glue joints (Figure 8b). The result of this measurement is projected in real time onto the hood itself: white dots indicate good adhesion quality, while red dots indicate poor quality (Figure 8c).

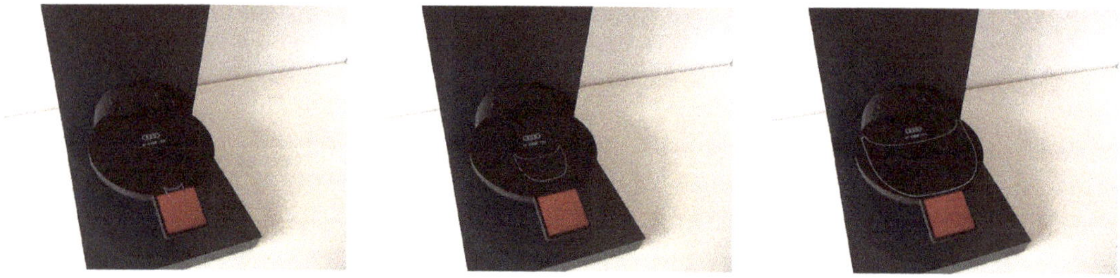

Figure 6. Hood icon slides from multi-touch display onto tabletop and grows to actual size.

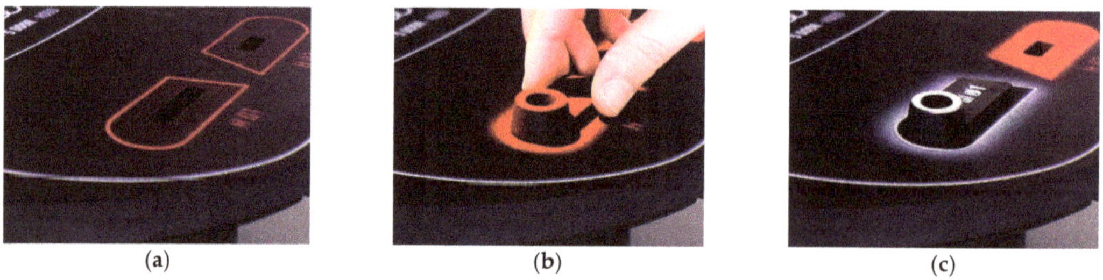

(a) (b) (c)

Figure 7. (**a**) Support contours are shown in red. (**b**) Operator places the left support. (**c**) Projection adapts to placed support and highlights its identification number and peg hole.

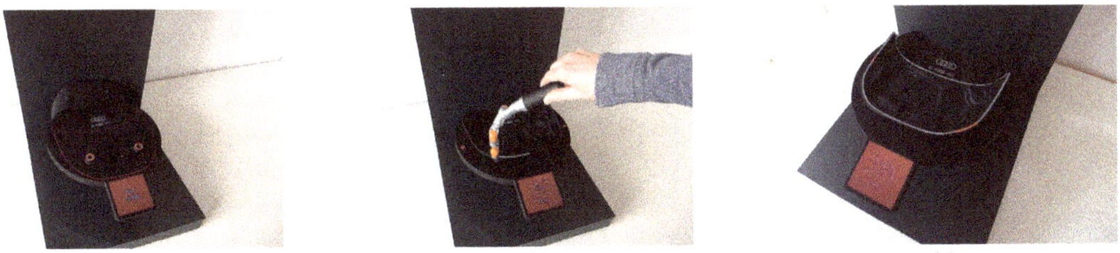

(a) (b) (c)

Figure 8. (**a**) Hood contour is projected in red. (**b**) Cobot tracks the hood contour. (**c**) Measurement result is projected onto hood.

3.4. Discussion

The Audi demonstrator combines two interaction styles: interaction with a multi-touch display, and interaction with a spatially augmented tabletop. We see novelty in the way the demonstrator couples these two interaction styles. We focus on two particular couplings, which are further developed in the demonstrators presented in the next two sections.

3.4.1. Coupling 1-On-Screen Event and Projected Event

The demonstrator is divided into two distinct working zones: the control unit with a multi-touch display, and the tabletop with SAR (Figure 4). This split-up determines the form semantics of the workplace. The square shape of the multi-touch display is adjacent to the circular shape of the SAR tabletop, forming a continuous surface. This continuity is further extended in the coupling of both on-screen and projected events. This coupling is clearly visible in the following moment. When the user selects the hood by touching the icon on the display, this icon moves upwards and slides onto the augmented tabletop (Figure 6). At this point, the projection takes over the on-screen imagery. The transition between the on-screen event and the projected event leads the operator from the control unit to the tabletop. In the design process, we relied on the rules of the Interaction Frogger Framework [3]. Both events are coupled on the aspects of time, location, and direction (Figure 9).

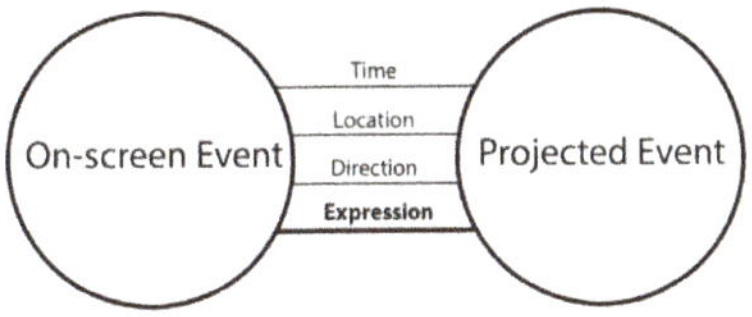

Figure 9. Coupling scheme representing movement of hood icon from on-screen environment to tabletop.

- They happen one after the other, thus in the same time span;
- Where the icon leaves the display, it is projected onto the tabletop; thus, both events have the same location;
- The direction of the icon's movement on the display is the same as the direction of its movement on the tabletop.

It is important to note that the three different couplings combine aspects that are already present in each event separately. Each event occurs at a particular time, on a particular location, and has a movement with a particular direction. Designing the coupling between them is clear and straightforward because the goal is clear: to ensure that the three aspects of both events are in line with each other. The benefit of these three couplings lies in the domain of ease of use. Unity of time, location, and direction between different events makes the events resonate with each other. These couplings create uniformity, coherence, and order in different movements and actions, and promote the naturalness of interaction, as already implied through the Interaction Frogger Framework. They make the functioning of the workplace logical and, as such, easy to read. They reveal the meaning of the workplace to the operator. We call them couplings of meaning.

However, there is another factor involved. The transition between the two different interaction styles, i.e., multi-touch interaction on a display and in situ projection on an augmented tabletop, is emphasized through the movement of the hood icon. The icon literally leaves the display and the GUI paradigm and enters the real world, where it grows into a full-scale work instruction. This transformation is not already present in each event separately. It emerges from the coexistence and interplay of the three other couplings. Its merit lies in the fact that it adds expression and engagement to the interaction, as it touches on aesthetics and emotional values. In former work, we referred to this phenomenon as the aesthetics of coupling. In this paper, we want to deepen our thinking. We state that both events show unity on a fourth aspect, next to time, location, and direction. We call this aspect expression (Figure 9). We call the coupling on this aspect coupling of expression.

We want to note that our definition of coupling of expression should not be confused with Wensveen's definition [3], in which the user expresses him- or herself during the interaction with a product. This expression is then reflected in the product's function. For example, cutting paper with a pair of scissors while feeling nervous and rushed will result in sloppy incisions.

3.4.2. Coupling 2: Physical Event and Projected Event

Another form of coupling occurs when the contours of both supports are projected onto the tabletop in real scale and all in white. At a certain point, the contour of the first support starts to pulsate slowly in red (Figure 7a). This event nudges the operator to place the first support. Once the operator has completed this task (Figure 7b), the projection changes color from red to white and adapts to the shape of the support (Figure 7c). The projection highlights the support's identification number, as well as its peg hole, in which the hood will eventually be positioned. The placement of the support by the operator is a physical event that is coupled to a projected event. This coupling occurs on two aspects: time and location (Figure 10).

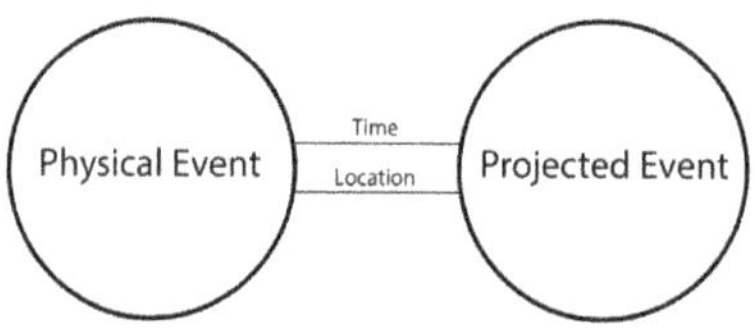

Figure 10. Coupling scheme representing placement of and projection on supports.

- The moment the operator positions the support on the tabletop, the projection appears;
- The projection appears on the support itself, not next to or near it.

3.4.3. Learnings for the Next Iteration

After building the demonstrator, and presenting it to people from Audi, Kuka and the Flemish make industry, we came to the conclusion that the demonstrator indeed elicited the aspects of coupling mentioned above, but did not exploit their full potential. We saw three angles to further explore the design space. Firstly, we wanted to build a real scale setup to enhance the sense of reality and immersion. Secondly, we wanted to create a real multi-touch display, rather than a simulated one, to increase the contrast between on-screen and projected events. Thirdly, we wanted to further explore coupling 2, i.e., the coupling between the supports and the projection. We decided to build a second installation, together with Kuka, in order to fulfil these intentions.

4. The Kuka Mockup

Demonstrator video can be seen in Supplementary Material: Video S3-KukaMockup.

4.1. Background

At this point in the research project, the plan rose to build a real demonstrator with a working cobot. In order to understand this idea, we decided to first construct a mockup out of wood, 3D printed components and spare parts. Since our focus was on interaction and coupling, rather than on force-based contour tracking and echolocation, we chose an artefact with a less complex 3D contour than the Audi A1 hood: a wooden longboard.

4.2. Description of the Mockup

The demonstrator was more an experience prototype or mockup than a full-blown demonstrator (Figure 11). It consisted of two ladders on which we mounted an aluminum beam with two projectors, and a full-scale wooden mockup of the Kuka LBR iiwa cobot. We built a wooden tabletop and a control unit with a working multi-touch display, and placed it under the beam structure. We designed specific supports for the longboard, which could be placed on the tabletop.

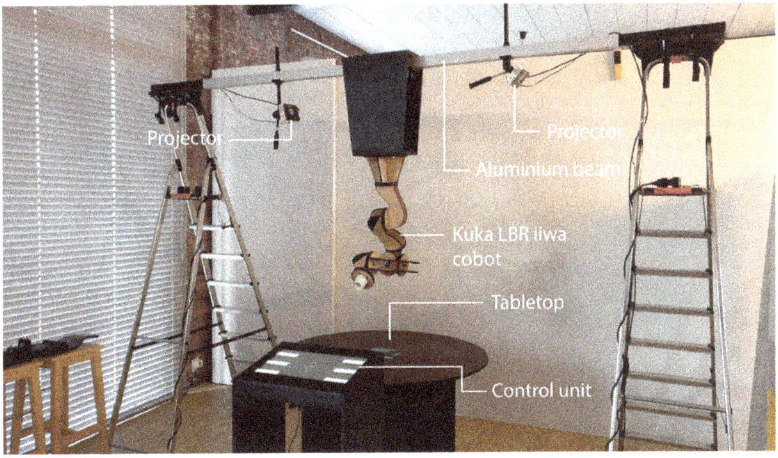

Figure 11. The Kuka mockup.

4.3. Description of the Interaction

The interaction routine is very similar to that of the Audi demonstrator. The idea is that the cobot measures the contour of a longboard instead of the Audi A1 hood and compares it with a reference contour. Another difference lies in the supports. Both longboard supports contain a dedicated projection surface, which is in the form of a pill-shaped cavity, printed on cardboard. Once the support is placed on the work surface, an identically shaped icon is projected into this cavity (Figures 12 and 13). The projected icon contains a slider that moves from one side of the printed cavity to the other, naturally guided via its boundary, informing the operator that the system has locked the support to the work surface (Figure 14). When the cobot has completed its measurement task, the zones with form deviations are indicated by a red light (Figure 15b), and the operator marks the zones with physical stickers (Figure 15c).

(a) (b) (c)

Figure 12. (**a**) Operator chooses longboard. (**b**) Longboard leaves display and slides on tabletop. (**c**) Longboard contour at full size.

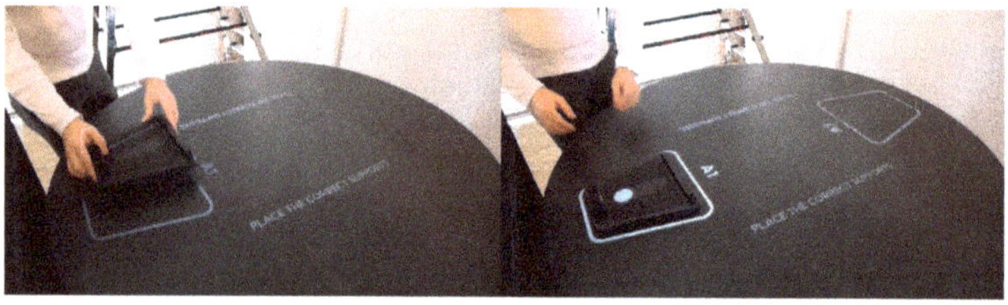

Figure 13. Operator places first support.

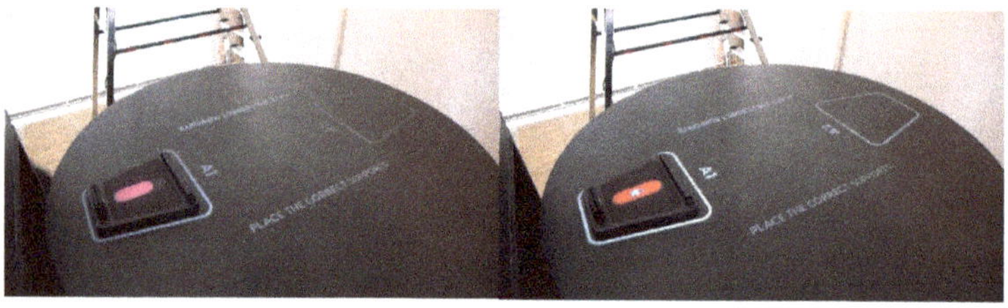

Figure 14. A moving slider is projected in support's cavity.

(a) (b) (c)

Figure 15. (**a**) Cobot tracks contour of longboard. (**b**) Deviating zones are marked with red light. (**c**) Operator marks zones with stickers.

4.4. Discussion

During the design of this demonstrator, we developed the two coupling themes that we had begun to explore in the Audi demonstrator.

4.4.1. Coupling 1–On-Screen Event and Projected Event

The transition from on-screen to projected event is realized literally through the presence of an operational multi-touch display on the control unit. The coupling between the two events is now a transition between two different media, which reinforces the transformative aspect of the interaction routine and, thus, the expressive power of the coupling. As soon as the icon disappears from the top of the display, it reappears as a projection on the tabletop, continuing the same movement (Figure 12). This process creates the effect of the icon literally crawling out of the display and into the real world. The sense of magic and surprise [3] that this process creates is the result of coupling of expression (Figure 9). We want to emphasize that, in our previous work, we stated that this form of aesthetic was only possible in the coupling between physical and digital events. In this demonstrator, both coupled events are digital in nature.

4.4.2. Coupling 2–Physical Event and Projected Event

We further explored Coupling 2 from the Audi demonstrator. The result of this exploration is visible in the placing of the longboard supports on the work surface (Figure 13). The manual placement of the support causes the appearance and movement of a projected element in the support's cardboard cavity, reflecting the status change in the support from unlocked to locked (Figure 14). With respect to the Audi demonstrator, we added coupling of direction to the concept, as the projected element follows the physical contour of the cavity. As a result of this design intervention, something remarkable happens: the support, which is a physical and inanimate object, suddenly has a moving part and seems to be brought to life through the projection. We consider this event to be coupling of expression (Figure 16).

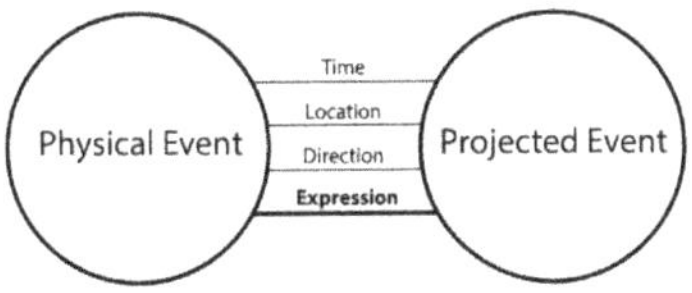

Figure 16. Coupling scheme representing placement of and projection on supports.

4.5. Learnings for the Next Iteration

As the cobot in the Kuka mockup was just a static wooden dummy, its movement capabilities remained underexposed, as do its coupling possibilities with other events. In our final demonstrator, we wanted to include a real, working cobot.

5. The Kuka Demonstrator

Demonstrator video can be seen in Supplementary Material: Video S4-KukaDemonstrator.

5.1. Background

Together with the people from Kuka, we designed and built a workplace around a limited set of cobot tasks, the most important of which was to measure the contour of a longboard deck and compare it to a reference contour. The envisaged workplace would contain a real Kuka LBR iiwa, real force-based contour tracking, and real-time projection of the measurement results onto the longboard itself.

5.2. Description of the Demonstrator

The demonstrator features a horizontal bridge of approximately 3m high (Figure 17). Beneath the bridge is a workbench containing two different work zones: a control unit with a display-button setup and a horizontal work surface. On the work surface, a longboard can be positioned and mounted by the operator using two supports. Above it, a Kuka LBR iiwa cobot hangs upside down from the bridge, with its movement envelope covering the entire work surface. On opposite sides of the cobot, two projectors are mounted on the bridge. We designed a special tool, which is mounted on the cobot itself, that allows it to physically touch and track the contour of the longboard, thereby sensing and processing the applied force. The two work zones involve different operator tasks. The zone with the control unit serves to select tasks through menu navigation. The work surface with the cobot and the projection is conceived to perform physical tasks, in cooperation with the cobot. Both work zones are connected via the large OK button below the control unit. This large button is located between the two different work zones and always remains accessible, whether the operator is working at the control unit or the work surface.

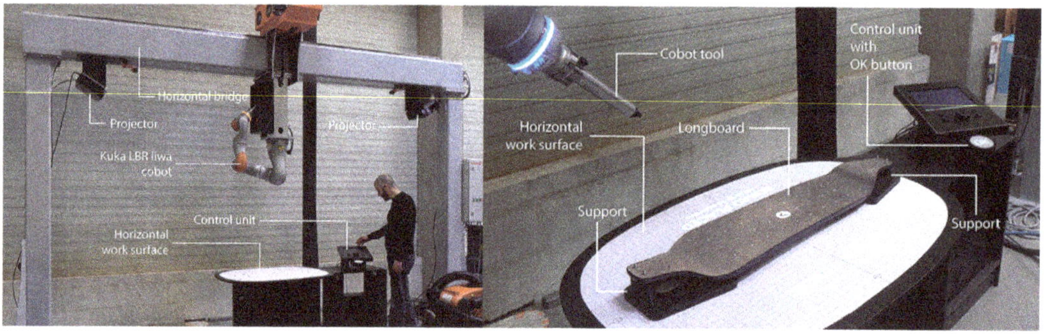

Figure 17. Main components of Kuka demonstrator.

5.3. Description of the Interaction

5.3.1. Interaction with the Control Unit

In a first phase, the cobot is in sleep mode (Figure 18a). We provided a box, attached to the bridge, into which the cobot can retreat, portraying a clear image of being at rest. The operator walks to the control unit and activates the system by pushing the slider button on the control unit to the left (Figure 18b). On-screen, a black curtain slides away together with the button, and the control unit is activated. As a result, the cobot above the work surface wakes up and moves towards the control unit. It adopts an attentive posture, as it seems to be looking at the display, together with the operator. We call this dialogue mode (Figure 19a). The operator navigates through the menus using a traditional rotary dial and push button interface (Figure 19b). As he/she turns the dial, the menus move horizontally.

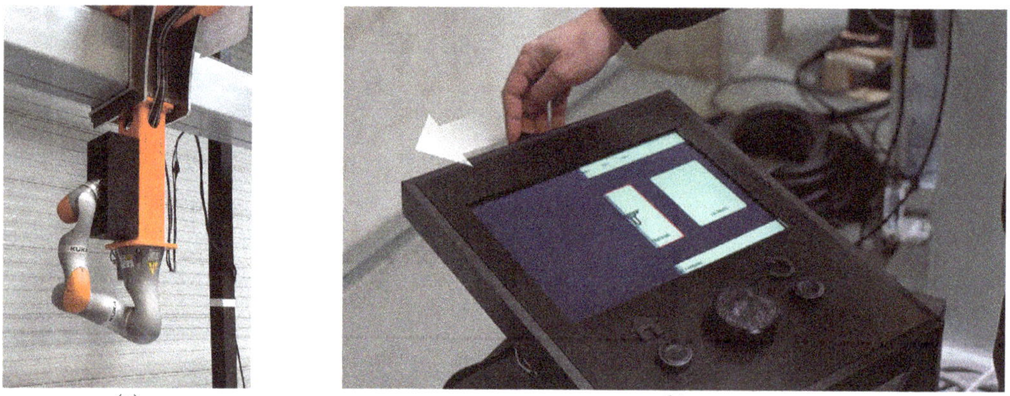

(**a**) (**b**)

Figure 18. (**a**) Cobot in sleep mode. (**b**) Operator activates system by pushing the slider.

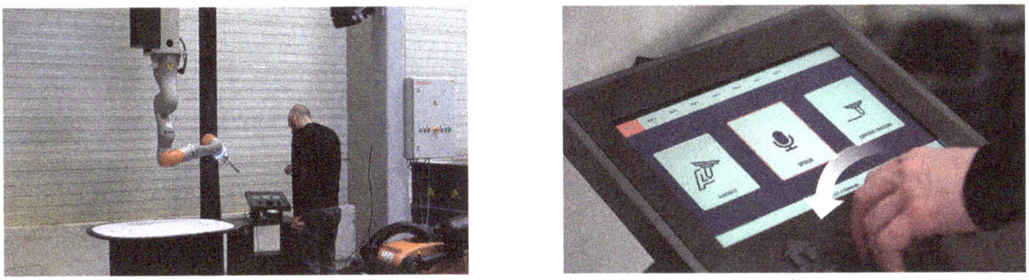

(**a**) (**b**)

Figure 19. (**a**) Cobot in dialogue mode. (**b**) Operator navigates through menus by turning the rotary dial.

5.3.2. Transition between Two Work Zones

The operator selects to perform a contour tracking task, and confirms this selection by pushing the rotary dial (Figure 20a). The cobot then moves away from the control unit towards the work surface. At this point, the on-screen images on the control unit's display move downwards, as if they flow onto the table below. At the same time, a projection is generated on the work surface, showing a sliding image that moves away from the control unit and fills the entire work surface. The cobot appears to "pull" the on-screen image out of the control unit onto the tabletop (Figure 20b). The operator is guided from one work zone towards the other based on the physical movements of the cobot and the movements of on-screen and projected images. The cobot is now looking at the work surface, and is in standby mode. Work instructions are projected onto the work surface (Figure 21).

(**a**) (**b**)

Figure 20. (**a**) Operator pushes rotary dial. (**b**) Cobot "pulls" images out of display and spreads them across work surface.

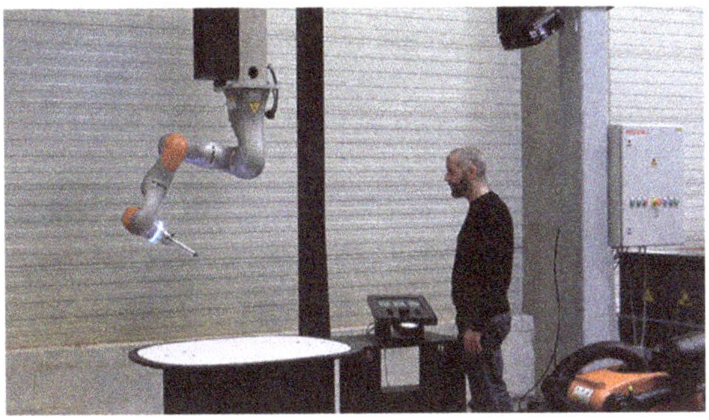

Figure 21. Cobot in standby mode.

5.3.3. Manual Mounting of the Supports and the Longboard

The operator follows the instructions on the work surface and mounts the supports (Figure 22a). After each support is mounted, the operator pushes the OK button (Figure 22b), and a moving icon is projected into a cavity on each support, indicating that the system has locked the support (Figure 23). The operator then places the longboard on the supports. This action is detected via the system, which responds with a projection on the longboard itself (Figure 24). The operator continues to follow the work instructions and manually bolts the longboard in place. When this task is completed, he/she pushes the OK button. The cobot approaches the longboard.

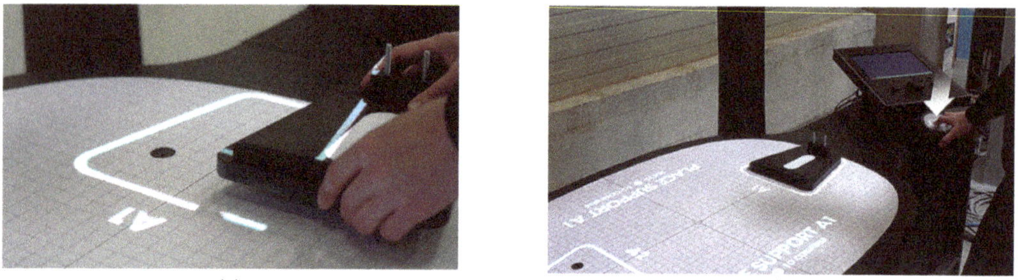

(**a**) (**b**)

Figure 22. (**a**) Operator manually mounts supports. (**b**) After mounting of each support, operator pushes OK button.

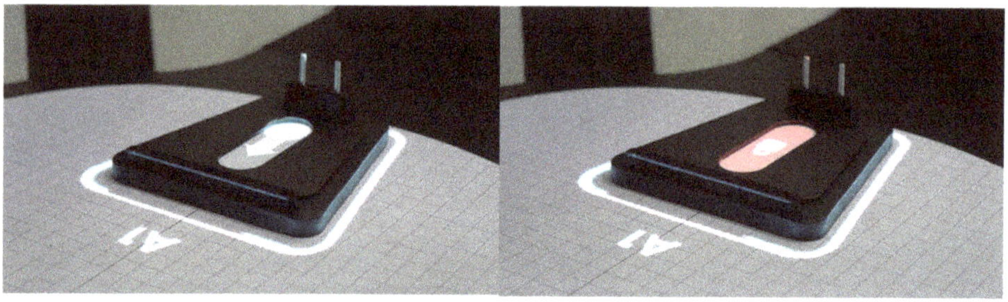

Figure 23. A slider is projected into physical cavity of mounted support and moves to other side of this cavity.

Figure 24. Operator places longboard on supports, and system reacts with a projection directly on longboard.

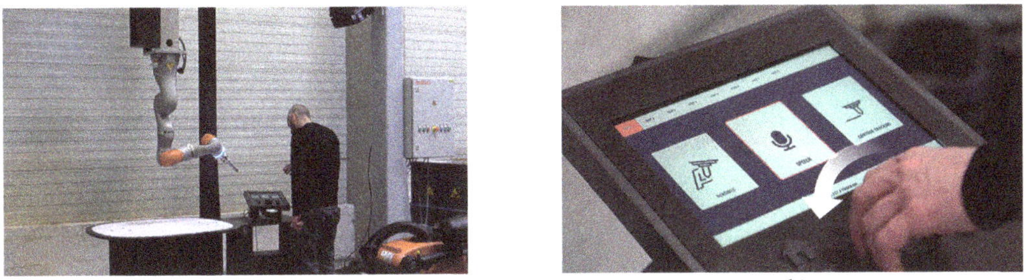

(**a**) (**b**)

Figure 19. (**a**) Cobot in dialogue mode. (**b**) Operator navigates through menus by turning the rotary dial.

5.3.2. Transition between Two Work Zones

The operator selects to perform a contour tracking task, and confirms this selection by pushing the rotary dial (Figure 20a). The cobot then moves away from the control unit towards the work surface. At this point, the on-screen images on the control unit's display move downwards, as if they flow onto the table below. At the same time, a projection is generated on the work surface, showing a sliding image that moves away from the control unit and fills the entire work surface. The cobot appears to "pull" the on-screen image out of the control unit onto the tabletop (Figure 20b). The operator is guided from one work zone towards the other based on the physical movements of the cobot and the movements of on-screen and projected images. The cobot is now looking at the work surface, and is in standby mode. Work instructions are projected onto the work surface (Figure 21).

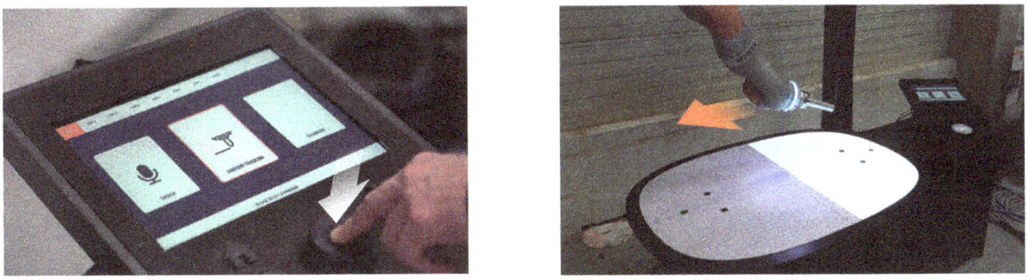

(**a**) (**b**)

Figure 20. (**a**) Operator pushes rotary dial. (**b**) Cobot "pulls" images out of display and spreads them across work surface.

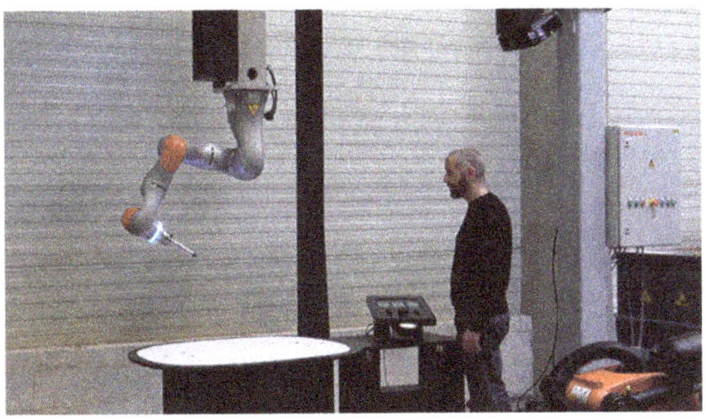

Figure 21. Cobot in standby mode.

5.3.3. Manual Mounting of the Supports and the Longboard

The operator follows the instructions on the work surface and mounts the supports (Figure 22a). After each support is mounted, the operator pushes the OK button (Figure 22b), and a moving icon is projected into a cavity on each support, indicating that the system has locked the support (Figure 23). The operator then places the longboard on the supports. This action is detected via the system, which responds with a projection on the longboard itself (Figure 24). The operator continues to follow the work instructions and manually bolts the longboard in place. When this task is completed, he/she pushes the OK button. The cobot approaches the longboard.

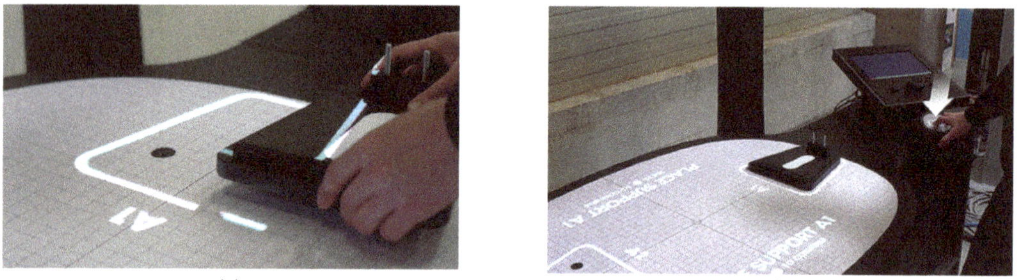

(**a**) (**b**)

Figure 22. (**a**) Operator manually mounts supports. (**b**) After mounting of each support, operator pushes OK button.

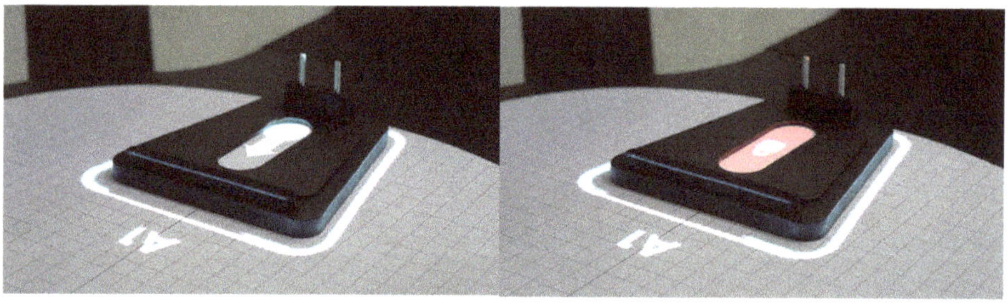

Figure 23. A slider is projected into physical cavity of mounted support and moves to other side of this cavity.

Figure 24. Operator places longboard on supports, and system reacts with a projection directly on longboard.

5.3.4. Force-Based Contour Tracking

The operator pushes the OK button, and the cobot begins to track the contour of the longboard. The tracked path is projected onto the longboard in real time (Figure 25a). The cobot is now in scan mode. In a first pass, the cobot recognizes the longboard contour and uses it as a reference for other longboards. When a longboard with a deviating contour is checked via the system (in the video, the deviation is added by the operator), the deviation is detected and marked with a red projection image on the longboard itself (Figure 25b). Correspondingly, the deviation value is projected on the work surface. After tracking, the operator pushes the OK button.

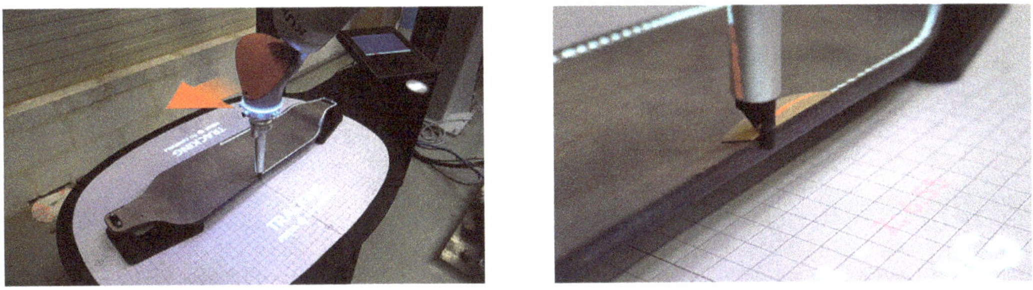

(**a**) (**b**)

Figure 25. (**a**) Scan mode. Cobot tracks contour of longboard, and measurement results are projected onto it in real time. (**b**) Cobot detects a deviation in contour of longboard, and marks it with red light.

5.3.5. Clearing the Work Surface and Shutting down the System

The operator removes the longboard and the supports, following the instructions on the work surface. When the work surface is empty, the OK button is pushed, and the cobot guides the projected images into the control unit with a physical movement (Figure 26a). The control unit is reactivated, and the cobot is back in dialogue mode. To deactivate the system, the operator moves the slider on top of the control unit to the right, and the on-screen image reacts accordingly (Figure 26b). The cobot returns to sleep mode (Figure 18a).

(**a**) (**b**)

Figure 26. (**a**) Cobot moves towards the control unit, takes projection with it, and "pushes" images in display. (**b**) Operator deactivates system by moving slider to right.

5.4. Discussion

The fact that we had the possibility to conceive and craft a semi-functional demonstrator on real scale, with a fully-functioning cobot, gave us the chance to further explore Coupling 1, which had already appeared in the two earlier demonstrators. Coupling 2 was refined, though its concept remained the same. In addition, we added a display-button setup as control unit, with specific couplings.

5.4.1. Coupling 1: On-Screen, Projected and Cobot Event

We considered the movements and postures of the cobot not only as a functional given, but also as a crucial part of the workplace's form semantics and affordances. We realized this intention by relying on the concept op Mode of Use Reflected in the Physical State (MURPS) [30]. This means that we designed the different postures of the cobot in such a way that they non-verbally express the state of the workplace's operating system. In this respect, we distinguish between sleep mode (Figure 18a), dialogue mode (Figure 19a), standby mode (Figure 21), and scan mode (Figure 25a). Moreover, the movements of the cobot are coupled to projected events. This coupling is most prominent in scan mode (Figure 25a), where the cobot tracks the contour of the longboard, and the result of the measurement is projected onto the longboard's surface.

We further explored Coupling 1 by enriching it with cobot movements. This approach led to a new insight. When the operator ends the dialogue mode by pushing the rotary dial (Figure 20a), three events are coupled: a cobot event, an on-screen event, and a projected event (Figure 27). The three events occur at the same time and at the same location, and share the same speed and direction. In addition, the couplings create the impression that the cobot is actually pulling the image out of the display and spreading it onto the tabletop (Figure 20b). A similar coupling occurs at the end of the contour tracking task. The projected images slide back from the work surface in the control unit (Figure 26a), seemingly being pushed by the cobot. This effect, where the cobot appears to pull images out of the display onto the tabletop and back, surpasses the three couplings of meaning and reinforces the coupling of expression. The cobot becomes an expressive medium that grasps intangible elements and moves them from one physical place to another.

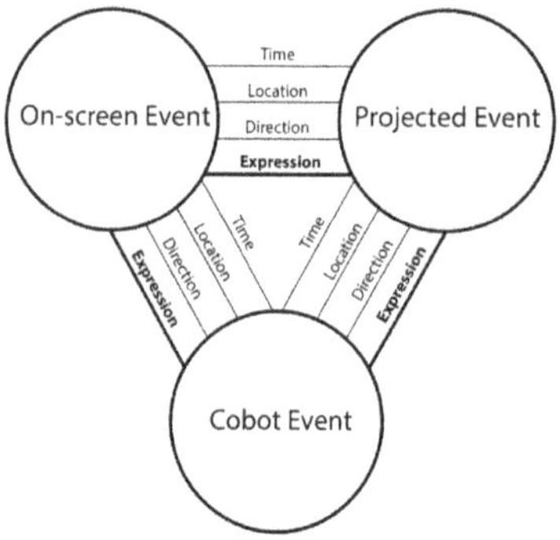

Figure 27. Coupling scheme representing cobot manipulating images on display and tabletop.

5.4.2. Additional Couplings of the Control Unit: Physical Event and On-Screen Event

What coupling of expression entails is clearly illustrated by the difference between two interaction routines on the control unit of the Kuka demonstrator. The first meaning is the turning of the rotary dial, which causes the different task icons on the display to move sideways, either to the left or to the right (Figure 19b). The rotary dial is positioned near the display with the task icons, its rotation is immediately translated into a lateral movement of the icons, and the icons move left or right, according to its direction of rotation. As such, there is coupling on the aspects of time, location, and direction (Figure 28a). However, the interaction feels rather plain and barely expressive. This outcome is because the coupling of direction is not very strong: the dial rotates, while the icons translate. The fact that a rotary dial allows this degree of randomness makes it a popular control element in many commercial GUIs, but it also makes its interaction standardized and generic. This method is very different to the second routine. This routine involves pushing the slider button at the top of the display, which causes a black curtain to slide across the display, activating or deactivating it (Figures 18b and 26b). Again, there is coupling on the aspects of time, location, and direction. In this case, however, the coupling of direction is strong. Both the slider button and the on-screen curtain translate in the same direction, over almost the same distance. For the user, it feels as if the slider button is directly attached to the on-screen curtain, as if he/she were dragging a physical curtain across the display. The specific design and coupling of the two events, the pushing of the slider button, and the movement of the on-screen curtain results in coupling of expression (Figure 28b).

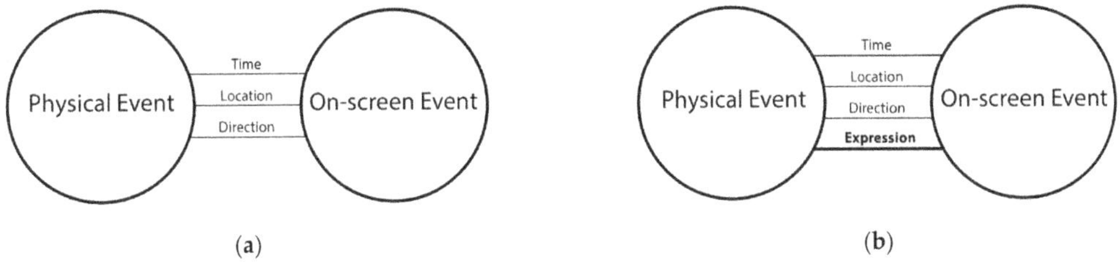

Figure 28. (**a**) Coupling scheme representing movement of rotary dial and its effect on on-screen menu. (**b**) Coupling scheme representing movement of slider button and on-screen curtain.

6. Discussion

In this section, we look back at the delivered work. We discuss the couplings that emerged across the three demonstrators, and reflect on how they have refined our understanding of coupling.

6.1. A Wealth of Events

In our previous work, we assumed two types of events—physical events and digital events—with coupling being the relationship between them. During the conception and creation of the three demonstrators, it became clear that this division was not sufficiently fine-grained. We encountered two types of digital events: on-screen events and projected events. It is to be expected that, as different forms of AR (tablet-based AR, SAR, head-mounted display-based AR) are adopted in workplaces, the number of digital event types will increase. In this scenario, we are thinking of holographic events, sound events, etc. Physical events can also be categorized in more detail. During our research, we already came across the cobot event; however, a further classification of physical events urges itself. Event classifiers include user movements, physical events performed with control elements, actuated events, etc. Digital events can be coupled to physical events, but also to other digital events. This fact clearly surfaced in our RtD process: in the three demonstrators, we defined couplings between on-screen and projected events, which were both digital in nature. Similarly, couplings between physical events are already commonplace. Any kitchen appliance that connects the pushing of a button to the activation of an electric motor couples two physical events.

In addition, coupling is not necessarily limited to two events. In the Kuka demonstrator, we coupled three events instead of two: an on-screen event, a projected event, and a cobot event. Four or five events may also be coupled together. Whether these events are physical or digital in nature is less important.

As digital technology evolves, and as digital phenomena break free from displays and enter the physical world in projected or holographic form, the dichotomy between the digital and the physical, which has been the blueprint for embodied interaction to date, becomes less dominant [31]. We want to open the door to an understanding of coupling that transcends the traditional bridging of the physical and the digital. Designers of digital products and systems need to be primarily concerned with the intuitive and engaging coupling of different events, regardless of their nature, rather than making interaction with digital phenomena more physical.

6.2. From Meaning to Expression

We defined four aspects of coupling. Three of them came directly from the Interaction Frogger Framework: time, location, and direction. We called them couplings of meaning. In the discussion about the Audi demonstrator, we stated that they are relatively easy for designers to understand and employ, as they only require the organization and alignment of aspects that are already present in each event separately.

The fourth aspect of coupling—expression—is more difficult to grasp, because it is not already present in each individual event. Instead, it emerges as a result of the other couplings and the design of the workplace in general. Coupling of expression stems from the aesthetics of coupling, which we discussed in Section 2.3. However, the aesthetics of coupling, as well as the MCRpd interaction model, relied on the dichotomy between the physical and the digital realms. If this dichotomy fades into the background, then how can we define coupling of expression and its relation to coupling of meaning?

Coupling of expression emerges as a consequence of, and as a contrast to, coupling of meaning [32]. In order to set the stage on which coupling of expression can perform, at least some realization of couplings of meaning is necessary. Couplings of meaning form, as it were, a reference, i.e., a background against which coupling of expression is perceived and felt by the operator. This stage or background has an orderly, logical, and natural character, as the various coupled events resonate with familiar, often Newtonian laws and common sense knowledge about the physical world [33]. However, when coupling of expression appears on the stage, this familiar character is challenged and transcended, as the coupling appears to violate the established laws. The resulting user experience enters the realm of magic, beauty, and surprise, and appeals to the operator's emotions, rather than to his/her reasoning. We give three examples:

- In the Kuka mockup, the movement of the on-screen longboard icon is adopted using the projected longboard icon; thus, it appears to leave the display and slide across the tabletop (Figure 12). Both icons appear to be one, which is an effect created via the couplings of time, location, and direction. The expressive appeal of the interaction routine lies in the fact that the operator clearly realizes that the two icons are not the same. They have inherently different aspects. In the Kuka demonstrator, this movement is reinforced through the cobot movements. The effect is that the cobot pulls the on-screen content out of the display and spreads it across the tabletop (Figure 20b).
- In the Kuka mockup and the Kuka demonstrator, the pill-shaped cavity of the support is filled with a projected slider (Figures 14 and 23). Couplings of time, location, and direction ensure that the slider carefully follows the contours of the physical cavity, as if it were a real slider. The expression comes into play when the operator realizes that the slider is not real.
- In the Kuka demonstrator, pushing the slider button on the control unit causes the on-screen curtain to slide simultaneously in the same direction at almost the same pace (Figures 18b and 26b). For the operator, it feels as if the curtain is physically attached to the slider, although he/she clearly sees that the curtain is only an on-screen representation, rather than a real one.

Why should designers bother to create couplings of expression? Apparently, unlike couplings of meaning, coupling of expression does not contribute to the ease of use or intuitive readability of an augmented workplace. Therefore, what is the point? We believe that coupling of expression enhances the quality perception of the workplace. It creates an interaction that is aesthetically pleasing, harmonious, and engaging for the operator, and as such heightens his/her appreciation of the workplace [34]. In other words, coupling of expression serves the operator's emotional well-being.

We see a parallel between our concept of coupling and Hassenzahl's theory of User Experience, which is specifically aimed at digital products and systems [35]. Couplings of meaning generate what Hassenzahl calls pragmatic quality. They concern the utility and usability of a digital product, and describe how well the product fulfils a particular function or completes a particular task. Coupling of expression generates hedonic quality [36], and determines how the operator feels when he/she performs a task in the workplace. The most radical aspect of Hassenzahl's theory, in our view, is the relationship between pragmatic and hedonic quality. Pragmatic quality, according to Hassenzahl, is never a goal in itself. It should be considered as an enabler of hedonic quality. The fact that a digital product fulfils the task for which it was designed is taken for granted by the user, and does not contribute to his/her well-being. A product that aims to provide pleasure and engagement should have hedonic quality, and its pragmatic quality is subordinate to this. This relationship between pragmatic and hedonic quality corresponds to how we position couplings of meaning in relation to coupling of expression. The former couplings are merely enablers of the latter type. The ultimate goal of the designer should be to design an augmented workplace in which the operator feels good and thrives. Coupling of expression is directly related to this goal. Couplings of meaning allow coupling of expression to flourish.

7. Conclusions

We started our investigation by formulating the following research question: what coupling possibilities emerge when a strong specific workplace is enriched with SAR? Our goal was to design a workplace as one holistic, integrated entity, combining physical and SAR components. During the conception and crafting of three demonstrators, several themes within the embodied interaction research agenda were addressed and explored.

Firstly, we believe that the traditional dichotomy between the physical and the digital is becoming less prominent as a driver in the design of digital products and systems. For years, this dichotomy prevailed in tangible interaction; the embodiment framework we briefly described in Section 2.1; and our own research on the aesthetics of coupling (Section 2.3). With the development of digital technology, the number of digital events in people's daily lives is increasing dramatically. Moreover, digital events are abandoning the traditional, detached display. Instead, they are taking on new forms, such as 2D projections or 3D holograms, which are better integrated into the physical world. As a result, the traditional distinction between the digital and the physical is fading and becoming less important. Together with this evolution, the concept of coupling is also changing. Coupling, which we previously defined as the connection between physical and digital events, can also occur between two digital events, for example between graphics on a display and projected images on a real object. Moreover, coupling should not be reduced to the connection between two events. Our research shows several action routines where coupling occurs between three events, and it is likely that the number of coupled events can be increased.

Secondly, we set the stage for a new taxonomy of couplings. As the amount of events in digital products and systems increases, the design of the coupling between these events, be they digital or physical, product or user-related, becomes more important. We propose to divide couplings into two groups by making the distinction between coupling of time, location, and direction on one hand, and coupling of expression on the other. The first three couplings, which we called couplings of meaning, are related to ease of use and pragmatic usability, while coupling of expression resides in the domain of psychological wellbeing. Bu writing this paper, we want to stress the importance of coupling in the practice of industrial and interaction design. It is our aim to establish the concept of coupling as a full-blown design theory, just like 2D and 3D composition, color theory, and affordance theory.

Lastly, from our point of view, the speculative view that we formulated in Section 1.2, i.e., the strong-specific workplace, opens up new possibilities for the design of spatially augmented workplaces. Throughout the three demonstrators, we designed a workplace that was fully tailored to a limited number of tasks. This allowed us to design the projected images in conjunction with the physical workplace itself. The potential benefits of this approach are best reflected in the design of the supports. In all three demonstrators, the design of both the supports and the projected images on them were created simultaneously by the same designer. As such, the physical shape of these supports and the projection onto them were allowed to influence each other, to the point where both were designed as a single system. As a result, the supports have multiple physical reference points that channel projected images. These physical reference points are persistent, meaning that they are always present and provide the operator with information about the projected images on the support. Even when there is no projected image present, the operator knows where on the support it will appear. The idea of imposing physical restrictions on projected content may seem counterintuitive, given the innate freedom of projection. However, we advocate this approach, because we believe that a workplace that physically channels its projected content makes that content more structured and predictable for the operator working within it. This approach might reduce the operator's chance of missing a projected message and contribute to his/her sense of control over the workplace.

8. Future Research

The work described in this paper opens the door to further research. In the demonstrators we built for Audi and Kuka, we encountered different types of events: physical, on-screen, projected, and cobot events. With the advent of head-mounted display-based AR, holographic objects emerge as a new event type. As holographic objects are not tied to a display or projection surface, their coupling possibilities with other event types offer a great deal of design freedom, and form a new and promising research space.

The dual approach to coupling that we established in this paper can be further elaborated. Where does meaning end, and where does expression begin? How can coupling of meaning support coupling of expression and the other way round?

Finally, this research is situated in the field of industrial workplaces. Further research is needed to show that the result of this work—the coupling framework—is relevant to a wider application area: digital products and systems in general.

Given the exploratory nature of this future research, we believe that Research through Design is a valuable method to tackle this research gap. We hope that this paper will inspire design researchers and design students to adopt this method and put it into practice.

Supplementary Materials: All images and videos can be can be downloaded at: https://www.mdpi.com/article/10.3390/designs7030069/s1. Video S1-NightLamp, Video S2-AudiDemonstrator, Video S3-KukaMockup, and Video S4-KukaDemonstrator.

Author Contributions: Conceptualization, L.V.C., W.V. and I.D.; methodology, L.V.C. and I.D.; validation, L.V.C. and W.V.; investigation, L.V.C. and W.V.; writing—original draft preparation, L.V.C.; writing—review and editing, L.V.C. and I.D.; visualization, L.V.C. and W.V.; supervision, L.V.C.; project administration, L.V.C.; funding acquisition, L.V.C. All authors have read and agreed to the published version of the manuscript.

Funding: This research was funded by VLAIO–AGENTSCHAP INNOVEREN EN ONDERNEMEN-BELGIUM, grant number HBC.2016.0461.

Data Availability Statement: The pictures presented in this study are available in the article. The presented videos are available in the supplementary material.

Acknowledgments: We want to thank the relevant people from Audi Brussels for offering us the opportunity to explore what embodied interaction could mean for their particular use case. Moreover, we want to thank the people from Kuka for their part in building the Kuka demonstrator.

Conflicts of Interest: The authors declare no conflict of interest.

References

1. Van Campenhout, L.; Frens, J.; Vaes, K.; Hummels, C. The Aesthetics of Coupling: An Impossible Marriage. *Int. J. Des.* **2020**, *14*, 1–16.
2. Ullmer, B.; Ishii, H. Emerging Frameworks for Tangible User Interfaces. *IBM Syst. J.* **2000**, *39*, 915–931. [CrossRef]
3. Wensveen, S.A.G.; Djajadiningrat, J.P.; Overbeeke, C.J. Interaction Frogger: A Design Framework to Couple Action and Function through Feedback and Feedforward. In Proceedings of the 2004 Conference on Designing Interactive Systems Processes, Practices, Methods, and Techniques-DIS '04, Cambridge, MA, USA, 1–4 August 2004; ACM Press: New York, NY, USA, 2004; p. 177. [CrossRef]

4. Zimmerman, J.; Stolterman, E.; Forlizzi, J. An Analysis and Critique of Research through Design: Towards a Formalization of a Research Approach. In Proceedings of the 8th ACM Conference on Designing Interactive Systems, Aarhus, Denmark, 16–20 August 2010.
5. van Dijk, J.; Moussette, C.; Kuenen, S.; Hummels, C. Radical Clashes: What Tangible Interaction Is Made of. In Proceedings of the 7th International Conference on Tangible, Embedded and Embodied Interaction-TEI '13, Barcelona, Spain, 10–13 February 2013; ACM Press: New York, NY, USA, 2013; p. 323. [CrossRef]
6. van Dijk, J. Designing for Embodied Being-in-the-World: A Critical Analysis of the Concept of Embodiment in the Design of Hybrids. *MTI* **2018**, *2*, 7. [CrossRef]
7. Gibson, J.J. *The Ecological Approach to Visual Perception: Classic Edition*, 1st ed.; Psychology Press: London, UK, 2014. [CrossRef]
8. Ishii, H.; Ullmer, B. Tangible Bits: Towards Seamless Interfaces between People, Bits and Atoms. In Proceedings of the ACM SIGCHI Conference on Human Factors in Computing Systems, Atlanta, GA, USA, 22–27 March 1997; ACM: New York, NY, USA, 1997; pp. 234–241. [CrossRef]
9. Fishkin, K.P. A Taxonomy for and Analysis of Tangible Interfaces. *Pers. Ubiquit. Comput.* **2004**, *8*, 347–358. [CrossRef]
10. Dourish, P. *Where the Action Is: The Foundations of Embodied Interaction*; The MIT Press: Cambridge, MA, USA, 2001. [CrossRef]
11. Hornecker, E.; Buur, J. Getting a Grip on Tangible Interaction: A Framework on Physical Space and Social Interaction. In Proceedings of the SIGCHI Conference on Human Factors in Computing Systems, Montreal, QC, Canada, 22–27 April 2006.
12. Van Campenhout, L.D.E.; Frens, J.; Hummels, C.; Standaert, A.; Peremans, H. Touching the Dematerialized. *Pers. Ubiquit. Comput.* **2016**, *20*, 147–164. [CrossRef]
13. Buxton, W. Less Is More (More or Less). In *The Invisible Future: The Seamless Integration of Technology in Everyday Life*; McGraw Hill: New York, NY, USA, 2001; pp. 145–179.
14. Follmer, S.; Leithinger, D.; Olwal, A.; Hogge, A.; Ishii, H. InFORM: Dynamic Physical Affordances and Constraints through Shape and Object Actuation. In Proceedings of the 26th Annual ACM Symposium on User Interface Software and Technology, St. Andrews, UK, 8–11 October 2013; ACM: New York, NY, USA, 2013; pp. 417–426. [CrossRef]
15. Nikolaidis, A. What Is Significant in Modern Augmented Reality: A Systematic Analysis of Existing Reviews. *J. Imaging* **2022**, *8*, 145. [CrossRef] [PubMed]
16. Palmarini, R.; Erkoyuncu, J.A.; Roy, R.; Torabmostaedi, H. A Systematic Review of Augmented Reality Applications in Maintenance. *Robot. Comput.-Integr. Manuf.* **2018**, *49*, 215–228. [CrossRef]
17. Nee, A.Y.C.; Ong, S.K.; Chryssolouris, G.; Mourtzis, D. Augmented Reality Applications in Design and Manufacturing. *CIRP Ann.* **2012**, *61*, 657–679. [CrossRef]
18. Xi, N.; Chen, J.; Gama, F.; Riar, M.; Hamari, J. The Challenges of Entering the Metaverse: An Experiment on the Effect of Extended Reality on Workload. *Inf. Syst. Front.* **2022**, *25*, 659–680. [CrossRef] [PubMed]
19. Reljić, V.; Milenković, I.; Dudić, S.; Šulc, J.; Bajči, B. Augmented Reality Applications in Industry 4.0 Environment. *Appl. Sci.* **2021**, *11*, 5592. [CrossRef]
20. Henderson, S.J.; Feiner, S. Evaluating the Benefits of Augmented Reality for Task Localization in Maintenance of an Armored Personnel Carrier Turret. In Proceedings of the 2009 8th IEEE International Symposium on Mixed and Augmented Reality, Orlando, FL, USA, 19–22 October 2009; IEEE: Piscataway, NJ, USA, 2009; pp. 135–144. [CrossRef]
21. Kipper, G.; Rampolla, J. *Augmented Reality: An Emerging Technologies Guide to AR*, 1st ed.; Syngress/Elsevier: Amsterdam, The Netherlands; Boston, MA, USA, 2013.
22. Boer, L.; Jenkins, T. Fostering Creative Confidence with SCD in Interaction Design Education. *IxD&A* **2021**, *51*, 270–302. [CrossRef]
23. Ellen, L. Stimulating Critical Thinking about the Self-Sustaining Network of Relations Which Reinforce Smartphone Use: A Critical Design Study. Master's Thesis, University of Twente, Enschede, The Netherlands, 2019.
24. Stienstra, J.; Bogers, S.; Frens, J. Designerly Handles: Dynamic and Contextualized Enablers for Interaction Designers. In Proceedings of the Conference on Design and Semantics of Form and Movement, Milano, Italy, 13–17 October 2015; pp. 86–94.
25. Djajadiningrat, T.; Wensveen, S.; Frens, J.; Overbeeke, K. Tangible Products: Redressing the Balance between Appearance and Action. *Pers. Ubiquit. Comput.* **2004**, *8*, 294–309. [CrossRef]
26. Khamaisi, R.K.; Prati, E.; Peruzzini, M.; Raffaeli, R.; Pellicciari, M. UX in AR-Supported Industrial Human–Robot Collaborative Tasks: A Systematic Review. *Appl. Sci.* **2021**, *11*, 10448. [CrossRef]
27. Uva, A.E.; Gattullo, M.; Manghisi, V.M.; Spagnulo, D.; Cascella, G.L.; Fiorentino, M. Evaluating the Effectiveness of Spatial Augmented Reality in Smart Manufacturing: A Solution for Manual Working Stations. *Int. J. Adv. Manuf. Technol.* **2018**, *94*, 509–521. [CrossRef]
28. Rupprecht, P.; Kueffner-McCauley, H.; Trimmel, M.; Schlund, S. Adaptive Spatial Augmented Reality for Industrial Site Assembly. *Procedia CIRP* **2021**, *104*, 405–410. [CrossRef]
29. Mengoni, M.; Ceccacci, S.; Generosi, A.; Leopardi, A. Spatial Augmented Reality: An Application for Human Work in Smart Manufacturing Environment. *Procedia Manuf.* **2018**, *17*, 476–483. [CrossRef]
30. Frens, J. Designing for Rich Interaction-Integrating Form, Interaction and Function. Ph.D. Thesis, Eindhoven University of Technology, Eindhoven, The Netherlands, 2006.
31. Scholz, R. Sustainable Digital Environments: What Major Challenges Is Humankind Facing? *Sustainability* **2016**, *8*, 726. [CrossRef]
32. Djajadiningrat, T.; Matthews, B.; Stienstra, M. Easy Doesn't Do It: Skill and Expression in Tangible Aesthetics. *Pers. Ubiquit. Comput.* **2007**, *11*, 657–676. [CrossRef]

33. Jacob, R.J.K.; Girouard, A.; Hirshfield, L.M.; Horn, M.S.; Shaer, O.; Solovey, E.T.; Zigelbaum, J. Reality-Based Interaction: A Framework for Post-WIMP Interfaces. In Proceedings of the Twenty-Sixth Annual CHI Conference on Human Factors in Computing Systems-CHI '08, Florence, Italy, 5–10 April 2008; ACM Press: New York, NY, USA, 2008; p. 201. [CrossRef]
34. Ross, P.R.; Wensveen, S.A.G. Designing Behavior in Interaction: Using Aesthetic Experience as a Mechanism for Design. *Int. J. Des.* **2010**, *4*, 3–13.
35. Hassenzahl, M.; Diefenbach, S.; Göritz, A. Needs, Affect, and Interactive Products–Facets of User Experience. *Interact. Comput.* **2010**, *22*, 353–362. [CrossRef]
36. Diefenbach, S.; Kolb, N.; Hassenzahl, M. The "hedonic" in Human-Computer Interaction: History, Contributions, and Future Research Directions. In Proceedings of the 2014 Conference on Designing Interactive Systems, Vancouver, BC, Canada, 14–18 June 2014; ACM: New York, NY, USA, 2014; pp. 305–314. [CrossRef]

designs

MDPI

Article

Predictive Maintenance in Industry 4.0 for the SMEs: A Decision Support System Case Study Using Open-Source Software

Mirjana Pejić Bach [1,*], Amir Topalović [2], Živko Krstić [2] and Arian Ivec [3]

1 Department of Informatics, Faculty of Economics and Business, University of Zagreb, 10000 Zagreb, Croatia
2 Aisma srl, 20122 Milan, Italy
3 Department of Physics, Faculty of Science, University of Zagreb, 10000 Zagreb, Croatia; ivec@phy.hr
* Correspondence: mpejic@net.efzg.hr; Tel.: +385-1-238-3333

Abstract: Predictive maintenance is one of the most important topics within the Industry 4.0 paradigm. We present a prototype decision support system (DSS) that collects and processes data from many sensors and uses machine learning and artificial intelligence algorithms to report deviations from the optimal process in a timely manner and correct them to the correct parameters directly or indirectly through operator intervention or self-correction. We propose to develop the DSS using open-source R packages because using open-source software such as R for predictive maintenance is beneficial for small and medium enterprises (SMEs) as it provides an affordable, adaptable, flexible, and tunable solution. We validate the DSS through a case study to show its application to SMEs that need to maintain industrial equipment in real time by leveraging IoT technologies and predictive maintenance of industrial cooling systems. The dataset used was simulated based on the information on the indicators measured as well as their ranges collected by in-depth interviews. The results show that the software provides predictions and actionable insights using collaborative filtering. Feedback is collected from SMEs in the manufacturing sector as potential system users. Positive feedback emphasized the advantages of employing open-source predictive maintenance tools, such as R, for SMEs, including cost savings, increased accuracy, community assistance, and program customization. However, SMEs have overwhelmingly voiced comments and concerns regarding the use of open-source R in their infrastructure development and daily operations.

Keywords: Industry 4.0; predictive maintenance; open source; R; machine learning; simulated data; recommender; case study

1. Introduction

Maintenance in general is critical to achieving operational priorities and strategies [1], especially within the context of smart factories, as indicated by Jerman et al. [2]. Mobley [3] points out that maintenance expenditures represent a large portion of the operating costs of a manufacturing or processing facility. Depending on the industry, maintenance costs can represent between 15% and 60% of the cost of goods produced.

A good maintenance mechanism helps reduce equipment and machinery downtime, improve efficiency, and increase productivity [1]. Since maintenance management is dynamic and complex and maintenance performance is difficult to evaluate and enumerate, choosing the right maintenance strategy is crucial [1], especially for the fulfillment of sustainability goals within the context of Industry 4.0, as indicated by Roblek et al. [4]. However, Moore and Starr [5] emphasize that inadequate maintenance leads to further unplanned plant and equipment failures, which are associated with a number of costs to the company, including labor, spare parts, rework, scrap, late order charges, and lost orders due to dissatisfied customers.

In this paper, we focus on SMEs. Implementing predictive maintenance can be difficult for SMEs due to the high cost of proprietary software and hardware solutions. However, by using open-source tools, SMEs can create predictive maintenance applications at a fraction of the cost of proprietary solutions. One of these is R. R is a robust programming language that is freely available for statistical computing and data analysis. It provides access to various statistical and machine learning methods required for predictive maintenance. R has a large community of users and developers who collaborate to further develop the programming language and the many packages it offers, making it easy to find support and other useful resources.

The paper consists of the following parts: After the introduction, in the second chapter, we outline the state of the art of predictive maintenance, provide an overview of the field, and discuss the benefits of predictive maintenance and its role in Industry 4.0. In the third chapter, we focus on IoT devices and sensors as a pillar of predictive maintenance. The fourth chapter presents the prototype of a DSS for predictive maintenance. It explains the architecture and purpose of the application, the machine learning engine, and the graphical user interface, and concludes with an example of how to use the platform. The last chapter contains a concluding remark.

2. State of the Art of Predictive Maintenance

2.1. *Predictive Maintenance Overview*

Although predictive maintenance has been around since 1940 [6], it has recently gained popularity and is being used by many companies [7,8]. Ahmad and Kamaruddin [9], (p. 140) even refer to predictive maintenance as "the most advanced and popular maintenance technique discussed in the literature". Many industries have adopted predictive maintenance, including those where reliability is critical, such as infrastructure, emergency services, transportation networks, power plants, and communication systems. The popularity of predictive maintenance is reflected in the Market Research Future Report forecast [10], which projects that the global predictive maintenance market will grow at a compound annual growth rate of 25.5% during the forecast period from 2019 to 2024, reaching 23 billion by 2025.

Turgis et al. [11] emphasize that predictive maintenance is often considered a subclass of preventive maintenance, which is defined by Chen and Trivedi [12] as an action performed on a system at predetermined intervals to minimize or eliminate cumulative degradation while maintaining satisfactory operation. Predictive maintenance as a maintenance strategy is driven by predictive analytics technologies whose solutions are installed to track and identify system faults or failures [10]. However, such solutions are activated only when a critical failure is possible. This contributes to the efficient use of available resources and capital, the maximization of system or equipment uptime, the improvement of quality and supply chain systems, and the overall satisfaction of all stakeholders [10].

It is important to note that some authors, such as Parpala and Iacob [13], understand condition-based maintenance as a precursor to predictive maintenance. In contrast, others, such as Ahmad and Kamaruddin [9], as mentioned earlier, equate these two terms. According to Parpala and Iacob [13], condition-based maintenance was a well-known method of industrial maintenance before the widely adopted IoT concept. In contrast, predictive maintenance includes the help of the IoT when it comes to maintenance. Jardine et al. [14] define condition-based maintenance as a maintenance program that provides maintenance recommendations and decisions based on the information obtained through condition monitoring and works in three steps, namely

- Data acquisition,
- Data processing, and
- Maintenance decision making.

It is important to note that some authors, such as Parpala and Iacob [13], understand condition-based maintenance as a precursor to predictive maintenance. In contrast, others, such as Ahmad and Kamaruddin [9], as mentioned earlier in this report, equate these two terms. According to Parpala and Iacob [13], condition-based maintenance was a well-known method of industrial maintenance before the IoT concept was widely adopted. In contrast, predictive maintenance includes the help of the IoT when it comes to maintenance. Jardine et al. [14] define condition-based maintenance as a maintenance program that provides maintenance recommendations and decisions based on the information obtained through condition monitoring and operates in three steps.

The predictive maintenance techniques market can be divided into traditional and advanced techniques, with advanced techniques additionally divided into two segments [10]:

- IoT and Big Data techniques and
- Machine learning-based techniques.

The Market Research Future Report [10] also shows that the traditional techniques segment held the largest market share in 2018. Nevertheless, the advanced techniques segment is expected to grow. In this context, Chuang et al. [15] state that most maintenance approaches rely on traditional routine maintenance to keep a system running smoothly. However, the cost of replacing components and oils during maintenance cannot be ignored. Therefore, more and more companies are turning to predictive maintenance, in which the system makes decisions individually and schedules the appropriate time for maintenance according to the requirements [15]. Chuang et al. [15] also point out that Big Data analytics and machine learning can help determine when assets are most likely to fail and when it is the right time to repair and replace components before they fail. IoT platforms also greatly support predictive maintenance as they can combine data from different machines and production processes [13].

Christiansen [7] argues that predictive maintenance will increase as technology costs decrease, mainly due to the shift from wired to wireless sensors.

The main component of a predictive maintenance strategy is the monitoring of machinery and equipment. This involves the use of traditional and advanced monitoring methods that enable the scheduling of machine maintenance before a malfunction occurs [10]. These monitoring methods include the use of various monitoring or testing tools such as infrared thermography, vibration monitoring, oil analysis, electrical isolation, ultrasonic leak detection, and temperature monitoring [7]. Based on the calculations provided by predictive maintenance and the monitoring and testing results with the appropriate tools and instruments, the company can run pre-programmed predictive algorithms to determine when a piece of equipment might fail so that repairs can be initiated only when a failure occurs [7]. In addition, predictive maintenance data provide diagnostic and prognostic information that indicates what is wrong, where the problem is, why the problem is occurring, whether it is a malfunction or just a failure, and when the failure will occur, if ever [6]. Similarly, Ahmad and Kamaruddin [9] argue that the main purpose of predictive maintenance is to review and assess the requirements and condition of devices, systems, and machines in real time to make maintenance decisions, resulting in less excessive maintenance and associated costs. Therefore, condition monitoring is the heart of predictive maintenance that leads to decision making [7]. In Figure 1, a graphical representation of the general process of predictive maintenance is proposed, showing that a control loop exists between the two elements of predictive maintenance.

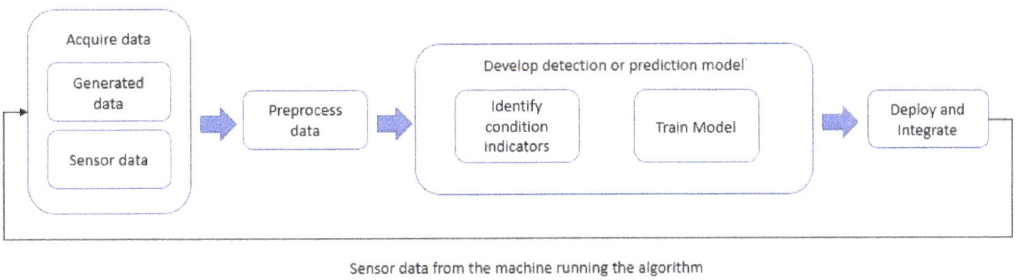

Figure 1. The process of applying predictive algorithms.

The Market Research Future Report [10] also shows that the traditional techniques segment held the largest market share in 2018. Nevertheless, the advanced techniques segment is expected to grow. In this context, Chuang et al. [15] state that most maintenance approaches rely on traditional routine maintenance to keep a system running smoothly. However, the cost of replacing components and oils during maintenance cannot be ignored. Therefore, more and more companies are turning to predictive maintenance, in which the system makes decisions individually and schedules the appropriate time for maintenance according to the requirements [15,16]. Big Data analytics and machine learning can help determine when assets are most likely to fail and when it is the right time to repair and replace components before they fail [15,17]. IoT platforms also greatly support predictive maintenance by:

1. Condition monitoring technology and predictive maintenance techniques;
2. Internet of Things (IoT) technology;
3. Predictive algorithms.

Condition monitoring and predictive maintenance technology refers to installed condition monitoring sensors that transmit real-time performance and system health data [7]. In this sense, Ahmad and Kamaruddin [9] explain that state monitoring is a tool used to specify the state of devices in a system with a dual purpose:

1. Collection of condition data on the equipment;
2. Increase in knowledge about failure reasons and causes;
3. About the effects and deterioration patterns of equipment.

Condition monitoring equipment and devices can be mounted or retrofitted in a variety of ways so that they can measure [9,18]: electrical currents, vibration, temperature, pressure, oil, noise, and corrosion levels.

Predictive maintenance is based on the data analyzed for condition monitoring, from which various patterns emerge that indicate that a machine or asset may be beginning to malfunction or deteriorate [18]. In addition, the analysis provided by condition monitoring allows the company to plan maintenance activities to avoid failures and prevent emergency breakdowns [18,19]. As argued in Hodge et al. [20], condition monitoring also helps the enterprise reduce the need for human inspection through automated monitoring, reduce repair costs by detecting defects early before they escalate, and increase protection and safety as well as efficiency and reliability. In addition, Christiansen [18] argues that another benefit of condition monitoring sensors is that they can provide an accurate picture of what is going on inside the plant without compromising efficiency, i.e., the plant does not have to be interrupted and disassembled to perform physical inspections.

Ahmad and Kamaruddin [9] argue that state monitoring can be performed in two ways:

- Online—refers to the monitoring process performed during the operational state of the machine, system, or equipment, i.e., during its running state.
- Offline—refers to the monitoring process performed during the phase when the machine, system, or equipment is not in operation.

Ahmad and Kamaruddin [9] also point out that condition monitoring with respect to time can be conducted in the following two ways:

- Periodic—refers to the monitoring process conducted at specified intervals, such as every 30 min, every hour, at the end of a work shift, etc., and conducted using portable indicators such as vibration pens, acoustic emission devices, handheld meters, etc.
- Continuous—refers to the monitoring process that is performed automatically and continuously based on specialized measuring devices, such as vibration and acoustic sensors.

There are two main limitations to continuous monitoring: (i) the high costs associated with continuous monitoring, which arise because many specialized machines are needed;

and (ii) the possibility of obtaining inaccurate information because the continuous flow of data leads to increased noise [14].

Based on the sensors used for condition monitoring and related testing, there are a variety of condition monitoring, i.e., predictive maintenance, methods and techniques that can be used, such as the following [6,7,9]:

1. vibration monitoring and dynamic analysis,
2. oil analysis and lubricant monitoring,
3. sound, ultrasonic, and acoustic monitoring,
4. motor circuit analysis,
5. different variations of thermography and thermal analysis,
6. electromagnetic measurements,
7. radiography and radiation analysis,
8. laser interferometry, and
9. different performance measurements.

Internet of Things (IoT) technology enables communication between machines and computers, software solutions, and cloud technologies, thus enabling the collection and analysis of large amounts of data [21–23]. IoT technology, in combination with the aforementioned sensors, helps collect and share data that predictive maintenance relies heavily on to connect assets to a central system that stores incoming information and from which assets (i.e., machine systems and devices) can communicate, perform data analysis, collaborate, recommend remedial actions, or take direct action, depending on how the systems are set up [7,21–23]. Christiansen [7] notes that such centralized systems are operated using cloud technology, WLAN, or LAN-based connectivity.

Predictive algorithms refer to the most important part of predictive maintenance: building predictive algorithms where all processed data are fed into predictive data models that then provide failure predictions, meaning that the model must take into account a large number of variables and how they interact and affect each other, with the goal of predicting system failures [7,24–26]. The more variables included in the models, the more reliable they are. Therefore, building predictive models is an iterative process [7,27,28]. Initial models would need to be based on equipment history in a Computerized Maintenance Management System or file cabinets, personal impressions, failure analysis, already usable internal sensors such as flow meters and accelerometers, and other related sources. In addition, it may be necessary to initially mount condition monitoring sensors and run them for a period of time to collect baseline data and complete initial predictive models [7].

The final product is a fully automated system, as shown in Figure 1, which:

1. monitors working conditions using installed sensors,
2. detects and predicts patterns characterized by data anomalies,
3. generates warnings when deviations from established thresholds occur.

With this in mind, the predictive maintenance workflow described above and explained in [7] has been illustrated in Figure 2.

2.2. Predictive Maintenance Advantages and Disadvantages

There are many different types of maintenance, i.e., methods of maintenance management, such as run-to-failure management [1,3,6,9], also called corrective maintenance, reactive maintenance, unplanned maintenance, failure-based maintenance, failure-controlled maintenance, fail-to-fix maintenance, or failure maintenance, which usually takes place in companies and is the most expensive method of maintenance management because it means that there is no maintenance approach in the company but actions are taken to restore the failed equipment. This is a reactive and unplanned occurrence of equipment failure, meaning the company does not spend a budget on maintenance until a system, machine, or device fails.

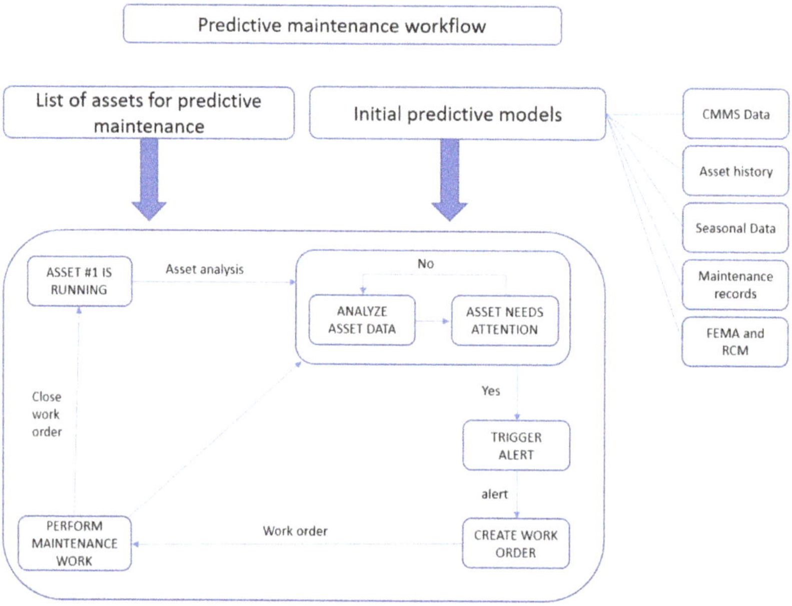

Figure 2. Predictive maintenance workflow.

Predictive maintenance has many benefits. For example, its implementation and application help companies identify product failures and customer dissatisfaction in advance so that they can be proactive and thus achieve significant cost savings, reduce product downtime, and improve customer service [29]. Moreover, predictive maintenance improves productivity, product quality, and the overall effectiveness of production and manufacturing organizations [3]. On the other hand, Christiansen [7] points out that the main advantage of predictive maintenance is the ability to plan and perform work based on the current condition of the plant and equipment.

Selcuk [6] summarizes the benefits of predictive maintenance in the following list:

- improved safety for workers and the environment,
- increased availability,
- increased reliability,
- reduced cost of parts and labor,
- improved product quality,
- reduced waste of raw materials and consumables such as lubricants,
- energy savings from quieter machines (e.g., alignment, which in some cases is reported as 3–5%; balancing, 1–2%).

In addition, Selcuk [6] also provide data on average industrial savings from predictive maintenance, as shown in Table 1.

Table 1. Industrial average savings. Adapted from [6].

Savings Area	Savings
Return on investment (ROI)	10 times
Reduction in maintenance costs	25–30%
Elimination of breakdowns	70–75%
Reduction in downtime	35–45%
Increase in production	20–25%

Christiansen [7] also provides a list of benefits of predictive maintenance, as follows:

- Minimizes the occurrence of unplanned downtime and maximizes the uptime of machines, equipment, and systems;
- Provides the company with a real-time view of the current condition of its plant, machinery, equipment, and systems;
- Ensures minimal disruption to productivity by allowing some predictive maintenance activities to be performed on running assets;
- Optimizes the amount of time the company spends on maintenance activities;
- Optimizes the use of spare parts;
- Improves equipment reliability.

The implementation of a predictive maintenance program plays an important role in the benefits that an organization can derive from predictive maintenance technologies [3]. If the predictive maintenance program is limited to avoiding catastrophic failures in specific plant systems, then this is the outcome that will result from such an implementation, but it also indicates that focusing solely on avoiding failures could lead to a significant increase in maintenance costs for the company [3]. In this vein, Mobley [3] argues that the benefits of predictive maintenance are virtually unlimited when properly applied and cautions that these benefits could be significantly limited if the scope of the program is artificially limited by the scope of work or if the business enforces certain constraints.

However, Christiansen [7] states that the main disadvantage of predictive maintenance is the complexity of its use and application, caused primarily by the high upfront costs and time required for implementation, the requirements for condition monitoring equipment, sensors, and software to be implemented and operated, and the expertise required for someone to understand and analyze the data collected by condition monitoring.

2.3. *Predictive Maintenance in Industry 4.0*

Considering the increasing importance of technology and the rapid development of Industry 4.0, as well as the main objective of asset management, which is to transform available data into valuable information, Demoly and Kiritsis [30] believe that asset management requires technology in two main areas:

- Data collection, storage, and analysis can be accomplished using powerful data servers and software; and
- Presenting and communicating the results of these analyses to decision-makers inside and outside the organization.

Using an IoT platform to collect usage data from various production equipment is the first step in developing a general cloud-based predictive maintenance system that facilitates factory maintenance [13]. In this sense, Noura et al. [31] argue that in recent years, many smart devices in the real world have become interconnected and communicate over established Internet networks, resulting in a massive network infrastructure known as the Internet of Things (IoT).

However, Parpala and Iacob [13] also emphasize that connectivity architecture is the key drawback in integrating and combining production systems with IoT platforms since most industrial communication protocols are not compatible with the current communication protocols introduced on IoT platforms. In this sense, they propose a modern and convenient approach for real-time control and maintenance of industrial machines, focusing on two features of networked manufacturing: (i) process monitoring that ensures consistent quality assurance, and (ii) condition monitoring that prevents unplanned downtime [13].

Gregersen [32] also states that while the current Internet framework is open to and can be used by all IoT devices, it is too bulky and power-hungry for most IoT applications, resulting in the HyperText Transfer Protocol (HTTP) being considered an inappropriate protocol for IoT. Moreover, IoT implementation requires advances in networking technologies necessary to realize the vision of reaching physical objects and connecting them to the Internet [33].

According to Ferretti and Schiavone [34], there is a very wide range of industrial IoT applications, with transportation and logistics being the central areas of IoT applications. Predictive maintenance is usually introduced in a company for one of the following reasons [3]:

1. As a maintenance management tool—when its use is limited to preventing spontaneous downtime and/or catastrophic failures;
2. As a business optimization tool—when its use relates to establishing best production practices and procedures for all critical production systems within an organization;
3. As a reliability improvement tool—when its use is to quantify even the smallest deviations from standard operating parameters. This enables company personnel (e.g., reliability engineers, repair planners) to prepare and plan for minor changes to prevent machine and equipment failures, thus avoiding extensive rebuilds and associated downtime.

Ahmad and Kamaruddin [9] point out that applications of predictive maintenance cover many problems in different fields, such as structures, industrial machines, and medical devices. On the other hand, Parpala and Iacob [13] argue that IoT technology was first applied to consumer devices such as washing machines, air conditioners, and smart home control systems that could connect directly to the Internet through smartphones or private networks. However, as the IoT has evolved, other products have been equipped with the necessary technology. Many companies are trying to integrate the IoT into existing production processes [13].

3. IoT Devices and Sensors as a Pillar of Predictive Maintenance

3.1. IoT Concept

Vermesan et al. [33] provide the simplest definition of IoT, defining it as a symbiotic interaction between the real/physical and digital/virtual worlds. The Internet of Things enables people and things to connect to anything and everything at any time and in any place, ideally over any path/network, or service. This requires that elements such as convergence, content, collections (repositories), computing, communication, and connectivity are addressed in a context where there is a seamless connection between people and things and/or between things and things, as graphically depicted in Figure 3.

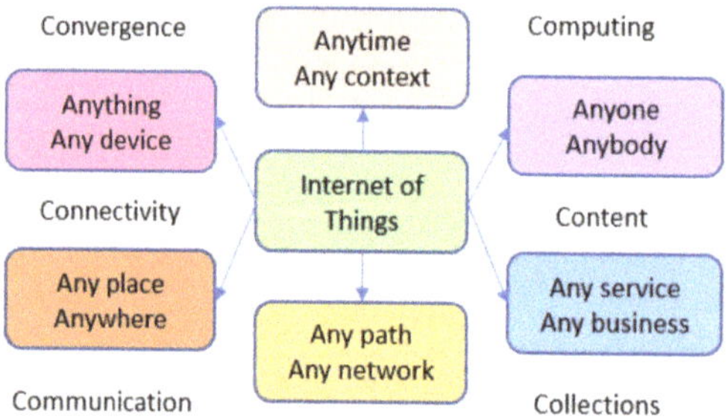

Figure 3. The IoT concept.

The Internet of Things (IoT) is a term that describes a modern environment where virtually all machines and devices used by humans are connected to a network. They could work together to accomplish difficult tasks that require high levels of intelligence. To achieve such a level of intelligence and interconnectivity, IoT systems contain integrated sensors, actuators, processors, and transceivers [35].

Noura et al. [31] argue that the IoT refers to a global network infrastructure of interconnected devices that communicate over the existing Internet infrastructure. However, interoperability issues arise when each solution has IoT networks, applications, APIs, and data formats. They also note that interoperability issues arise for several critical reasons, such as vendor lock-in, the inability to develop IoT applications that expose cross-platform and/or cross-domain data, and the complexity of integrating non-interoperable IoT devices into separate IoT systems, which ultimately hinders the widespread adoption of IoT technology [31]. In this sense, Vermesan et al. [33] point out that protection, scalability, and cross-platform connectivity between different networked systems are critical in the IoT. In this context, the network infrastructure must have cost-effective solutions that allow virtually anyone to connect to the network. This ubiquitous connectivity can change the way information is handled.

Moreover, Noura et al. [31] emphasize that academia and practitioners have highlighted the value of the IoT interoperability problem. In this sense, practitioners are trying to solve IoT interoperability problems through standardization [31]. Moreover, Noura et al. [31] explain that there are different perspectives on IoT interoperability (Figure 4), such as:

- device interoperability
- network interoperability
- syntactic interoperability
- semantic interoperability
- platform interoperability

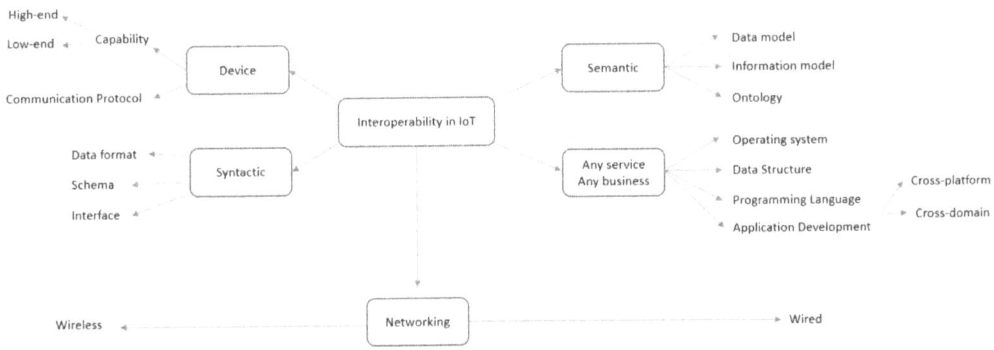

Figure 4. The IoT taxonomy and its different perspectives.

There are many approaches and technologies to improve the state of IoT interoperability [31]:

- Adapters and gateways—they take care of interoperability by developing a method known as a mediator to increase interoperability between IoT devices. Among other things, they aim to establish a connection between different specifications, data, standards, and middleware;
- Virtual networks or overlay-based solutions—the main idea is to build a virtual network on top of a physical network that can communicate with other types of devices, such as sensor nodes. The main goal is to seamlessly connect sensors and actuators, as well as other smart IP objects, to the Internet to enable end-to-end communication that is possible within each virtual network using different protocols;
- Network technologies—include various network technologies and protocols such as IP-based approaches, software-defined networking (SDN), network functions virtualization, and Fog computing;
- Open API—refers to an interface provided by service providers that exposes functions or data to an application written in a high-level language;
- Service-Oriented Architecture (SOA)—interaction with the end operations of various wireless devices is divided into different service components, and application layer software can access resources provided by the devices as services;
- Semantic Web Technologies—refers to the Semantic Web of Things (swot) paradigm, which is proposed to integrate the semantic web with the Web of Things (wot), with the further goal of achieving a common understanding of the various entities that make up the IoT;

- Open standard—nowadays numerous standardization bodies, consortia, and alliances are trying to find a solution to the IoT standard problems, such as IPSO, OIC, seen Alliance, etc.

Protocols for IoT data exchange are communication and security protocols that serve as forms of communication and ensure that data exchanged between connected devices is protected as much as possible [36]. Gregersen [32] emphasizes that the IoT infrastructure stack is incomplete without IoT protocols, as IoT hardware would be considered useless. Gregersen [32] further explains how IoT protocols enable hardware to communicate with each other so that end users can derive valuable information from this exchanged data.

Typically, IoT devices are connected to the Internet via an Internet Protocol (IP) network. However, they can also be connected locally via Bluetooth and radio frequency identification (RFID), which leads to differences in the range, power, and storage of such IoT devices [36]. Uppalapati [36] points out that while range is not a problem, connections over IP networks are more difficult and require more memory and power from IoT devices. In contrast, non-IP networks require less power and memory but have a shorter range. In this sense, there are two main categories of IoT protocols and standards [32,36]:

1. IoT network protocols—used to connect devices over a network, usually the Internet;
2. IoT data protocols—used to connect to low-power IoT devices by allowing users to communicate with hardware over a cellular or wired network without requiring an Internet connection.

All IoT technology requires the handling of immense data volumes to realize its full potential [37,38]. To deal with this, increasingly complex systems and frameworks are being constructed, and new approaches are constantly being developed to tackle this open problem [37,39–41]. The key issue is not the quantity but the quality of the data used for model input, i.e., how to successfully clean and preprocess data from ever more abundant data sources [42–46]. For example, Alharam et al. [47] discuss one facet of IoT complexity in the healthcare industry, and Pappas et al. [48] in the tourist accommodation industry.

3.2. Analysis and Identification of IoT Devices and Sensors Available in the Market

According to Gartner [49], IoT is one of "the "most hyped technologies that could transform the way businesses operate"." Therefore, this chapter analyzes and identifies the IoT devices and sensors available on the market and defines the protocols for IoT data exchange.

Nowadays, many IoT devices and sensors are available in the global market. According to the Arm Glossary [50], examples of IoT devices are sensors, actuators, hardware, equipment, gadgets, or devices designed for specific purposes that can transmit data over the Internet or other networks and can be integrated into a variety of products, including cell phones, industrial machines, medical devices, home appliances, environmental sensors, and more.

Sharma [51] provides a list of the 15 most common types of IoT sensors used in various industries, namely:

1. temperature sensors,
2. proximity sensors,
3. pressure sensors,
4. water quality sensors,
5. chemical sensors,
6. gas sensors,
7. smoke sensors,
8. infrared sensors,
9. level sensors,
10. image sensors,
11. motion detection sensors,
12. accelerometers,

13. gyroscope sensors,
14. humidity sensors,
15. optical sensors.

Temperature sensors measure the amount of thermal energy and convert the data for a device or consumer to detect a physical temperature difference from a specific source [51]. Proximity sensors refer to devices that can sense the absence or presence of a surrounding object or its characteristics and convert the information into a signal that can be easily read by a consumer or simple electronic instrument without coming into contact with it [51]. Pressure sensors sense pressure and convert it into an electrical signal, with the amount of pressure determining the amount applied. A more detailed overview of sensors can be found in [51].

4. Prototyping a Decision Support System (DSS) for Predictive Maintenance

4.1. Case Study Description

The DSS was developed for the case study as a proof of concept for predictive maintenance of industrial air conditioning systems. These systems operate 24/7 throughout the year and collect various types of data that we can use thanks to IoT and sensors. The SME has several storage facilities, which are connected in a network using innovative IoT technologies. The following information is collected for the monitoring and maintenance of these facilities: location, device, energy, temperature, humidity, region, age, and cost. The study presents an improved predictive maintenance DSS for Industry 4.0. The proposed DSS analyzes data from multiple sensors. It uses machine learning and artificial intelligence algorithms to report deviations from the ideal process and correct them to the best settings, directly or indirectly through operator involvement or self-collection. The program also predicts certain variables over time, such as energy consumption.

4.2. Architecture of the Application

The architecture of the application was developed following the predictive maintenance framework [7]. Figure 5 shows the DSS architecture. Data is collected from different devices at different locations. The DSS is fed by the stream of IoT data that is analyzed using the machine learning application for predictive maintenance. This consists of the IoT data database, the data preparation module, the machine learning for predicting IoT data, and the predictive maintenance dashboard.

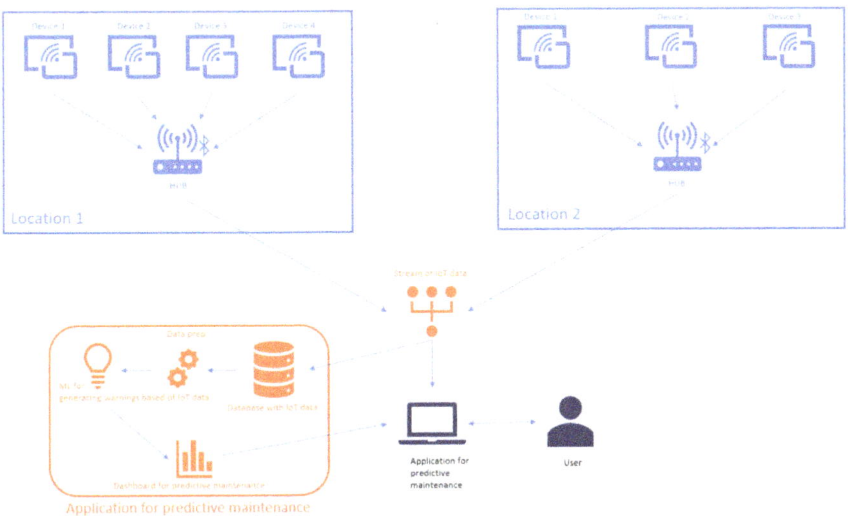

Figure 5. DSS architecture; Source: Authors' work.

The application was developed using the open-source software R [52]. The packages and libraries used in the development are:

- Tidyverse—The tidyverse is a collection of R packages designed for data science. All packages share a common design philosophy, grammar, and data structures [53,54].
- Shiny—Shiny is an open-source R package that provides an elegant and powerful framework for building web applications with R [55].
- shinyWidgets—the family of pre-built widgets in the Shiny package, each created with a transparently named R function.
- Shinycustomloader—a custom file for loading the screen in the R Shiny package.
- bs4Dash—the R package for developing modern dashboards in R Shiny.
- echarts4r—a package for interactive charts.
- echarts4r.maps—a dataset with the latitude and longitude of all cities used for interactive charts.
- Reactable—creates a table from tabular data with default sorting and pagination. The data table is an HTML widget that can be used in R Markdown documents and Shiny applications or displayed via an R console.
- Fresh—used for Custom 'Bootstrap" themes in Shiny.

4.3. *Machine Learning Deployment*

Machine learning has been used for three purposes:

- Dataset creation—The small company for which the DSS was developed chose to keep its data anonymous. Therefore, we created an illustrative dataset.
- Prediction of data—The DSS predicts the observed data for the selected time period in the future.
- Warnings for devices—The DSS generates warnings for the devices in the future based on the observed data in the past and the predicted data.

4.3.1. Dataset Generation

To complete our case study, we simulated the dataset to be collected by networking storage facilities based on innovative IoT technologies. The dataset used was simulated based on the information on the indicators measured as well as their ranges collected by the in-depth interviews, as is common in the field [56,57]. For data simulation and prediction, the following R function was used: rnorm function—for generating the simulated dataset using the normal distribution for a given integer sequence (depending on the metric) for a synthetic site/device string. A normal distribution has been selected since the synthetic dataset has been used for demonstration purposes. Such an approach has been often used in machine learning research, as indicated by Bolón-Canedo et al. [58]. More specifically, normal Gaussian distributions have been used by Camacho [59] and Panagiotakis et al. [60] for generating synthetic datasets for recommender systems. Simulated datasets that mimic real datasets well can either augment or entirely remove the need for training on real sensor data [61–65].

The complete list of variables included in the dataset, as well as their format and modalities as nominal variables, can be found in Table 2.

Table 2. Variables—Formats and Modality.

Variable	Description of the Variable	Format	Modality of (Nominal) Variables
Location	Location of the device	Numeric	
Device	Device number id	Numeric	
Energy	Energy consumption of the device	Numeric	
Temperature	The temperature of the device	Numeric	
Humidity	The humidity of the device	Numeric	

Table 2. *Cont.*

Variable	Description of the Variable	Format	Modality of (Nominal) Variables
Region	Region of the device	Nominal	South, Central, West, East
Age	Device age	Numeric	
Cost	The average cost of maintenance for the device	Numeric	Home Office, Corporate, Consumer...
Warnings	Detected warnings on the device	Numeric	

The simulated dataset was created using the normal distribution for a given integer sequence (depending on the metric) for a synthetic site/device chain. The dataset includes nine variables, including device location, device ID, energy, temperature, warnings, etc.

4.3.2. Forecasting Data

Second, the aggregate daily averages of the data are used to make predictions using the Prophet algorithm implemented in R, a method for predicting time series data based on an implemented additive model. This model fits non-linear trends with annual, weekly, and daily seasonality and vacation influences. It works best when applied to time series with significant seasonal effects and historical data from many seasons. The Prophet algorithm is resilient to missing data and fluctuations in trends, and it usually copes well with outliers.

The program uses the historical simulated data from the last 100 days to predict the next 100 days. The data were forecasted using the Prophet algorithm. Prophet is a forecasting technique that can be implemented in R and Python. It is fast and provides automated predictions, although data scientists and analysts can still post-process the predictions by hand if they wish. Prophet is a technique for predicting time series data based on an additive model and involves fitting nonlinear trends with annual, monthly, and daily seasonality and vacation influences. It works most effectively for time series with significant seasonal influences and historical data that spans many seasons. The Prophet algorithm is resilient to missing data and fluctuations in trends, and it usually handles outliers well. Prophet has been used for forecasting in many fields, such as energy [66], sales [67], and the stock market [68].

4.3.3. Warning for Devices

To generate warnings about devices that need urgent maintenance, we apply the following Recommenderlab algorithm: The package is used to apply the Recommender algorithm, more precisely the collaborative filtering algorithm, which generates predictions for each device based on observed patterns and predicted data.

The data must be sufficiently prepared to apply the collaborative filtering algorithm in R using the "Recommender" function from the "recommenderlab" package. This package provides a research platform for developing and evaluating collaborative filtering and recommender algorithms [69]. This includes a sparse representation of user-element matrices, many well-known methods, top-N recommendations, and cross-validation.

First, we transform the dataset into a matrix useful for machine learning using the Devic X metric function. We performed cleanup and preparation steps and transformed the matrix into a binarized matrix for faster computation. The matrix was split into a training set and a test set. Using the training set and the test set, we generate the recommendation algorithm for device alerts using item-based collaborative filtering [70]. The recommendation algorithm is trained to generate warnings for the future based on the warnings in the past. In this illustrative example, energy, temperature, and humidity were input variables while the warning was issued. The portion of the training data that was used as the basis for the prediction is shown in Table 3.

Table 3. Training data.

no	ds	Energy	Temperature	Humidity	Warnings
<int>	<date>	<int>	<int>	<int>	<dbl>
1	11 February 2022	32.5	13	67.6	0
2	12 February 2022	16.9	10.4	70.2	0
3	13 February 2022	14.3	7.8	78	0
4	14 February 2022	13	7.8	67.6	0
5	15 February 2022	24.7	5.2	76.7	0
6	16 February 2022	32.5	13	58.5	0
7	17 February 2022	36.4	7.8	67.6	1
8	18 February 2022	16.9	2.6	59.8	0
9	19 February 2022	36.4	6.5	63.7	0
10	20 February 2022	26	6.5	65	0

Note: Data types are presented in the second row of the table.

Recommendations are generated for each device. The recommendations are displayed in the graphical user interface on the Analysis Table.

4.4. Graphical User Interface

The purpose of the graphical user interface is to provide insight into device behavior based on past data and discover patterns that improve device maintenance by predicting the future state of the devices.

The graphical user interface for the Predictive Maintenance application is shown below (Figure 6).

Figure 6. Application Overview opened on the Real-time tab.

5. Use Case of the DSS Platform Usage

The following use case demonstrates the validation of the platform and data collection functions through the various sensors and operational actions.

An example of how to use the platform is presented in the following steps, which illustrate the functionalities of the application:

1. Step 1: Location selection
2. Step 2: Checking devices based on energy consumption
3. Step 3: Checking devices based on temperature
4. Step 4: Overview of key performance indicators
5. Step 5: Inspecting devices by age and cost

6. Step 6: Inspecting devices by age and cost
7. Step 7: Forecasting data
8. Step 8: User feedback

5.1. Step 1: Location Selection

First, the user selects the location to be monitored. Once a location is selected, all charts for that specific location are filtered. We obtain the three most important Key Performance Indicators (KPIs) for that location: number of devices, energy consumption (kWh—expected), and alerts (today), as shown above in Figure 7. You can filter the entire dashboard using a synthetic site group. Depending on the selection, different devices and readings will be displayed to simulate the real situation.

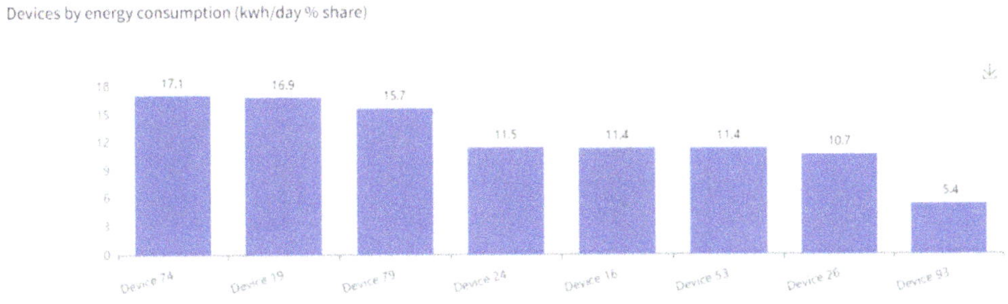

Figure 7. Real-time energy consumption of the devices for the selected location.

There are five active devices at this site with a current expected energy consumption of 5.93 kWh and one warning (Figure 8).

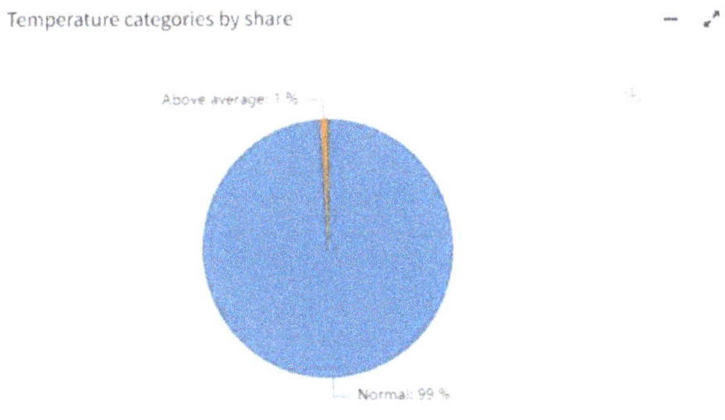

Figure 8. Temperature categories by share of daily occurrences.

5.2. Step 2: Checking Devices Based on Energy Consumption

Below the KPIs is a graph showing the energy consumption of the devices, sorted by percentage of total energy consumption for the selected site. The real-time energy consumption for units 29, 46, 42, 36, and 22 at the time of writing is shown in Figure 7. When you hover over a particular device, a pop-up window appears with the device name and the percentage of total energy consumption for the selected location. In this particular example, the real-time energy consumption at the time of writing was:

- Device 29—29.4 kWh/day % share

- Device 46—20.9 kWh/day % share
- Device 42—20.7 kWh/day % share
- Device 36—19.6 kWh/day % share
- Device 22—9.5 kWh/day % share

5.3. Step 3: Checking Devices Based on Temperature

On the right, under Temperature Categories by Proportion of Daily Occurrence, is a doughnut chart showing the detection of anomalies in temperature and all warnings in a day, as shown in Figure 8, which represents the warnings for Site 8.

5.4. Step 4: Overview of Key Performance Indicators

The final table is the key performance indicator table, which lists the devices with three main graphs for each metric: energy consumption, temperature, and humidity. The final table is shown in Figure 9.

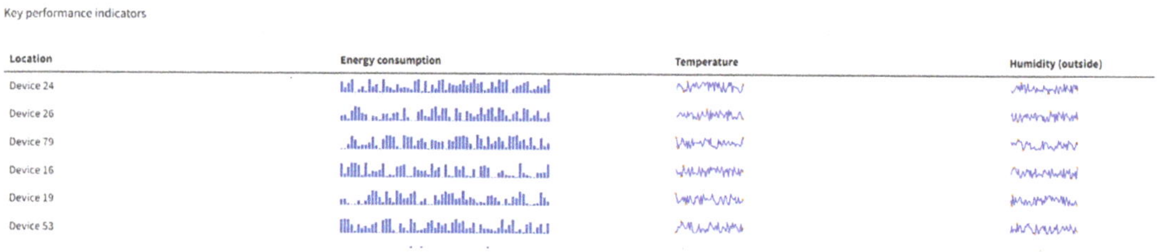

Figure 9. Key performance indicators data.

5.5. Step 5: Inspecting Devices by Age and Cost

The Devices by Cost and Age chart allows us to look at the correlation between age and cost for each device to see possible patterns, as shown in Figure 10. In this example, no interesting patterns are observed, which is to be expected since simulated data were used. Hovering over a specific point on the graph will display the exact data for that device.

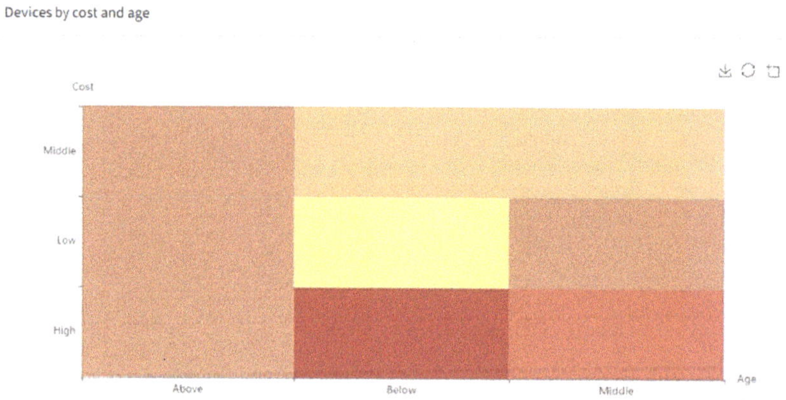

Figure 10. Devices by cost and age; UI is in eye-saving night mode.

5.6. Step 6: Inspecting Devices by Regions

Energy consumption by region is a doughnut chart that shows us the energy consumption of the different regions in the dataset. In this case, these are the west, central, south, and east regions. The exact energy consumption data (% share) for each region in this example is as follows:

- West—45 (30.82%)
- Central—45 (30.82%)
- South—34 (23.29%)
- East—21 (14.38%)

The Energy consumption by region doughnut chart can be viewed in Figure 11.

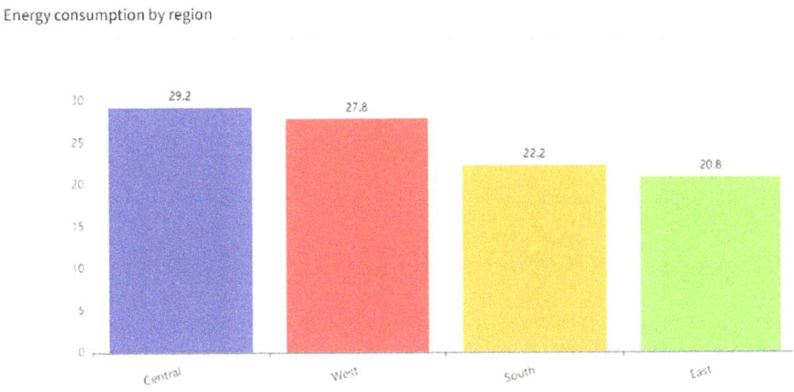

Figure 11. Energy consumption by region.

5.7. Step 7: Forecasting Data

To demonstrate the use of the DSS for scenario planning, we employ the prediction algorithm, the Prophet algorithm [71]. The advantage of this approach is that it avoids the opaque nature of other common machine learning algorithms [72]. The data to be predicted are the aggregate average data for each day. The algorithm uses historical simulated data from the last 100 days to forecast 100 days into the future.

Once 100 days of data are generated for each metric, the data are sent to the ML function, which receives the data, cleans and processes it, and trains it with the Prophet algorithm to predict 100 days into the future.

In this example, the user has selected the energy forecast for site 10. As we can see in Figure 12, a slight increase in energy consumption is expected over the next 100 days.

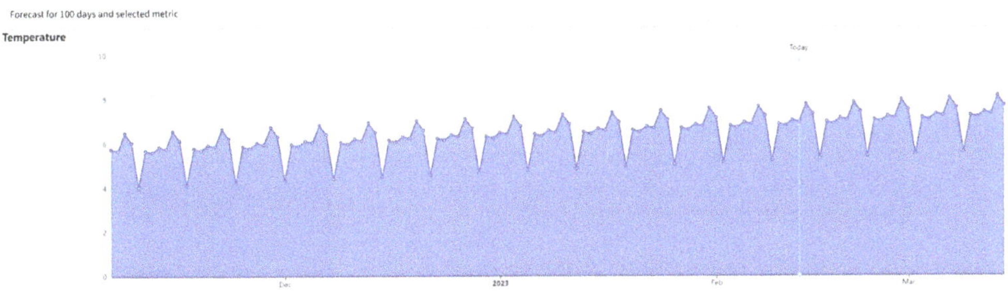

Figure 12. Forecast for 100 days for the selected metric (Energy).

5.8. Step 8: Generating Warnings for Devices

Finally, the recommendation algorithm is used to generate the device warnings. Future alerts are generated based on the trained algorithm and predicted data. Devices by number of alerts is a bar chart with multiple alerts for each device to identify potential outliers and problems. The bar chart can be seen in Figure 13. In this example, we have four devices with one warning.

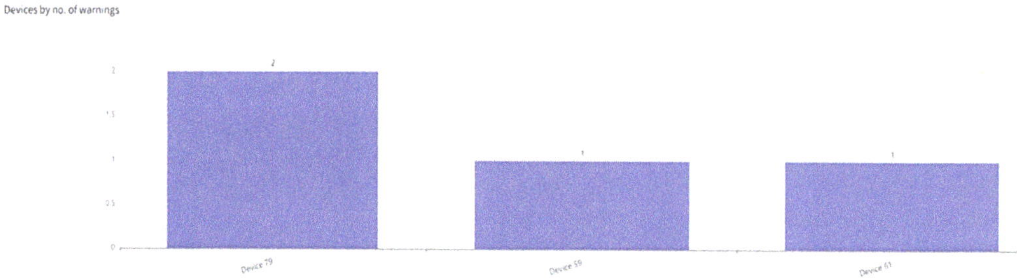

Figure 13. Devices by number of warnings. The y-axis shows the number of warnings, and the x-axis is the device's name.

5.9. Step 8: User Feedback

The application was tested with four potential users: SMEs in the manufacturing sector. The respondents were in the positions of Chief Technology Officer and Chief Information Officer or involved in monitoring air conditioning and heating systems.

The positive comments highlighted that using open-source predictive maintenance software, such as R, can be beneficial to SMEs, including cost savings, greater accuracy, community support, and the ability to customize the program. The following positive comments were made:

- "... looking for a cost-effective and efficient solution, and R was able to deliver just that ... helped us save a significant amount of money compared to proprietary alternatives."
- "...... a wide range of statistical and machine learning methods, we are now able to make more accurate predictions about equipment failure and significantly reduce downtime ......"
- "...The open-source community, of which R is a part, is an extremely useful resource..."
- "... the integration of the predictive maintenance solution into our existing infrastructure, which allows us to make it a seamless part of our business processes ......"

Negative comments referred to the following issues users may encounter when using open-source software such as R:

- "... People who have no experience in programming or statistics may find the learning curve to be quite high..."
- "... is constantly evolving and changing, which can be both beneficial and annoying... the solution became more difficult to maintain and keep up to date ......"

In summary, SMEs overwhelmingly expressed comments and concerns about the use of open-source R in their infrastructure development and day-to-day operations. These conclusions are consistent with the findings of McCredie et al. [73], who emphasized that flexible support, licensing cost savings, community support, and functionality are the positive drivers and complexity is the negative driver for SME adoption of open source software within the theory-based framework.

6. Open-Source Predictive Maintenance Playbook for SMEs

We suggest that emerging SMBs adopt open-source predictive maintenance solutions using the following steps:

1. Define the problem—Determine which devices or systems need predictive maintenance and clearly define the maintenance goals.
2. Data collection—Identify the necessary data sources for the system and set up a data collection infrastructure to collect real-time data.

3. Data pre-processing—Remove outliers, missing numbers, and discrepancies from the data and normalize or scale the data to ensure consistency and comparability between features. Develop new features.
4. Flagging and anomaly detection—Define criteria for flagging errors and maintenance events. Label historical data to indicate normal operation or specific failure scenarios. Identify aberrant data points.
5. Model selection—Evaluate different machine learning algorithms suitable for predictive maintenance, considering tradeoffs between accuracy, interpretability, scalability, and real-time capabilities.
6. Model Training—Split the labeled dataset into two parts: Training and Validation. Train the selected machine learning model on the labeled data, modifying hyperparameters and the model architecture as necessary.
7. Model evaluation—Select metrics to evaluate the model, such as F1 score, accuracy, precision, recall, or similar, and evaluate the model.
8. Integration and deployment—Integrate the trained model into the Industry 4.0 ecosystem by creating an inference API to obtain real-time predictions.
9. Monitoring and maintenance—Continuously monitor the performance of the predictive maintenance system, including data quality, model accuracy, and false positive/negative rates. Retrain models as needed using updated data.
10. Continuous Improvement—Analyze feedback from the installed predictive maintenance system, including maintenance activities performed and their consequences. Incorporate findings into future iterations of the system to increase accuracy, efficiency, and effectiveness.

7. Conclusions

The paper demonstrates the DSS we developed and the results of a predictive maintenance case study using a simulated IoT climate dataset. The DSS allows users to make these predictions without advanced knowledge of the underlying machine learning algorithms via its easy-to-use graphical user interface. The paper discusses technologies relevant to the horizontal approach to predictive maintenance. The authors consider the application of the methodology used in this research relevant, especially for SMEs. We have presented the use of the DSS step by step so that it is relevant to other industrial applications due to the benefits of an appropriate predictive maintenance strategy.

This work has important implications for business management, particularly in relation to the adoption of predictive maintenance using the open-source software R in SMEs, which provides advanced features that can be used for predictive maintenance in the context of Industry 4.0. Users concluded that R has an intuitive user interface and is easily customizable, allowing the predictive maintenance solution to be tailored to their individual needs. In addition, SMEs can further streamline their operations by integrating the application with their existing systems and tools, such as their enterprise resource planning (ERP) system. Negative impacts of open-source software, such as a lack of support, should also be considered and remedial actions planned.

The study presented here has some limitations because the hypothetical case was developed to test the DSS prototype and used a limited number of machine algorithms and a use case based on only one measurement device. These limitations call for future research directions. First, real datasets should be collected from multiple facilities in the field so that multiple algorithms can be tested. Such real-world cases would also allow further testing of the theoretically-based framework for explaining the adoption of open-source software. Second, future research should apply multiple machine learning algorithms to real-world datasets to recommend which algorithms would be best suited for predictive maintenance. Carvalho et al. [74] suggest the following algorithms as best suited for predictive maintenance: Random Forest [75], Artificial Neural Networks [63,64,74,76–78], Support Vector Machines [79,80], and k-means [81], as well as others such as Decision Trees [82].

Therefore, in future efforts to test DSS for predictive maintenance, these algorithms should be tested with both simulated and real datasets and then compared to the approach proposed by Nardo [83,84] for their effectiveness. Third, since the use case was developed only for the measurement devices, future use cases should focus on the upstream functions relevant to Industry 4.0, as indicated by Irsa et al. [85], e.g., self-optimizing processes, productivity improvements, the development of novel services, and additive manufacturing.

Author Contributions: Conceptualization, M.P.B. and A.T.; methodology, M.P.B. and Ž.K.; software, Ž.K.; validation, A.T., M.P.B. and Ž.K.; formal analysis, A.T.; investigation, Ž.K.; resources, A.T. and A.I.; data curation, Ž.K. and A.I.; writing—original draft preparation, A.I., Ž.K., and M.P.B.; writing—review and editing, A.I. and M.P.B.; visualization, A.I. and Ž.K.; supervision, A.T.; project administration, A.T. All authors have read and agreed to the published version of the manuscript.

Funding: This research received no external funding.

Conflicts of Interest: The authors declare no conflict of interest.

References

1. Shyjith, K.; Ilangkumaran, M.; Kumanan, S. Multi-criteria decision-making approach to evaluate optimum maintenance strategy in textile industry. *J. Qual. Maint. Eng.* **2008**, *14*, 375–386. [CrossRef]
2. Jerman, A.; Bertoncelj, A.; Erenda, I. The influence of critical factors on business model at a smart factory: A case study. *Bus. Syst. Res. Int. J. Soc. Adv. Innov. Res. Econ.* **2019**, *10*, 42–52.
3. Mobley, R.K. *An Introduction to Predictive Maintenance*; Elsevier: Amsterdam, The Netherlands, 2002.
4. Roblek, V.; Thorpe, O.; Bach, M.P.; Jerman, A.; Meško, M. The fourth industrial revolution and the sustainability practices: A comparative automated content analysis approach of theory and practice. *Sustainability* **2020**, *12*, 8497. [CrossRef]
5. Moore, W.; Starr, A. An intelligent maintenance system for continuous cost-based prioritisation of maintenance activities. *Comput. Ind.* **2006**, *57*, 595–606. [CrossRef]
6. Selcuk, S. Predictive maintenance, its implementation and latest trends. *Proc. Inst. Mech. Eng. Pt. B J. Eng. Manuf.* **2017**, *231*, 1670–1679. [CrossRef]
7. Christiansen, B. A Complete Guide to Predictive Maintenance. Limble CMMS. Available online: https://limblecmms.com/predictive-maintenance/ (accessed on 21 March 2023).
8. Nunes, P.; Santos, J.; Rocha, E. Challenges in predictive maintenance—A review. *CIRP J. Manuf. Sci. Technol.* **2023**, *40*, 53–67. [CrossRef]
9. Ahmad, R.; Kamaruddin, S. An overview of time-based and condition-based maintenance in industrial application. *Comput. Ind. Eng.* **2012**, *63*, 135–149. [CrossRef]
10. Market Research Future. Global Predictive Maintenance (PdM) Market Research Report. 2020. Available online: https://www.marketresearchfuture.com/reports/predictive-maintenance-market-2377 (accessed on 21 March 2023).
11. Turgis, F.; Auder, P.; Coutadeur, Q.; Verdun, C. Industrialization of condition based maintenance for complex systems in a complex maintenance environment, example of NAT. In Proceedings of the 12th World Congress on Railway Research, Tokyo, Japan, 28 October 2019.
12. Chen, D.; Trivedi, K.S. Closed-form analytical results for condition-based maintenance. *Reliab. Eng. Syst. Saf.* **2002**, *76*, 43–51. [CrossRef]
13. Parpala, R.C.; Iacob, R. Application of IoT concept on predictive maintenance of industrial equipment. In Proceedings of the MATEC Web of Conferences, Bucharest, Rumania, 9 August 2017; p. 02008.
14. Jardine, A.K.; Lin, D.; Banjevic, D. A review on machinery diagnostics and prognostics implementing condition-based maintenance. *Mech. Syst. Signal Process.* **2006**, *20*, 1483–1510. [CrossRef]
15. Chuang, S.-Y.; Sahoo, N.; Lin, H.-W.; Chang, Y.-H. Predictive maintenance with sensor data analytics on a Raspberry Pi-based experimental platform. *Sensors* **2019**, *19*, 3884. [CrossRef]
16. Pech, M.; Vrchota, J.; Bednář, J. Predictive maintenance and intelligent sensors in smart factory. *Sensors* **2021**, *21*, 1470. [CrossRef]
17. Kanawaday, A.; Sane, A. Machine learning for predictive maintenance of industrial machines using IoT sensor data. In Proceedings of the 2017 8th IEEE international conference on software engineering and service science (ICSESS), Beijing, China, 24–26 November 2017; pp. 87–90.
18. Christiansen, B. Complete List of Condition Monitoring Techniques. Machinery and Equipment MRO. Available online: https://www.mromagazine.com/features/complete-list-of-condition-monitoring-techniques (accessed on 21 March 2023).
19. Davies, A. *Handbook of Condition Monitoring: Techniques and Methodology*; Springer Science & Business Media: Berlin/Heidelberg, Germany, 2012.
20. Hodge, V.J.; O'Keefe, S.; Weeks, M.; Moulds, A. Wireless sensor networks for condition monitoring in the railway industry: A survey. *IEEE Trans. Intell. Transp. Syst.* **2014**, *16*, 1088–1106. [CrossRef]

21. Mukhopadhyay, S.C.; Suryadevara, N.K. *Internet of things: Challenges and Opportunities*; Springer: Berlin/Heidelberg, Germany, 2014; pp. 2347–4718.
22. Khalil, N.; Abid, M.R.; Benhaddou, D.; Gerndt, M. Wireless sensors networks for Internet of Things. In Proceedings of the 2014 IEEE ninth international conference on Intelligent sensors, sensor networks and information processing (ISSNIP), Singapore, Singapore, 21–24 April 2014; pp. 1–6.
23. Kocakulak, M.; Butun, I. An overview of Wireless Sensor Networks towards internet of things. In Proceedings of the 2017 IEEE 7th annual computing and communication workshop and conference (CCWC), Las Vegas, NV, USA, 9–11 January 2017; pp. 1–6.
24. Civerchia, F.; Bocchino, S.; Salvadori, C.; Rossi, E.; Maggiani, L.; Petracca, M. Industrial Internet of Things monitoring solution for advanced predictive maintenance applications. *J. Ind. Inf. Integr.* **2017**, *7*, 4–12. [CrossRef]
25. Dong, L.; Mingyue, R.; Guoying, M. Application of internet of things technology on predictive maintenance system of coal equipment. *Procedia Eng.* **2017**, *174*, 885–889. [CrossRef]
26. Passlick, J.; Dreyer, S.; Olivotti, D.; Grützner, L.; Eilers, D.; Breitner, M.H. Predictive maintenance as an internet of things enabled business model: A taxonomy. *Electron. Mark.* **2021**, *31*, 67–87. [CrossRef]
27. Chehri, A.; Jeon, G. The industrial internet of things: Examining how the IIoT will improve the predictive maintenance. In Proceedings of the Innovation in Medicine and Healthcare Systems, and Multimedia: Proceedings of KES-InMed-19 and KES-IIMSS-19 Conferences, Berlin/Heidelberg, Germany, 6 June 2019; pp. 517–527.
28. Lade, P.; Ghosh, R.; Srinivasan, S. Manufacturing analytics and industrial internet of things. *IEEE Intell. Syst.* **2017**, *32*, 74–79. [CrossRef]
29. Grambau, J.; Hitzges, A.; Otto, B. Predictive Maintenance in the Context of Service. In Proceedings of the 20th International Conference on Enterprise Information Systems (ICEIS 2018), Dortmund, Germany, 21–24 March 2018; pp. 223–230.
30. Demoly, F.; Kiritsis, D. Asset optimization and predictive maintenance in discrete manufacturing industry. *IFAC Proc. Vol.* **2012**, *45*, 1–6. [CrossRef]
31. Noura, M.; Atiquzzaman, M.; Gaedke, M. Interoperability in internet of things: Taxonomies and open challenges. *Mob. Netw. Appl.* **2019**, *24*, 796–809. [CrossRef]
32. Gregersen, C. A Complete Guide to IoT Protocols & Standards in 2021. Available online: https://www.nabto.com/guide-iot-protocols-standards/ (accessed on 21 March 2023).
33. Vermesan, O.; Friess, P.; Guillemin, P.; Gusmeroli, S.; Sundmaeker, H.; Bassi, A.; Jubert, I.S.; Mazura, M.; Harrison, M.; Eisenhauer, M. Internet of things strategic research roadmap. In *Internet of Things-Global Technological and Societal Trends from Smart Environments and Spaces to Green ICT*; River Publishers: Aalborg, Denmark, 2022; pp. 9–52.
34. Ferretti, M.; Schiavone, F. Internet of Things and business processes redesign in seaports: The case of Hamburg. *Bus. Process Manag. J.* **2016**, *22*, 271–284. [CrossRef]
35. Sethi, P.; Sarangi, S.R. Internet of things: Architectures, protocols, and applications. *J. Electr. Comput. Eng.* **2017**, *2017*, 9324035.
36. Uppalapati, K. How IoT Protocols and Standards Support Secure Data Exchange in the IoT Ecosystem? Available online: https://www.kelltontech.com/kellton-tech-blog/internet-of-things-protocols-standards (accessed on 21 March 2023).
37. Alkhalil, A.; Ramadan, R.A. IoT data provenance implementation challenges. *Procedia Comput. Sci.* **2017**, *109*, 1134–1139. [CrossRef]
38. Udoh, I.S.; Kotonya, G. Developing IoT applications: Challenges and frameworks. *IET Cyber-Phys. Syst. Theory Appl.* **2018**, *3*, 65–72. [CrossRef]
39. Cheruvu, S.; Kumar, A.; Smith, N.; Wheeler, D.M.; Cheruvu, S.; Kumar, A.; Smith, N.; Wheeler, D.M. *IoT frameworks and complexity. Demystifying Internet of Things Security: Successful IoT Device/Edge and Platform Security Deployment*; Springer: Berlin/Heidelberg, Germany, 2020; pp. 23–148.
40. Gil, D.; Johnsson, M.; Mora, H.; Szymański, J. Review of the complexity of managing big data of the internet of things. *Complexity* **2019**, 4592902. [CrossRef]
41. Zhang, K.; Han, D.; Feng, H. Research on the complexity in Internet of Things. In Proceedings of the 2010 International Conference on Advanced Intelligence and Awarenss Internet (AIAI 2010), Beijing, China, 23–25 October 2010; pp. 395–398.
42. Song, S.; Zhang, A. IoT data quality. In Proceedings of the 29th ACM International Conference on Information & Knowledge Management, Virtual Event Ireland, 19 October 2020; pp. 3517–3518.
43. Ding, X.; Wang, H.; Li, G.; Li, H.; Li, Y.; Liu, Y. IoT data cleaning techniques: A survey. *In Intelligent and Converged Networks* **2022**, *3*, 325–339. [CrossRef]
44. Jane, V. Survey on iot data preprocessing. *Turk. J. Comput. Math. Educ.* **2021**, *12*, 238–244.
45. Chu, X.; Ilyas, I.F.; Krishnan, S.; Wang, J. Data cleaning: Overview and emerging challenges. In Proceedings of the 2016 international conference on management of data, San Francisco, CA, USA, 26 June–1 July 2016; pp. 2201–2206.
46. Khare, S.; Totaro, M. Big data in IoT. In Proceedings of the 2019 10th International Conference on Computing, Communication and Networking Technologies (ICCCNT), Kanpur, India, 6–8 July 2019; pp. 1–7.
47. Alharam, A.K.; Elmedany, W. Complexity of cyber security architecture for IoT healthcare industry: A comparative study. In Proceedings of the 2017 5th international conference on future internet of things and cloud workshops (FiCloudW), Prague, Czech Republic, 21–23 August 2017; pp. 246–250.
48. Pappas, N.; Caputo, A.; Pellegrini, M.M.; Marzi, G.; Michopoulou, E. The complexity of decision-making processes and IoT adoption in accommodation SMEs. *J. Bus. Res.* **2021**, *131*, 573–583. [CrossRef]

49. Gartner. Internet of Things: Unlocking True Digital Business Potential. Available online: https://www.gartner.com/en/information-technology/insights/internet-of-things (accessed on 21 March 2023).
50. Arm. Arm Glossary. IoT devices. Available online: https://www.arm.com/glossary/iot-devices (accessed on 21 March 2023).
51. Sharma, R. Top 15 Sensor Types Being Used Most by IoT Application Development Companies. Available online: https://www.finoit.com/blog/top-15-sensor-types-used-iot/ (accessed on 21 March 2023).
52. Ihaka, R.; Gentleman, R. R: A language for data analysis and graphics. *J. Comput. Graph. Stat.* **1996**, *5*, 299–314.
53. Wickham, H.; Averick, M.; Bryan, J.; Chang, W.; McGowan, L.D.A.; François, R.; Grolemund, G.; Hayes, A.; Henry, L.; Hester, J. Welcome to the Tidyverse. *J. Open Source Softw.* **2019**, *4*, 1686. [CrossRef]
54. Wickham, H.; Wickham, M.H. Package Tidyverse, Easily Install and Load the 'Tidyverse, Version 1.2.0. Available online: https://tidyverse.tidyverse.org/ (accessed on 21 March 2023).
55. Chang, W.; Cheng, J.; Allaire, J.; Xie, Y.; McPherson, J. Package Shiny, Version 1.7.4.1. Available online: https://cran.r-project.org/web/packages/shiny/index.html (accessed on 21 March 2023).
56. Anderson, J.W.; Kennedy, K.E.; Ngo, L.B.; Luckow, A.; Apon, A.W. Synthetic data generation for the internet of things. In Proceedings of the 2014 IEEE International Conference on Big Data (Big Data), Washington, DC, USA, 27–30 October 2014; pp. 171–176.
57. Kannan, S. Synthetic time series data generation for edge analytics. *F1000Research* **2022**, *11*, 67. [CrossRef]
58. Bolón-Canedo, V.; Sánchez-Maroño, N.; Alonso-Betanzos, A. A review of feature selection methods on synthetic data. *Knowl. Inf. Syst.* **2013**, *34*, 483–519. [CrossRef]
59. Camacho, V.T. Synthetic dataset generation methodology for Recommender Systems using statistical sampling methods, a Multinomial Logit model, and a Fuzzy Inference System. *arXiv* **2022**, arXiv:14350.2212.
60. Panagiotakis, C.; Papadakis, H.; Fragopoulou, P. Unsupervised and supervised methods for the detection of hurriedly created profiles in recommender systems. *Int. J. Mach. Learn. Cybern.* **2020**, *11*, 2165–2179. [CrossRef]
61. La Russa, F.M.; Santagati, C. An AI-based DSS for preventive conservation of museum collections in historic buildings. *J. Archaeol. Sci. Rep.* **2021**, *35*, 102735. [CrossRef]
62. Li, D.-C.; Lin, L.-S.; Peng, L.-J. Improving learning accuracy by using synthetic samples for small datasets with non-linear attribute dependency. *Decis. Support Syst.* **2014**, *59*, 286–295. [CrossRef]
63. Agarwal, P.; Gao, B.; Huo, S.; Reddy, P.; Dechu, S.; Obeidi, Y.; Muthusamy, V.; Isahagian, V.; Carbajales, S. A Process-Aware Decision Support System for Business Processes. In Proceedings of the 28th ACM SIGKDD Conference on Knowledge Discovery and Data Mining, Washington, DC, USA, 14 August 2022; pp. 2673–2681.
64. Jabbari, M.; Sheikh, S.; Rabiee, M.; Oztekin, A. A collaborative decision support system for multi-criteria automatic clustering. *Decis. Support Syst.* **2022**, *153*, 113671. [CrossRef]
65. Piri, S.; Delen, D.; Liu, T. A synthetic informative minority over-sampling (SIMO) algorithm leveraging support vector machine to enhance learning from imbalanced datasets. *Decis. Support Syst.* **2018**, *106*, 15–29. [CrossRef]
66. Forootan, M.M.; Larki, I.; Zahedi, R.; Ahmadi, A. Machine learning and deep learning in energy systems: A review. *Sustainability* **2022**, *14*, 4832. [CrossRef]
67. Jha, B.K.; Pande, S. Time series forecasting model for supermarket sales using FB-prophet. In Proceedings of the 2021 5th International Conference on Computing Methodologies and Communication (ICCMC), Erode, India, 8–10 April 2021; pp. 547–554.
68. Saiktishna, C.; Sumanth, N.S.V.; Rao, M.M.S.; Thangakumar, J. Historical Analysis and Time Series Forecasting of Stock Market using FB Prophet. In Proceedings of the 2022 6th International Conference on Intelligent Computing and Control Systems (ICICCS), Madurai, India, 25–27 May 2022; pp. 1846–1851.
69. Schafer, J.B.; Frankowski, D.; Herlocker, J.; Sen, S. Collaborative filtering recommender systems. In *The Adaptive Web: Methods and Strategies of Web Personalization*; Springer: Berlin/Heidelberg, Germany, 2007; pp. 291–324.
70. Deshpande, M.; Karypis, G. Item-based top-n recommendation algorithms. *ACM Trans. Inf. Syst.* **2004**, *22*, 143–177. [CrossRef]
71. Gong, F.; Han, N.; Li, D.; Tian, S. Trend analysis of building power consumption based on prophet algorithm. In Proceedings of the 2020 Asia Energy and Electrical Engineering Symposium (AEEES), Chengdu, China, 29–31 May 2020; pp. 1002–1006.
72. Pearl, J. The limitations of opaque learning machines. *Possible Minds* **2019**, *25*, 13–19.
73. D Macredie, R.; Mijinyawa, K. A theory-grounded framework of Open Source Software adoption in SMEs. *Eur. J. Inf. Syst.* **2011**, *20*, 237–250. [CrossRef]
74. Carvalho, T.P.; Soares, F.A.; Vita, R.; Francisco, R.d.P.; Basto, J.P.; Alcalá, S.G. A systematic literature review of machine learning methods applied to predictive maintenance. *Comput. Ind. Eng.* **2019**, *137*, 106024. [CrossRef]
75. Paolanti, M.; Romeo, L.; Felicetti, A.; Mancini, A.; Frontoni, E.; Loncarski, J. Machine Learning approach for Predictive Maintenance in Industry 4.0. In Proceedings of the 2018 14th IEEE/ASME International Conference on Mechatronic and Embedded Systems and Applications (MESA), Oulu, Finland, 2–4 July 2018; pp. 1–6.
76. Wu, S.-j.; Gebraeel, N.; Lawley, M.A.; Yih, Y. A neural network integrated decision support system for condition-based optimal predictive maintenance policy. *IEEE Trans. Syst. Man Cybern. -Part A Syst. Hum.* **2007**, *37*, 226–236. [CrossRef]
77. Kellner, D.; Lowin, M.; von Zahn, M.; Chen, J. Towards designing a user-centric decision support system for predictive maintenance in SMEs. *INFORMATIK 2021* **2021**, 1255–1260. [CrossRef]

78. Carnero, M.C. Selection of diagnostic techniques and instrumentation in a predictive maintenance program. A case study. *Decis. Support Syst.* **2005**, *38*, 539–555. [CrossRef]
79. Susto, G.A.; Schirru, A.; Pampuri, S.; Pagano, D.; McLoone, S.; Beghi, A. A predictive maintenance system for integral type faults based on support vector machines: An application to ion implantation. In Proceedings of the 2013 IEEE International Conference on Automation Science and Engineering (CASE), Madison, WI, USA, 17–20 August 2013; pp. 195–200.
80. Chaudhuri, A. Predictive maintenance for industrial iot of vehicle fleets using hierarchical modified fuzzy support vector machine. *arXiv* **2018**, arXiv:1806.09612.
81. Yoo, J.-H.; Park, Y.-K.; Han, S.-S. Predictive maintenance system for wafer transport robot using k-means algorithm and neural network model. *Electronics* **2022**, *11*, 1324. [CrossRef]
82. Arena, S.; Florian, E.; Zennaro, I.; Orrù, P.F.; Sgarbossa, F. A novel decision support system for managing predictive maintenance strategies based on machine learning approaches. *Saf. Sci.* **2022**, *146*, 105529. [CrossRef]
83. Di Nardo, M.; Murino, T.; Osteria, G.; Santillo, L.C. A New Hybrid Dynamic FMECA with Decision-Making Methodology: A Case Study in An Agri-Food Company. *Appl. Syst. Innov.* **2022**, *5*, 45. [CrossRef]
84. Nardo, M.; Converso, G.; Castagna, F.; Murino, T. Development and implementation of an algorithm for preventive machine maintenance. *Eng. Solid Mech.* **2021**, *9*, 347–362. [CrossRef]
85. Irsa, W.; Dalaqmeh, N. Impact of Additive Manufacturing on the Strategic Alignment of Business Processes in the Logistics Industry in Europe. *ENTRENOVA-ENTerprise REsearch InNOVAtion* **2022**, *8*, 188–199.

designs

Article

Evaluating the Ranking of Performance Variables in Flexible Manufacturing System through the Best-Worst Method

Anthony Bagherian [1,*], Gulshan Chauhan [2], Arun Lal Srivastav [3] and Rajiv Kumar Sharma [4]

1 International School of Management, 17, Boulevard Raspail, 75007 Paris, France
2 Geeta University, Panipat 132145, India; pvc@geetauniversity.edu.in
3 Chitkara University School of Engineering and Technology, Chitkara University Himachal Pradesh, Chandigarh 175103, India; arun.srivastav@chitkarauniversity.edu.in
4 National Institute of Technology, Hamirpur 177005, India; rksfme@nith.ac.in
* Correspondence: anthony6@gmx.de

Abstract: Flexible Manufacturing Systems (FMSs) provide a competitive edge in the ever-evolving manufacturing landscape, offering the agility to swiftly adapt to changing customer demands and product lifecycles. Nevertheless, the complex and interconnected nature of FMSs presents a distinct challenge: the evaluation and prioritization of performance variables. This study clarifies a conspicuous research gap by introducing a pioneering approach to evaluating and ranking FMS performance variables. The Best-Worst Method (BWM), a multicriteria decision-making (MCDM) approach, is employed to tackle this challenge. Notably, the BWM excels at resolving intricate issues with limited pairwise comparisons, making it an innovative tool in this context. To implement the BWM, a comprehensive survey of FMS experts from the German manufacturing industry was conducted. The survey, which contained 34 key performance variables identified through an exhaustive literature review and bibliometric analysis, invited experts to assess the variables by comparing the best and worst in terms of their significance to overall FMS performance. The outcomes of the BWM analysis not only offer insights into the factors affecting FMS performance but, more importantly, convey a nuanced ranking of these factors. The findings reveal a distinct hierarchy: the "Quality (Q)" factor emerges as the most critical, followed by "Productivity (P)" and "Flexibility (F)". In terms of contributions, this study pioneers a novel and comprehensive approach to evaluating and ranking FMS performance variables. It bridges an evident research gap and contributes to the existing literature by offering practical insights that can guide manufacturing companies in identifying and prioritizing the most crucial performance variables for enhancing their FMS competitiveness. Our research acknowledges the potential introduction of biases through expert opinion, delineating the need for further exploration and comparative analyses in diverse industrial contexts. The outcomes of this study bear the potential for cross-industry applicability, laying the groundwork for future investigations in the domain of performance evaluation in manufacturing systems.

Keywords: Flexible Manufacturing System; FMS; performance variables; quality; productivity; flexibility; BWM

1. Introduction

Flexible Manufacturing Systems (FMSs) have recently emerged as a focal point in the manufacturing domain, primarily due to their exceptional adaptability to dynamic production demands and their capacity to boost efficiency and productivity levels significantly. FMSs seamlessly amalgamate an array of machinery, equipment, and computer-controlled systems to automate manufacturing processes, enabling swift reconfiguration for producing diverse product ranges. The ultimate significance of assessing the efficacy of FMSs is emphasized by the potential they hold for optimizing system efficiency, pinpointing areas for refinement, and guiding well-informed decisions concerning system design and operation.

Research in this domain, as noted in [1], defines the flexibility of a manufacturing system as its "capacity to efficiently and effectively adapt to changes in the product mix, volume, or timing of activities" [2]. Furthermore, ref. [3] characterizes FMSs as a manufacturing approach that "employs programmed machines, computer systems, and/or robotics for processing and assembling raw parts" [4]. In the context of modern digital manufacturing, FMSs assume an imperative role, vividly portrayed in Figure 1, by bolstering productivity, elevating quality standards, and boosting responsiveness to changes while simultaneously curbing time, effort, and operational costs, even in the face of proliferating product variations. It is essential to recognize that, as emphasized by [5], the quest to achieve flexibility in conjunction with productivity and quality stands as a substantial challenge confronting numerous manufacturers [6]. However, it is noteworthy that the flexibility of an FMS hinges on a multitude of factors, including its components, capabilities, interconnections, and mode of operation and control [2].

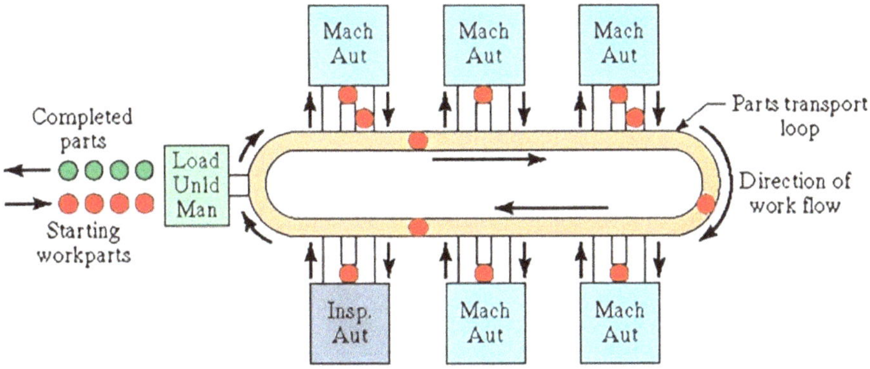

Figure 1. FMS loop layout [7].

In recent years, there has been a surge in studies aimed at investigating the performance variables within FMSs and formulating suitable methodologies for evaluating their effectiveness. Researchers have attempted to adopt diverse methodologies, ranging from mathematical modeling and simulation to statistical assessment, to gauge the efficiency of FMSs through key performance indicators (KPIs) such as productivity, throughput, flexibility, quality, and cost-effectiveness. A recent study by [6] has identified three paramount parameters closely associated with FMS performance: productivity, flexibility, and quality. These performance factors within FMSs are delineated in Figure 2.

Researchers have employed a variety of methodologies to evaluate FMS performance, including conventional approaches like the Analytical Hierarchy Process (AHP) [8] or fuzzy logic [9], as well as contemporary methodologies such as data envelopment analysis [10] and machine learning algorithms [11]. In this study, our goal is to make a significant contribution to the current body of knowledge related to FMS performance evaluation by introducing an innovative approach based on the Best-Worst Method (BWM). The BWM, which has gained popularity in recent years, excels at systematically and intuitively capturing the comparative ranking of various criteria used in decision-making processes.

Through the implementation of the BWM, we aim to provide a comprehensive and reliable assessment of performance variables in FMS, enabling a deeper understanding and optimization of FMS performance in modern manufacturing environments. Furthermore, our approach incorporates insights gained from an extensive literature review and consultation with industry professionals and experts. As a result, we have identified a total of 34 variables that significantly influence FMS performance.

The subsequent sections of this article are organized as follows: Section 2 offers a comprehensive overview of the literature related to FMS performance evaluation, including a discussion of relevant studies that have employed various methodologies to assess FMS performance variables. Section 3 outlines the methodology and provides the specifics of the proposed BWM-based approach for evaluating FMS performance. Section 4 presents the outcomes and analyzes the performance evaluation using our proposed approach. Finally, in Section 5, we draw conclusions from the implications of the results and outline future research directions in the field of FMS performance evaluation.

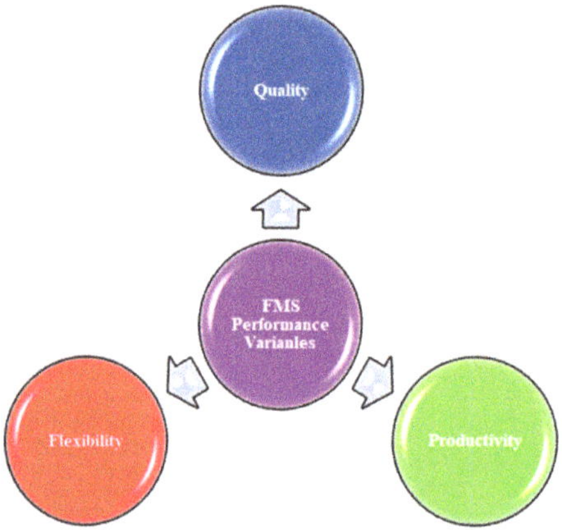

Figure 2. FMS performance variables.

2. Review of Literature

This literature review introduces an overview of relevant research conducted on the performance variables that influence the outcomes of FMSs. The authors conducted a comprehensive investigation of the literature, reviewing a total of 272 scientific publications. While the 'Dimensions' database spanning the years 2013 to 2023 conveys a comprehensive selection of recent research, it is essential to note that a substantial portion of the literature review incorporates seminal works published before 2000. It is imperative to include older publications in comprehending the essential concepts and theoretical development in the domain of FMSs. It allows us to trace the historical evolution of FMS research and recognize the enduring principles that continue to shape current investigations. The scrutinizing of the literature comprises three primary sections. The first part summarizes prior research on the factors that affect FMS performance. The second section discusses the research approach of MCDM in relation to FMSs. Finally, the third part identifies gaps in the current literature related to FMS performance.

By acknowledging the historical development of FMS research, we gain valuable insights into the origins and evolution of the field, conveying a comprehensive perspective that combines both essential principles and contemporary findings. This holistic approach enriches our interpretation of the complexity of FMS performance.

2.1. Literature Review on FMS Performance, Accompanied by a Bibliometric Overview

The literature offers various definitions of flexibility in the context of manufacturing. Ref. [12] presents flexibility principles, while [13] proposes additional types, including material handling flexibility, program flexibility, and industry flexibility. Ref. [14] identified four additional dimensions of flexibility: automation flexibility, labor flexibility, modern design flexibility, and distribution flexibility. Numerous studies have explored diverse performance indicators and research methods to determine the most influential factors in FMSs. According to [15], earlier research on FMSs focused solely on investigating the systems' performance from a single perspective. For example, certain research has focused on the productivity dimension; nevertheless, other studies have investigated time flow as a single metric or dimension.

Ref. [4] classified 15 performance variables into three groups—quality, productivity, and flexibility—in order to analyze FMSs. Ref. [1] defined FMS capacity as the ability to accommodate changes in product mix, volume, or timing of activities. Ref. [2] further emphasizes that FMS adaptability is contingent on its components, capabilities, interconnections, and mode of operation and control. Additionally, ref. [5] emphasizes the significant challenge manufacturers face in achieving flexibility alongside productivity and quality.

Researchers have utilized FMS in conjunction with various frameworks and methods to model productivity variables and evaluate FMSs. Ref. [15] employed Data Envelope Analysis (DEA) for FMS evaluation. Ref. [16] quantified the advantages of FMS implementation by employing the Multiple Attribute Decision Making (MADM) framework, specifically the Analytic Hierarchy Process (AHP). Ref. [17] utilized the AHP strategy to assess advanced technologies. Ref. [18] employed MADM frameworks such as MOORA (Multi-Objective Optimization by Ratio Analysis) and PSI (Preferential Similarity Index) to rank FMS performance variables. Refs. [19,20] applied FMSs to assess machine workload balance. Ref. [18] utilized FMSs for modeling parameters influencing flexibility. Ref. [14] applied FMSs as a case study for evaluating performance parameters in a printed circuit board manufacturing plant. Ref. [19] utilized FMSs to assess loading and routing influences. Refs. [20–23] disclose formulations and methods for resolving system-loading issues through the FMS framework.

Ref. [24] conducted an evaluation and prioritization of the Industry 4.0 challenges pertaining to Indian automotive industry utilizing the BWM approach.

The literature also identifies common and uncommon indicators affecting FMS quality, flexibility, and productivity. Common variables include production lead time, scrap percentage, automation, unit labor cost, and setup time. On the other hand, uncommon variables, such as training, tool inventory, customer satisfaction, and rejection reduction, are statistically insignificant to FMS performance variables. Additionally, bibliometric analysis quantitatively evaluates the interconnections among published papers based on FMS performance variables. By reviewing a wide array of literature from the Dimensions database from 2014–2023, about 50 articles were identified in the context of FMSs in their titles, abstracts, and keywords (see Figure 3). Utilizing VOSViewer software, the interactions among parameters in the publications are visually represented in a diagram (refer to Figure 4).

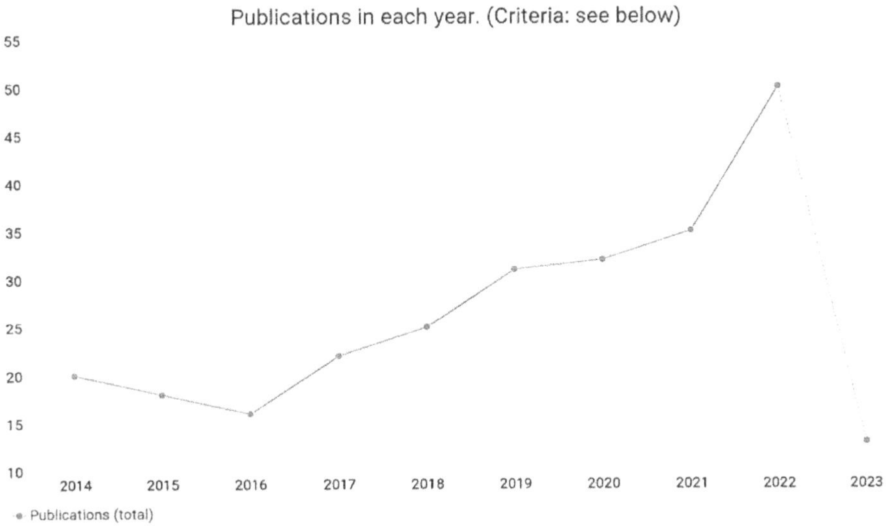

Figure 3. The number of scientific articles published in academic journals. Source: extractions were conducted by the authors with VOSViewer based on the data extracted in the Dimensions Database [Dimensions Database, www.dimensions.ai, accessed on 20 September 2023].

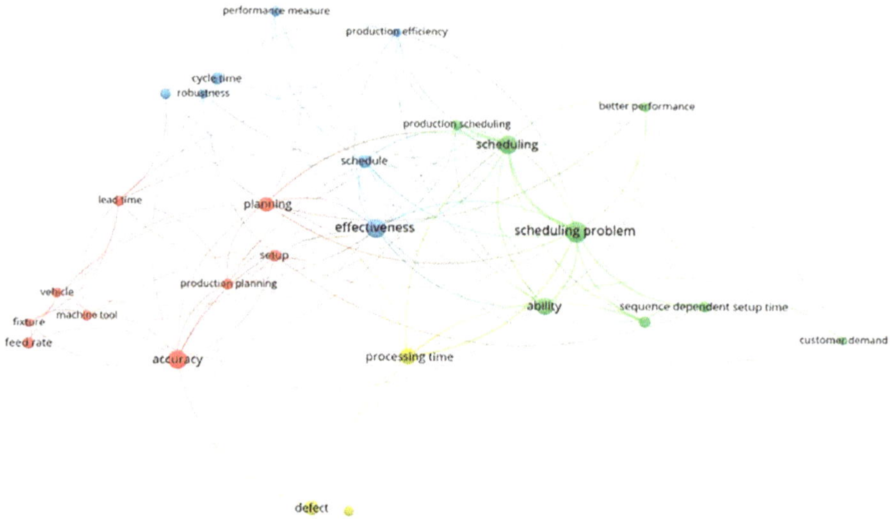

Figure 4. A graphical representation of connections among academic studies related to FMS performance variables. Source: extractions were conducted by the authors with VOSViewer based on the data extracted in the Dimensions Database [Dimensions Database, www.dimensions.ai, accessed on 20 September 2023].

Furthermore, multiple studies deployed a variety of performance variables and research methods to spot the most effective FMS factors. The contributions of previous authors are depicted in Table 1.

Table 1. Contributions of previous authors.

No	Author	Publication	Methodology	Performance Variable
1	[2]	2011	Simulation modelling, Fuzzy logic	1. Routing variety; 2. Routing efficiency; 3. Routing versatility.
2	[4]	2018	Exploratory Factor Analysis (EFA), Confirmatory Factor Analysis (CFA), Absolute Fit Indices, Incremental Fit Indices	1. Automation; 2. Capacity to handle new product; 3. Flexible fixturing; 4. Automation; 5. Increase machine utilization; 6. Flexibility in the design of the production system; 7. Use of automated material handling devices; 8. Ability to manufacture a variety of products; 9. Manufacturing lead time and setup time reduction; 10. Quality consciousness; 11. Speed of response; 12. Combination of operation; 13. Reduced WIP inventories; 14. Reduction in material flow; 15. Reduction in scrap; 16. Use of a reconfigurable machine tool.

Table 1. *Cont.*

No	Author	Publication	Methodology	Performance Variable
3	[6]	2014	ISM, SEM, GTMA	1. Effect of tool life; 2. Training; 3. Financial incentive; 4. Unit labor cost; 5. Customer satisfaction; 6. Reduction in scrap percentage; 7. Reduction of rejection; 8. Reduction in rework percentage; 9. Equipment utilization; 10. Trained worker; 11. Manufacturing lead time and setup time; 12. Unit manufacturing cost; 13. Setup cost; 14. Throughput time; 15. Automation; 16. Use of automated material handling devices; 17. Reduction in material flow; 18. Reduced work in process inventory; 19. Ability to manufacture a variety of products; 20. Capacity to handle new product.
4	[25]	2016	Effectiveness Index, ISM	1. Machine flexibility; 2. Setup or changeover time; 3. Tool magazine or tool current capacity; 4. Availability of technical know-how; 5. Skills and versatility of workers in the system; 6. Type of machine; 7. Max. No. of tools available; 8. Variety of parts to be handled by the machine; 9. Space availability; 10. Max. no. of operations available; 11. Number of machines available in the system; 12. Common tooling available; 13. Similarities of parts in the system; 14. Tool changing time of the machine; 15. Design changes required in the product; 16. Flexibility of material handling system; 17. Similarity of workstations; 18. Variety of products; 19. No of existing part families matching the new product design; 20. Type of operations to be performed on the machine; 21. Maximum number of routes available; 22. Offline part programming preparation facility.

Table 1. *Cont.*

No	Author	Publication	Methodology	Performance Variable
5	[19]	2018	Interpretive Structural Modelling (ISM), Structural Equation Modelling (SEM), Graph Theory, Matrix Approach (GTMA)	1. Effect of tool life; 2. Unit manufacturing cost; 3. Unit labor cost; 4. Manufacturing lead time; 5. Throughput time; 6. Setup cost; 7. Scrap percentage; 8. Rework percentage;; 9. Automation; 10. Use of automated material handling devices; 11. Equipment utilization; 12. Ability to manufacture a variety of product; 13. Capacity to handle new product; 14. Setup time; 15. Reduced work in process inventory.
6	[26]	2016	Total Interpretive Structural Modelling (TISM)	1. Capacity to handle new products; 2. Ability to manufacture a variety of product; 3. Flexibility to design production system; 4. Combination of operation; 5. Automation; 6. Flexible fixturing; 7. Use of automated material handling devices; 8. Increased machine utilization; 9. Use of reconfigurable machine tool; 10. Speed of response; 11. Reduced work in progress; 12. Manufacturing lead time and setup time reduction; 13. Quality consciousness; 14. Reduction in material flow; 15. Reduction in scrap.
7	[21]	1991	Identification of flexibilities, Fishbone diagram	1. Minimize machine to machine movements; 2. Balance workload per machine for equal size machine; 3. Unbalance workload per machine for unequal size machine; 4. Balance machine processing time; 5. Maximize the number of operation assignments; 6. Operation processing time variation; 7. Tool inventory.

Table 1. *Cont.*

No	Author	Publication	Methodology	Performance Variable
8	[27]	2018	MOORA Approach, Ratio System Approach	1. Increased machine utilization; 2. Automation; 3. Use of automated material handling devices; 4. Manufacturing lead time and setup time; 5. Flexible fixturing; 6. Scrap percentage.
9	[19]	2018	MOORA Approach, Ratio System Approach	1. Effect of tool life; 2. Unit manufacturing cost; 3. Unit labor cost; 4. Manufacturing lead time; 5. Setup cost; 6. Scrap percentage; 7. Throughput time; 8. Rework percentage; 9. Setup time; 10. Equipment utilization; 11. Automation; 12. Ability to manufacture variety of products; 13. Use of automated material handling devices; 14. Reduced work in process inventory; 15. Training; 16. Capacity to handle new product; 17. Financial incentive; 18. Customer satisfaction; 19. Reduction of rejection; 20. Reduction in material flow; 21. Trained worker; 22. Flexibility in the design of production system 23. Flexible fixturing; 24. Use of reconfigurable/machine tool; 25. Speed of response; 26. Quality consciousness; 27. Combination of operation.

Table 1. *Cont.*

No	Author	Publication	Methodology	Performance Variable
10	[28]	2019	TISM, Fuzzy logic	1. Unit manufacturing cost; 2. Unit labor cost; 3. Manufacturing lead time; 4. Effect of tool life; 5. Throughput time; 6. Setup cost; 7. Scrap percentage; 8. Setup time; 9. Rework percentage; 10. Equipment utilization; 11. Automation; 12. Ability to manufacture a variety of product; 13. Use of automated material handling devices; 14. Capacity to handle new product; 15. Reduced work in process inventory.
11	[29]	2012	COPRAS approach	1. Increased machine utilization; 2. Automation; 3. Use of automated material handling devices; 4. Flexible fixturing; 5. Scrap percentage; 6. Manufacturing lead time and set up time.

2.2. Review of Multi-Criteria-Decision-Making (MCDM) Approaches

The multicriteria decision-making (MCDM) approach is commonly used to address complex problems [30]. There are four main MCDM pathways for constructing structural networks: Decision Making Trial and Evaluation Laboratory (DEMATEL), Fuzzy Cognition Map (FCM), and Interpretative Structural Modeling (ISM) [31]. Based on a systematic literature review of MCDM approaches, it was found that the DEMATEL and FCM approaches have limitations compared to ISM [32]. Particularly, the DEMATEL approach lacks consideration of all criteria and the aggregation of relative weights from experts for group decisions [33]. On the other hand, FCM requires rigorous optimization and convergence of membership functions, which can be cumbersome [34]. In contrast, ISM overcomes these limitations by effectively identifying interrelationships among factors and is considered a reliable approach for developing hierarchical structural models [31]. Combining ISM with MICMAC yields favorable results for decision makers and researchers [34]. Additionally, integrating SEM (Structural Equation Modeling) enables the estimation and testing of interactions among both measured and latent factors in the developed structural network [30], and it allows for the validation of the proposed network fitness based on expert responses [31]. Notably, BWM outperforms AHP in terms of consistency, minimal violation, total deviation, and conformity, as demonstrated in studies by [35–37]. Thus, the BWM framework is known for producing consistent results and has been extensively utilized in various domains, including manufacturing, supplier selection, risk assessment, biology, automotive, air freight transportation, R&D performance evaluation, banking services, communication technologies, and logistics. These approaches provide a systematic and quantitative means to evaluate the relative importance and impact of different factors on FMS performance, aiding decision makers in making informed decisions.

2.3. Gap Analysis

Drawing from the extensive literature assessment discussed in the preceding section, subsequent gaps have been pinpointed:

1. **Lack of Consistent Labeling**: While numerous researchers have defined the weights of performance variables in various studies pertaining to Flexible Manufacturing Systems (FMSs), only a few have classified them into dimensions based on "Quality (Q)", "Productivity (P)", and "Flexibility (F)", which would encompass the manufacturing system and technological methods [30]. This research introduces a novel classification system based on these dimensions, conveying a structured framework to assess FMS performance.
2. **Limited Deployment of a Novel MCDM Approach**: Although FMS performance variables have been considered in studies using various approaches, such as ISM, SEM, Exploratory Factor Analysis (EFA), Confirmatory Factor Analysis (CFA), and others, no single study has employed a novel Multi-Criteria Decision Making (MCDM) approach comparable to BWM for assessing the significance (weight) of these variables. BWM offers enhanced consistency, minimal violation, total deviation, and conformity.
3. **Inclusion of More Variables** [14]: This research has incorporated 34 key performance variables and three factors extracted from the manufacturing industry, encompassing a larger number of variables compared to other studies, indicating a more comprehensive approach.
4. **Empirical Validation**: There is a need for more empirical studies that validate the findings from conceptual frameworks and propose practical solutions for enhancing FMS performance [2]. This study bridges this gap by presenting a comprehensive empirical analysis based on a broad literature review, consultations with industry experts, and the BWM approach.
5. **Exploring Technological Advancements**: Additionally, there is limited research on the implication of technological advancements, such as Industry 4.0, on FMS performance, indicating a potential research gap in this area [4,30]. This study acknowledges this gap and, through the BWM methodology, explores the implications of these advancements on FMS performance.
6. **Scarcity of Case Studies in Europe and the USA Context**: There is a lack of case studies on FMS implementation not only in India [37] but also in German manufacturing firms, which hinders a precise understanding of the outcomes and implications of performance variables in the German context. This research draws attention to this gap and, by providing a case study, offers valuable insights into FMS performance in distinct geographical contexts.

2.4. Contributions of the Study

This study makes numerous significant contributions to the field of FMS performance evaluation. The key contributions are summarized below in Table 2:

Table 2. Key contributions of this study to FMS performance evaluation.

1.	Introduction of an innovative approach based on BWM for evaluating FMS performance.
2.	Incorporation of insights gained from an extensive literature review and consultation with industry professionals.
3.	Comprehensive overview and analysis of 272 scientific publications.
4.	Identification of research gaps in the literature.
5.	Development of a novel classification for FMS performance variables based on "(Q)", "(P)", and "(F)" dimensions.
6.	Enhancement of consistency, minimal violation, total deviation, and conformity compared to other existing approaches such as AHP.
7.	Identification of the need for further empirical studies.

Table 2. *Cont.*

8.	Highlighting the limited research on the implications of technological advancements, such as Industry 4.0, on FMS performance.
9.	Integration with other smart production system components.
10.	Consideration of sustainability.
11.	Generalizability to other industries.

3. Research Methodology

The BWM is a favored MCDM approach employed through research to evaluate and rank alternatives based on their relative strengths and weaknesses. The research process for applying the BWM method encompasses several key steps, including defining the research problem, identifying criteria and alternatives, developing the BWM survey, collecting and analyzing data, interpreting the results, validating the findings, and presenting the outcomes in a comprehensive and systematic manner. A visual representation of the research method is depicted in Figure 5, illustrating the flowchart of the further steps involved.

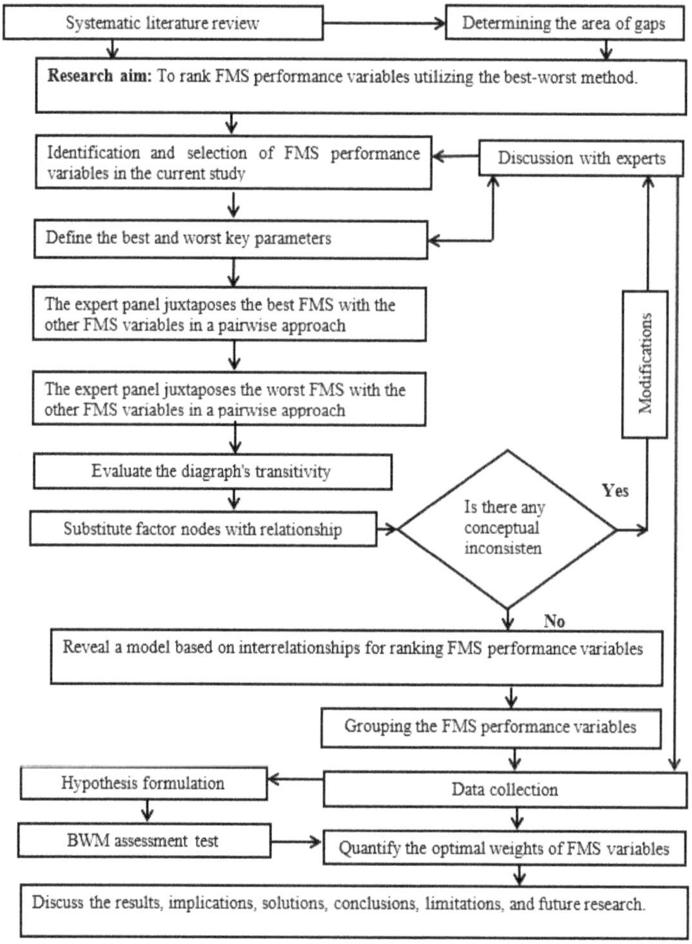

Figure 5. Research methodology flowchart.

In this investigation, a panel comprising thirteen experts was convened to determine the FMS variables that would be included in the analysis (see Table 3).

Table 3. List of decision panel experts.

Expert	Designation	Education	Exp. (in Years)
Expert 1	Academician	Ph.D. Supply Chain Management	13
Expert 2	Production Manager	M. Tech Production Engineering	9
Expert 3	Production Manager	M. Tech Industrial Engineering	13
Expert 4	Manufacturing Manager	B. E. Mechanical	9
Expert 5	Plant Manager	M. Industrial Management	14
Expert 6	Operations Manager	M. Tech Mechanical	13
Expert 7	Operation Management	Ph.D. Operational Management	4
Expert 8	Manufacturing Manager	Ph.D. Operational Management	10
Expert 9	Production Engineer	M. Industrial Management	7
Expert 10	Quality Engineer	M. Industrial Management	8
Expert 11	Production Management	M. Operational Management	10
Expert 12	Operation Management	Ph.D. Operational Management	9
Expert 13	Manufacturing Manager	Ph.D. Operational Management	13

The authors conducted in-person visits to the case organization and held brainstorming sessions with the organization's experts to present the study's framework and procedures. A concise survey was designed to collect input from the participants, and the experts were asked to complete it. Throughout the data study, the authors personally engaged with the experts to gather their viewpoints and inputs at each stage.

Based on a comprehensive literature review and expert analysis, Table 4 presents details about the chosen FMS elements. These selected FMS variables were then classified into three primary criteria: quality (Q), productivity (P), and flexibility (F).

Table 4. FMS performance variables.

Major Factor	Sub-Factor	References
Quality (Q)	Defect rate (Q1)	[37]
	Automation (Q2) *	[37,38]
	Scrap rate (Q3)	[4,38,39]
	Process capability (Q4)	[37,40,41]
	Conformance to specification (Q5)	[42]
	Effect of tool life (Q6)	[43]
	Rework percentage (Q7)	[4,43]
	First-pass yield (FPY) (Q8)	[44]
	Customer satisfaction (Q9)	[26,45]
	Rejection percentage (Q10)	[4,46]
	Takt time (Q11)	[47]

Table 4. *Cont.*

Major Factor	Sub-Factor	References
Productivity (P)	Machine utilization (P1)	[48]
	Unit labor cost (P2)	[4,49,50]
	Unit manufacturing cost (P3)	[4,43]
	Production rate (P4)	[45]
	Manufacturing lead time (P5)	[43,51]
	Work-in-progress (WIP) inventory (P6)	[4]
	Setup time (P7)	[45,48]
	OEE (Overall Equipment Effectiveness) (P8)	[52,53]
	Throughput time (P9)	[4,43]
	Labor productivity (P10)	[54]
	Setup cost (P11)	[4,48]
	Cycle time (P12)	[55]
Flexibility (F)	Changeover time (F1)	[56]
	Equipment utilization (F2)	[4,57]
	Volume flexibility (F3)	[28]
	Routing flexibility (F4)	[48]
	Product mix (F5)	[4]
	Use of automated material handling device (F6)	[25,58]
	Reduced work in process inventory (F7)	[4,43]
	Redundancy (F8)	[58]
	Use of reconfigurable machine tool (F9)	[38]
	Flexible fixturing (F10)	[59–61]
	Machine reconfiguration time (F11)	[53]

Automation (Q2) * in this study refers to the level of automation in the manufacturing process itself, not the automation of quality inspection.

3.1. Overview of BWM Approach

"Rezaei introduced the Best-Worst Method (BWM) in 2015 as a pairwise evaluation methodology used for multicriterion decision-making (MCDM) issues [36]". Its goal is to achieve consistent comparison outcomes while minimizing the number of pairwise comparisons required. BWM replaces the complete pairwise matrix with two vectors, which enables analysts to make decisions with less data. Furthermore, BWM employs a single integer scale ranging from 1 to 9 for ease of contemplation. Due to its ability to obtain quality outcomes with fewer pairwise comparisons, BWM is well suited for assessing the ranking of 34 FMS (Flexible Manufacturing System) indicators in this study.

In BWM, multiple variables are evaluated and consolidated into a single standard variable, which is then used to calculate the values of other items based on the preferences of the most effective element and the most severe element. The most effective or best parameter is compared to all other chosen elements, while the most severe or worst parameter is compared to all other parameters. BWM follows a step-by-step approach that involves five stages of implementation. In this article, to determine the pairwise comparison of the components throughout the BWM process, the Excel Solver program was employed.

3.1.1. Phases of the BWM Process

To convey a clearer understanding of how BWM works, it is essential to outline the specific phases of the methodology:

Phase 1: Determine the performance variables of the FMS.

The first step in the BWM process involves identifying the performance metrics of the FMS. An expert panel comprising manufacturing experts, denoted as {C1, C2, . . ., Cn}, is engaged to finalize the chosen FMS criteria. These criteria play an essential role in clarifying the industry's specific requirements and maximizing its capabilities. After extensive consultation with manufacturing experts, an array of 34 FMS components was validated for ranking and assessment.

Phase 2: Determining the Best and Worst FMS Components.

In this phase, the most favorable and least favorable components of the FMS are identified with input from manufacturing experts. The best criterion is considered the most imperative, preferable, or indispensable, while the worst criterion is recognized as the weakest, least essential, least preferred, or of lesser value.

Phase 3: Expert Panels' Pairwise Comparison.

In this phase, the expert panel employs a pairwise approach to compare the best parameter with the remaining elements. The panel assigns values ranging from one to nine to convey the preference for the best variable over each of the other variables. The preference of the superior variable *B* over variable j, as assessed by the expert panel, is denoted as

$$AB = (aB1, aB2, \ldots, aBn) \quad (1)$$

Here, aBj signifies the preference of the superior variable *B* over variable j, as assessed by the expert panel.

Phase 4: Expert Panels' Pairwise Comparison of the Worst Parameter.

During this phase, the expert panels utilize a pairwise approach to compare the worst parameter with the remaining variables. The panels assign values ranging from one to nine to express the preference of other variables over the suboptimal variable. The worst performance of variable *W* over variable j is denoted as:

$$AW = (a1w, a2w, \ldots, anw)T \quad (2)$$

where ajw reveals the favor of parameter j over the suboptimal variable *W*, which was determined by the expert panel.

The value of aww = 1 is constant, signifying an equal preference between the worst variable and itself.

Phase 5: Estimation of Weights for the Optimal FMS Components.

In this critical step, the previously described issue is mathematically transformed into a linear programming framework. This framework presents a structured approach to optimizing the assigned weights while considering various constraints. The primary objective is to minimize the maximum value among the expressions {$|WB/Wj - aBj| \leq \xi$ for every 'j' and $|Wj/Ww - ajW| \leq \xi$ for every 'j'. Here is a detailed explanation of this essential step:

- Minimizing Maximum Discrepancies: The core objective is to minimize the maximum absolute discrepancy between two expressions for each variable 'j'. The first expression is $|WB/Wj - aBj|$, which measures the extent to which the assigned weight WB deviates from the normalized weight aBj. The second expression is $|Wj/Ww - ajW|$, assessing the extent to which Wj differs from the variable's weight relative to the normalized.

- Balancing Variables: The linear programming framework aims to balance these variables and minimize their discrepancies. The objective is to find an optimal solution where the differences between the assigned weights (Wj) and the benchmark-based weights (aBj and ajW) are as small as possible.
- Accounting for Constraints: The framework considers specific constraints to ensure a feasible solution. These constraints include:
    - Non-negativity constraints: ensuring that all weights (Wj) are non-negative ($Wj \geq 0$ for every 'j').
    - Weight sum condition: maintaining the sum of all weights equal to 1, indicating that the weights encompass all evaluated components ($\sum j\ Wj = 1$).

This linear programming approach has a single unique solution and offers a systematic method for balancing the weights while considering the constraints and normalized weight:

$$|WB/Wj - aBj| \leq \xi \text{ for all j}, |Wj/Ww - ajW| \leq \xi \text{ for every j}, \sum j\ Wj = 1, Wj \geq 0 \text{ for all j} = 1 \quad (3)$$

The outcome of this optimization process provides not only the optimal weights (W1*, W2*, W3*, . . ., Wn*) for each variable but also the optimal value of ξ, denoted as ξ*. This value reflects the degree of balance achieved in assigning weights to the FMS performance variables, with a lower ξ* indicating a more balanced and consistent allocation of weights based on the chosen normalized weight. This comprehensive approach ensures that the relative significance of these components is carefully and consistently evaluated in the FMS assessment process.

3.1.2. Consistency Ratio and Interpretation

An important aspect of the BWM methodology is assessing consistency. To evaluate how reliable the rankings are, we calculate a consistency ratio employing the consistency index.

The consistency ratio can be calculated using Equation (4):

$$\text{Consistency Ratio} = \xi^*/(\text{Consistency Index}) \quad (4)$$

The closer the ratio is to 0, the more consistent the outcome. A consistency ratio closer to 1 would imply a less reliable ranking.

3.2. Case Explanation

3.2.1. Introduction to the Subject Company

Our case study focuses on a long-standing manufacturing firm specializing in automotive component production. With nearly 40 years of experience in the automobile industry, a workforce of 1300 employees, and prestigious certifications in TS 16949 and ISO 14001, this company has maintained its position but aspired to achieve market leadership.

3.2.2. Motivation for BWM Implementation

Despite their significant achievements, the organization faces increasing competition and a need to boost manufacturing performance. To respond more effectively to customer demands for product diversity, demand, and quality in real time, the leadership decided to implement a comprehensive strategy.

After careful consideration, they selected BWM as the ideal approach to evaluate and prioritize essential performance variables for successful FMS implementation. The stakeholders and management teams were convinced by the method's analytical robustness and its systematic approach to measuring the weights of FMS variables. This adoption reflects a commitment to data-driven decision making and strategic manufacturing improvements.

3.3. Analysis of Weight Ranking

3.3.1. The Expert Panel

In an earlier phase, a panel of thirteen experts representing diverse organizations was responsible for evaluating FMS performance variables. Their task was to identify the most effective-to-others (Table 5) and other-to-most severe (Table 6) relationships for each major element. The experts' evaluations were based on their collective industry knowledge and experience. These evaluations led to the creation of best-to-others and other-to-most severe vectors, which were essential data for the subsequent analysis.

Table 5. Best compared to other variables.

Expert	Best	Q	P	F
Expert 1	Q	1	3	5
Expert 2	Q	1	8	6
Expert 3	P	2	1	3
Expert 4	Q	1	8	7
Expert 5	Q	1	2	3
Expert 6	Q	1	3	2
Expert 7	F	7	8	1
Expert 8	F	4	3	1
Expert 9	Q	1	6	4
Expert 10	P	2	1	7
Expert 11	Q	1	4	7
Expert 12	Q	1	4	7
Expert 13	Q	1	7	6

Table 6. Others compared to the worst variables.

Expert	Worst	Q	P	F
Expert 1	P	8	1	4
Expert 2	F	4	6	1
Expert 3	F	8	9	1
Expert 4	P	9	1	7
Expert 5	F	8	7	1
Expert 6	F	3	5	1
Expert 7	P	3	1	2
Expert 8	P	5	1	6
Expert 9	F	2	4	1
Expert 10	Q	1	3	4
Expert 11	F	7	4	1
Expert 12	F	7	4	1
Expert 13	P	9	1	6

3.3.2. Calculating Average Weights

After the expert panel's assessment, the individual weights assigned by each expert for Quality (Q), Productivity (P), Flexibility (F), and the Ksi* index were combined to generate average weights (see Table 7). These average weights represent a consensus of expert opinions.

Table 7. Weightage of major factors.

Expert	Q	P	F	Ksi*
Expert 1	0.6790	0.1234	0.1975	0.3086
Expert 2	0.7636	0.1454	0.0909	0.4000
Expert 3	0.3958	0.5208	0.0833	0.2708
Expert 4	0.8000	0.0555	0.1444	0.3555
Expert 5	0.5750	0.3000	0.1250	0.3250
Expert 6	0.5278	0.3611	0.1111	0.1944
Expert 7	0.6786	0.1786	0.1429	0.3929
Expert 8	0.1667	0.6852	0.1481	0.3519
Expert 9	0.7083	0.2083	0.0833	0.1250
Expert 10	0.1750	0.1000	0.7250	0.3250
Expert 11	0.7589	0.0714	0.1696	0.2589
Expert 12	0.6607	0.2143	0.1250	0.4107
Expert 13	0.7083	0.2083	0.0833	0.1250
Final weight	0.5844	0.2240	0.1715	0.2956

The formula to calculate the Ksi* (consistency index) in the Best-Worst Method (BWM) is as follows: Ksi* = $(\sum(\omega i - w\bar{\imath})2)/(n(n-1))$; Where: Ksi* is the consistency index; ωi is the BWOR (Best-to-Worst Ratio) for performance variable I; $w\bar{\imath}$ is the average BWOR ratio across all performance variables; n is the number of performance variables.

3.3.3. Ensuring Consistency

The AHP methodology was employed for this analysis, and it was imperative to assess the consistency of expert responses. The median consistency ratio weight, shown in Table 7, was calculated and found to be 0.2956. This value, close to zero, indicates a high level of consistency in expert comparisons, reinforcing the reliability of the outcomes.

3.3.4. Determining Global Weights

The major weights for each group of FMS performance variables (Quality, Productivity, and Flexibility) were calculated using their average weights. While detailed calculations are beyond the scope of this section, these weights are derived from the expert assessments in Table 8. The final rankings in Table 8 represent the global weighting for each performance variable, determined by multiplying the major weights and local weights and ranking them based on their relative importance. The local weights were assessed using Formula (3).

Table 8. Final ranking of FMS performance variables.

Major Factor	Major Weight	Sub-Factor Element	Local Weighting	Global Weighting	Ultimate Rank
Quality (Q)	0.5544	Q1	0.0942	0.0522	4
		Q2	0.0969	0.0537	3
		Q3	0.0815	0.0452	10
		Q4	0.0724	0.0401	11
		Q5	0.0933	0.0517	5
		Q6	0.0889	0.0493	7
		Q7	0.0933	0.0517	5
		Q8	0.0981	0.0544	2
		Q9	0.0863	0.0478	8
		Q10	0.1093	0.0606	1
		Q11	0.0858	0.0476	9

Table 8. *Cont.*

Major Factor	Major Weight	Sub-Factor Element	Local Weighting	Global Weighting	Ultimate Rank
Productivity (P)	0.2240	P1	0.0792	0.0177	20
		P2	0.0902	0.0202	15
		P3	0.0844	0.0189	18
		P4	0.0889	0.0199	16
		P5	0.0728	0.0163	27
		P6	0.0658	0.0147	31
		P7	0.0959	0.0215	12
		P8	0.0764	0.0171	22
		P9	0.0922	0.0207	14
		P10	0.0759	0.0170	24
		P11	0.0857	0.0192	17
		P12	0.0926	0.0207	13
Flexibility (F)	0.1715	F1	0.0790	0.0135	32
		F2	0.0912	0.0156	29
		F3	0.0975	0.0167	25
		F4	0.1069	0.0183	19
		F5	0.0892	0.0153	30
		F6	0.0924	0.0158	28
		F7	0.0750	0.0129	33
		F8	0.1005	0.0172	21
		F9	0.0958	0.0164	26
		F10	0.0733	0.0126	34
		F11	0.0992	0.0170	23

3.3.5. Final Ranking of FMS Performance Variables

In this phase of our research, the BWM analysis resulted in the assignment of weights to each major FMS factor and its subfactors. These weights indicate the relative significance of each element in the overall evaluation of FMS performance. The outcomes are presented in Table 7 for major factors and Table 8 for the final ranking of FMS performance variables based on their global weights.

To ensure the credibility of our results, we calculated the outgoing median consistency ratio, which was found to be 0.2956, a value close to zero. This proximity indicates a high level of consistency in our comparisons, reinforcing the reliability of our findings.

Table 8 provides a concise summary of the weights assigned to the major FMS factors. These factors are pivotal in evaluating FMS performance, and their weights were determined through expert evaluation and the robust BWM methodology.

Table 8 presents the global weightings of various FMS performance variables. These variables underwent a rigorous evaluation process following the BWM methodology. Due to space limitations, we provide a concise overview of the results in this section. The table includes the final rankings, scores, and relative influence of each extracted FMS performance variable.

To highlight the significance of these outcomes, let us focus on the three primary performance variables:

- Quality (Q): This factor carries a weight of 0.5544, indicating its paramount role in the FMS performance assessment.
- Productivity (P): With a weight of 0.1715, productivity is a notable, although secondary, factor in the assessment.

- Flexibility (F): Flexibility is assigned a weight of 0.2208, signifying its essential yet less influential role in the assessment process.

3.4. Findings and Discussion

The primary objective of this study was to investigate and prioritize the critical performance variables and qualitative attributes, along with their respective weightings, within German manufacturing companies utilizing the BWM method. In this study, the chief factors of FMSs that influence their performance were taken into account, i.e., quality, productivity, and flexibility, and 34 attributes that affect these parameters were also considered. To enhance the level of flexibility in manufacturing systems, organizations must therefore prioritize these performance variables based on their importance or weight. The case study's findings disclosed that among the essential parameters, the quality performance variable had the highest weight (0.5542), followed by productivity (0.2247) and flexibility (0.2208).

Quality (Q): This factor, bestowed with a weight of 0.5544, is paramount in the FMS performance hierarchy. Its high weightage signifies its vital role in the overall assessment. The highest weighted variable in terms of quality is the rejection percentage (Q10), with a weight of 0.0606. Ref. [47] argue that rejection might be caused by malfunctioning equipment and tools, workers' low levels of skill, or errors in technical working instruction and control. Nonetheless, rejected parts could be recycled; however, the effects of part rejection might also be varied and are generally grouped into two classifications: "ROI losses and operational disruptions" [43].

Nevertheless, rejection might be resolved by the following strategies: process improvement, training and skill development, feedback loops, supplier evaluation, and continuous improvement. The First-pass yield was ranked second as a quality factor, with a global weight of 0.0544. First-pass yield and rejection percentage have comparable sources of defects. Ref. [61] suggests that First-pass yield could be maintained through robust design and optimization, preventive maintenance, and supplier qualification.

Automation (Q2) came in third place in the rankings, embracing a global weight of 0.0537. According to [37,62], automation and technological improvements reduce annual labor costs while increasing productivity and flexibility in the manufacturing system. Furthermore, "a higher level of automation increases this flexibility, partly as a result of both lower machine setup costs and lower variable costs" [62].

Defect Rate (Q1) resulted in the fourth position with a global weight of 0.0522. The sources of defect rate are design, manufacturing defects, or inspection errors, which lead to increased costs, lost sales, and customer dissatisfaction. According to the study by [36], the defect in manufacturing could vary significantly depending on the industry, the product being manufactured, and the manufacturing process employed. Ref. [37] assert that the defect rate in manufacturing has been declining over time due to numerous factors, including the adoption of novel technologies, the implementation of quality improvement initiatives, and an enhanced awareness of the importance of quality. Process capability (Q4) had a global weight of 0.0401 and was the lowest ranking among the Quality (Q) segments as performance variables. Process capability is a statistical measure that quantifies the ability of a process to consistently produce output within specified limits or tolerances of customer requirements [41]. Process capability is typically evaluated by a process capability index of Cp or Cpk.

Flexibility (F): In the dimension of flexibility, F4 (routing flexibility), F8 (redundancy), F11 (machine reconfiguration time), F3 (volume flexibility), and F9 (use of reconfigurable machine tools) have the most global weight and thus have the greatest influence on FMSs [49]. The F4 (routing flexibility) had the highest global weight of 0.0236. Routing flexibility is a vital element that enables manufacturers to produce a wide range of products with varying specifications and requirements. It enhances production agility, reduces lead time, and increases overall efficiency in the manufacturing process. F8 (redundancy) is in the second position with a global weight of 0.0222. Ref. [63] define it as the presence of duplicate equipment within the FMS that can be used to take over for a machine that has failed or is undergoing maintenance. Higher redundancy can increase the FMS's flexibility [53].

Productivity (P): Among the Productivity (P) factors, Setup time (P7), Cycle time (P12), Throughput time (P9), and Unit labor cost (P2) have the highest performing effects on FMS. Setup time (P7) has a global weight of 0.0215 and refers to the time required to prepare the machines, equipment, tools, and materials for a specific production run [60]. Furthermore, Cycle time (P12) has a final weight of 0.0208, and this is the time required to finish a part from beginning to completion [55].

In summary, this study investigated a total of 34 parameters, emphasizing the importance of effective management of these flexibilities by decision makers. Ultimately, the analysis highlights that the following elements—Rejection percentage (Q10), First-pass yield (Q8), Automation (Q2), and Defect Rate (Q1)—stand out as the key factors exerting a substantial implication on FMS performance.

4. The Implications of this Study

The Experts and researchers in the manufacturing industry can derive practical value from the study's findings:

- Strategic Prioritization: Manufacturing professionals can employ the insights to strategically prioritize FMS elements. By recognizing the essential role of quality, they can focus on enhancing quality management practices and minimizing rejection rates.
- Operational Improvements: The findings on productivity and automation suggest opportunities for operational enhancement. Implementing automation and optimizing cycle times could lead to cost reductions and increased efficiency.
- Flexibility Enhancements: Manufacturing experts can leverage the interpretation of flexibility dynamics to boost their production agility in real time. Strategies such as routing flexibility and redundancy can improve overall efficiency and responsiveness in the context of Industry 4.0.
- Research and Innovation: Researchers in the field could build on this study's findings to explore related topics further. Future research might delve into specific strategies for implementing prioritized factors in real manufacturing settings.

In conclusion, this study's implications are intended to motivate practitioners and academics to develop diverse strategies for prioritizing and managing FMS variables in their respective domains of work and drive improvements within their manufacturing processes.

4.1. Theoretical Contributions

This study breaks new ground in the realm of Flexible Manufacturing Systems (FMSs) research. While several prominent manufacturing firms, such as Toyota and General Motors, have delved into the examination of FMS performance variables, none have approached the task with the comprehensive and structured methodology employed. Our research develops a novel theoretical framework that harmonizes queuing theory with decision theory [57], creating a model that unravels the intricate dynamics of FMS performance variables within the stochastic production environment. Queuing theory, a discipline rooted in mathematics, opens a gateway to understanding and analyzing an array of systems, from manufacturing to transportation and communication [64]. It allows for an in-depth exploration of how different factors, such as arrival rates, service rates, and queueing disciplines, influence these systems' performance. By merging queuing theory with decision theory, we introduce an innovative theoretical framework that affords the ability to weigh the trade-offs between distinct performance objectives like productivity, flexibility, and quality. This empowers us to pinpoint optimal strategies for operating FMSs within diverse production contexts.

Additionally, our study identifies and validates a number of significant performance variables, including operator skill level, system flexibility, and organizational culture, as the primary determinants of FMS effectiveness within real-world manufacturing environments.

According to [43], an FMS pertains to a combined computer-controlled system of digitally controlled machinery and automated material handling parts and tools that are capable of processing medium-sized volumes of various part kinds sequentially. These insights offer essential guidance to managers, enabling them to prioritize these factors when implementing and managing FMS.

Furthermore, our research serves as an empirical validation for the resource-based view (RBV) theory, emphasizing its relevance in elucidating the competitive advantage of firms that have embraced FMSs. This quantitative analysis, conducted across multiple firms, supports the applicability of the RBV concept in the context of FMSs, enriching the field of strategic management theory.

Moreover, we conveyed a lucid and comprehensive definition of FMSs, accompanied by a classification framework that classifies FMSs into diverse types based on their operational characteristics. This lucidity ensures consistency and clarity within the existing body of knowledge regarding FMS the concept and classification of FMS.

The contributions of this research extend beyond the academic realm, shedding light on the roles that require fortification and anticipating prospective enhancements. The unparalleled application of the BWM approach to assess the potential weights of FMS performance parameters, particularly within the context of 34 variables, sets this research apart in the FMS domain. Prior studies have primarily examined a smaller array of performance variables, making this comprehensive analysis a noteworthy addition to the study of FMS in the manufacturing industry.

4.2. *Practical Implications*

The research on FMS performance variables utilizing the BWM method offers two distinct advantages. Firstly, it requires smaller pairwise relations compared to other MCDM strategies, simplifying weight assessment for experts. Additionally, due to the reduced number of pairwise comparisons, the BWM approach generates more consistent outcomes. These reliable results and the efficiency of the BWM strategy encourage professionals and decision makers to adopt FMSs, facilitating the transformation of traditional operations into sustainable business practices.

By employing the BWM approach to analyze an array of 34 qualitative performance variables, which have not previously been explored in the field of FMSs, this study becomes a valuable tool for clarifying complex problems involving the selection of significant factors from a large array of variables.

While the study was conducted at a manufacturing firm, its findings could serve as a catalyst for further exploration and application of the BWM strategy in diverse manufacturing industries, such as steel and iron, aviation, and others. Given the growing concerns in the manufacturing industry about diminishing environmental implications, this research could be essential for companies and researchers seeking to boost sustainable frameworks within manufacturing firms. Identifying the most significant performance variables through the BWM strategy can enable companies facing limitations in personnel or resources. By prioritizing and focusing on key performance variables, companies can indirectly clarify other variables; solutions for significant variables might positively affect related ones. This approach allows for a more efficient allocation of resources and efforts to the most influential performance variables.

Furthermore, the outcomes of this investigation open up novel research avenues for academics. Future studies could expand the scope of performance variables by grouping them based on other conceptual dimensions, such as the three major dimensions of business (human, economic, and environmental). This broader application of the BWM methodology to investigate other aspects within the domain of FMS can lead to a deeper interpretation and analysis of performance variables in various contexts and perspectives.

The comprehensive approach of the FMS model, encompassing challenges in applicability and efficiency in batch system planning and design, while incorporating the job-shop layout, ensures that key system parameters and design considerations, such as work center and lot dimensions, storage management, and planning processes, are considered. This approach leads to an enhanced applicability and efficiency in batch system planning and design, particularly in the context of job-shop layouts.

5. Discussion and Conclusions

In this study, the BWM was effectively harnessed to manage the assessment of FMS performance metrics, taking into account qualitative characteristics. This method distinctly captures the qualitative characteristics of pairwise comparisons using the BWM framework, thus empowering researchers to pinpoint variables with the highest loading values. The selection of key FMS parameters was a complex process, initiated with an extensive literature review to ensure comprehensive FMS variable prioritization. Subsequently, the identified performance metrics for the study were thoughtfully curated with the invaluable input of experts from both academic and industrial backgrounds.

Through consultation with these experts, the identified FMS parameters were thoughtfully classified into three major categories: Quality, Productivity, and Flexibility. Leveraging the BWM methodology, this research assessed and assigned appropriate weightings to these finalized parameters. As a result, the study not only enables professionals to have essential interpretations of FMSs but also equips them with the tools to analyze the individual implications of each variable. This, in turn, enables deeper explorations of the most vital parameters, fostering the development of systematic strategies for their optimization.

A distinctive facet of this research is its utilization of BWM for ranking and rating, a choice that yields notably more consistent results compared to other MCDM methods. These findings will prove instrumental for professionals seeking to grasp the nuanced dynamics within their organizations, as each FMS performance variable considered in this study mirrors the influence of one of the key stakeholders—employees, suppliers, and clients. Consequently, this investigation transcends the limitations of singular, one-dimensional analyses and delves into the multifaceted aspects of FMS.

Additionally, by spotting underperforming functions within an organization, this research enables companies to proactively devise strategies for enhancing their performance by focusing on the selected key variables. The insights gathered throughout this study suggest that the performance variables it identifies, along with their respective influence powers, will serve as invaluable tools for professionals looking to discern and address the most pivotal elements for effective FMS deployment.

In addition to these merits, it is essential to note that this research makes a significant contribution in comparison to existing studies in the field. Unlike numerous prior works that concentrate on the singular dimensions of FMS, this study casts a wider net by comprehensively exploring distinct dimensions of FMS performance. Its robust methodology and inclusion of qualitative characteristics for pairwise comparisons ensure a more holistic comprehension of the subject. This not only boosts its value for experts and researchers but also solidifies its position as a significant reference point for future explorations in the realm of FMS performance assessment.

6. Limitations and Future Research

Although the current study classified 34 performance variables into three groups, it is imperative to acknowledge that other parameters might continue to influence the adoption of FMSs. These additional parameters could be further organized into logically coherent groups for more in-depth analysis.

Furthermore, the outcomes of the study heavily rely on the opinions and perspectives of experts, making a thorough evaluation of expert input essential.

It is also important to note that the current investigation focuses on the manufacturing industry in Germany, and the findings might not be directly applicable to other sectors such as aviation, construction, services, etc. Nonetheless, they could still hold value for manufacturing industries in other emerging economies such as France, the UK, Italy, and others. Conducting a large-scale survey in the manufacturing sector and comparing and verifying the findings with other research results could provide more accurate insights.

Additionally, the results of this study are specific to the case analyzed and cannot be generalized to the entire manufacturing industry in all domains. Further analysis could investigate the interrelationships between FMS performance variables using fuzzy BWM with a different set of performance variables. The findings could also be compared with other fuzzy MCDM strategies such as fuzzy PROMETHEE, fuzzy TOPSIS, fuzzy VIKOR, and fuzzy ELECTRE.

Finally, investigating the implications of external factors, such as regulatory changes, technological advancements, and market dynamics, on FMS adoption could reveal a more comprehensive interpretation of the topic.

In conclusion, the authors believe that this study makes a significant contribution to the adoption of FMSs by prioritizing performance variables. Nonetheless, additional research and analysis are needed to validate and generalize the findings to different contexts and explore other fuzzy MCDM approaches for a more comprehensive understanding of FMS adoption.

Author Contributions: Conceptualization, A.B. and G.C.; methodology, A.B. and A.L.S.; validation, A.L.S. and R.K.S.; formal analysis, A.B.; investigation, A.B.; resources, A.B.; data curation, G.C.; writing—original draft preparation, A.B. and A.L.S.; writing—and review and editing, A.B. and G.C.; visualization, R.K.S.; supervision, A.B.; project administration, A.L.S.; funding acquisition, Not Required. All authors have read and agreed to the published version of the manuscript.

Funding: This research received no external funding.

Data Availability Statement: Data are contained within the article.

Conflicts of Interest: The authors declare no conflict of interest.

References

1. Chan, F.T.S. The state-of-the-art of flexible manufacturing systems. *Int. J. Adv. Manuf. Technol.* **2003**, *21*, 534–541. [CrossRef]
2. Joseph, R.; Sridharan, R. Performance measurement of flexible manufacturing systems: A review. *Int. J. Oper. Res. Inf. Syst.* **2011**, *2*, 21–42.
3. Santuka, R.; Mahapatra, S.S.; Dhal, P.R.; Mishra, A. An Improved Particle Swarm Optimization Approach for Solving Machine Loading Problem in Flexible Manufacturing System. *J. Adv. Manuf. Syst.* **2015**, *14*, 167–187. [CrossRef]
4. Jain, V.; Raj, T. Identification of performance variables which affect the FMS: A state-of-the-art review. *Int. J. Process Manag. Benchmarking* **2018**, *8*, 470. [CrossRef]
5. Cordero, R. *Flexible Manufacturing Systems: The Future of Manufacturing*; CRC Press: Boca Raton, FL, USA, 1997.
6. Jain, V.; Raj, T. Flexible manufacturing systems: A review of literature. *Int. J. Eng. Technol. Res.* **2014**, *2*, 87–90.
7. Latest Quality. How to Implement a Flexible Manufacturing System—Latest Quality. 2018. Available online: https://www.latestquality.com/flexible-manufacturing-system (accessed on 19 August 2021).
8. Chang, D.-Y. Applications of the extent analysis method on fuzzy AHP. *Eur. J. Oper. Res.* **1996**, *95*, 649–655. [CrossRef]
9. Suresh, N.C.; Yang, B. A review on performance evaluation of flexible manufacturing system. *Int. J. Eng. Technol.* **2018**, *7*, 6–9.
10. Ghosh, S.; Gagné, C. Flexible manufacturing system performance measurement and analysis: A review. *Int. J. Prod. Econ.* **2019**, *209*, 160–177.
11. Kumar, A.; Bhattacharya, A. Evaluation of performance measures for a flexible manufacturing system: A case study. *Comput. Mater. Contin.* **2017**, *53*, 117–136.
12. Gupta, Y.P.; Goyal, S. Flexibility of manufacturing systems: Concepts and measurements. *Eur. J. Oper. Res.* **1989**, *43*, 119–135. [CrossRef]
13. Sethi, A.K.; Sethi, S.P. Flexibility in manufacturing: A survey. *Int. J. Flex. Manuf. Syst.* **1990**, *2*, 289–328. [CrossRef]
14. Mishra, R.; Pundir, A.K.; Ganapathy, L. Conceptualizing sources, key concerns and critical factors for manufacturing flexibility adoption: An exploratory study in Indian manufacturing firms. *J. Manuf. Technol. Manag.* **2016**, *27*, 379–407. [CrossRef]
15. Pasha, N.; Mahdiraji, H.A.; Hajiagha, S.H.R.; Garza-Reyes, J.A.; Joshi, R. A multi-objective flexible manufacturing system design optimization using a hybrid response surface methodology. *Oper. Manag. Res.* **2023**, 1–17. [CrossRef]
16. Talluri, S.; Whiteside, M.M.; Seipel, S.J. A nonparametric stochastic procedure for FMS evaluation. *Eur. J. Oper. Res.* **2000**, *124*, 529–538. [CrossRef]
17. Wabalickis, R.N. Justification of FMS with the analytic hierarchy process. *J. Manuf. Syst.* **1988**, *7*, 175–182. [CrossRef]
18. Kuei, C.-H.; Lin, C.; Aheto, J.; Madu, C.N. A strategic decision model for the selection of advanced technology. *Int. J. Prod. Res.* **1994**, *32*, 2117–2130. [CrossRef]

19. Jain, V. Application of combined MADM methods as MOORA and PSI for ranking of FMS performance factors. *Benchmarking Int. J.* **2018**, *25*, 1903–1920. [CrossRef]
20. Shanker, K.; Tzen, Y.-J.J. A loading and dispatching problem in a random flexible manufacturing system. *Int. J. Prod. Res.* **1985**, *23*, 579–595. [CrossRef]
21. Chen, I.J.; Chung, C.-H. Effects of loading and routeing decisions on performance of flexible manufacturing systems. *Int. J. Prod. Res.* **1991**, *29*, 2209–2225. [CrossRef]
22. Mukhopadhyay, S.K.; Singh, M.; Srivastava, R. FMS machine loading: A simulated annealing approach. *Int. J. Prod. Res.* **1998**, *36*, 1529–1547. [CrossRef]
23. Nayak, G.K.; Acharya, D. Part type selection, machine loading and part type volume determination problems in FMS planning. *Int. J. Prod. Res.* **1998**, *36*, 1801–1824. [CrossRef]
24. Wankhede, V.A.; Vinodh, S. Analysis of Industry 4.0 challenges using best worst method: A case study. *Comput. Ind. Eng.* **2021**, *159*, 107487. [CrossRef]
25. Gothwal, S.; Raj, T. Analyzing the factors affecting the flexibility in FMS using weighted interpretive structural modeling (WISM) approach. *Int. J. Syst. Assur. Eng. Manag.* **2016**, *8*, 408–422. [CrossRef]
26. Zijm, H. Manufacturing systems. In *Operations, Logistics and Supply Chain Management*; Springer: Berlin/Heidelberg, Germany, 2019; pp. 75–95.
27. Chakraborty, S. Applications of the MOORA method for decision making in manufacturing environment. *Int. J. Adv. Manuf. Technol.* **2011**, *54*, 1155–1166. [CrossRef]
28. Yadav, A.; Jayswal, S. Modelling of flexible manufacturing system: A review. *Int. J. Prod. Res.* **2018**, *56*, 2464–2487. [CrossRef]
29. Jain, V.; Raj, T. Modeling and analysis of FMS flexibility factors by TISM and fuzzy MICMAC. *Int. J. Syst. Assur. Eng. Manag.* **2015**, *6*, 350–371. [CrossRef]
30. Ghosh, S.; Chakraborty, T.; Saha, S.; Majumder, M.; Pal, M. Development of the location suitability index for wave energy production by ANN and MCDM techniques. *Renew. Sustain. Energy Rev.* **2016**, *59*, 1017–1028. [CrossRef]
31. Yap, J.Y.L.; Ho, C.C.; Ting, C.-Y. A systematic review of the applications of multi-criteria decision-making methods in site selection problems. *Built Environ. Proj. Asset Manag.* **2019**, *9*, 548–563. [CrossRef]
32. Bagherian, A.; Gershon, M.; Kumar, S.; Mishra, M.K. Analyzing the Relationship between Digitalization and Energy Sustainability: A Comprehensive ISM-MICMAC and DEMATEL Approach. *Expert Syst. Appl.* **2023**, *236*, 121193. [CrossRef]
33. Tzeng, G.-H.; Chen, W.-H.; Yu, R.; Shih, M.-L. Fuzzy decision maps: A generalization of the DEMATEL methods. *Soft Comput.* **2010**, *14*, 1141–1150. [CrossRef]
34. Sheng-Li, S.; Xiao-Yue, Y.; Hu-Chen, L.; Zhang, P. DEMATEL technique: A systematic review of the state-of-the-art literature on methodologies and applications. *Math. Probl. Eng.* **2018**, *2018*, 3696457.
35. Shah, R.; Goldstein, S.M. Use of structural equation modeling in operations management research: Looking back and forward. *J. Oper. Manag.* **2006**, *24*, 148–169. [CrossRef]
36. Khaba, S.; Bhar, C. Analysing the barriers of lean in Indian coal mining industry using integrated ISM-MICMAC and SEM. *Benchmarking Int. J.* **2018**, *25*, 2145–2168. [CrossRef]
37. Talib, F.; Asjad, M.; Attri, R.; Siddiquee, A.N.; Khan, Z.A. A road map for the implementation of integrated JIT-lean practices in Indian manufacturing industries using the best-worst method approach. *J. Ind. Prod. Eng.* **2020**, *37*, 275–291. [CrossRef]
38. Jain, V.; Soni, V.K. Modeling and analysis of FMS performance variables by fuzzy TISM. *J. Model. Manag.* **2019**, *14*, 2–30. [CrossRef]
39. Liao, K.; Tu, Q. Leveraging automation and integration to improve manufacturing performance under uncertainty: An empirical study. *J. Manuf. Technol. Manag.* **2007**, *19*, 38–51. [CrossRef]
40. Gupta, Y.P. Organizational issues of flexible manufacturing systems. *Technovation* **1988**, *8*, 255–269. [CrossRef]
41. Boer, H.; Hill, M.; Krabendum, K. FMS implementation management: Promises and performance. *Int. J. Oper. Prod. Manag.* **1990**, *10*, 5–20. [CrossRef]
42. Modgil, S.; Sharma, S. Total productive maintenance, total quality management and operational performance: An empirical study of Indian pharmaceutical industry. *J. Qual. Maint. Eng.* **2016**, *22*, 353–377. [CrossRef]
43. Wu, Z.; Huang, N.; Zheng, X.; Li, X. Cyber-physical avionics systems and its reliability evaluation. In Proceedings of the 4th Annual IEEE International Conference on Cyber Technology in Automation, Control and Intelligent [Preprint], Hong Kong, China, 4–7 June 2014.
44. Montgomery, D.C. *Introduction to Statistical Quality Control*; John Wiley & Sons: Hoboken, NJ, USA, 2019.
45. Jain, V.; Raj, T. Modeling and analysis of FMS performance variables by ISM, SEM and GTMA approach. *Int. J. Prod. Econ.* **2016**, *171*, 84–96. [CrossRef]
46. Ward, S. Understanding first pass yield. *Quality* **2006**, *45*, 26.
47. Löthgren, M.; Tambour, M. Productivity and customer satisfaction in Swedish pharmacies: A DEA network model. *Eur. J. Oper. Res.* **1999**, *115*, 449–458. [CrossRef]
48. Anderson, E.W.; Fornell, C.; Rust, R.T. Customer satisfaction, productivity, and profitability: Differences between goods and services. *Mark. Sci.* **1997**, *16*, 129–145. [CrossRef]
49. Cheng, H.C.; Chan, D.Y.K. Simulation optimization of part input sequence in a flexible manufacturing system. In Proceedings of the 2011 Winter Simulation Conference—(WSC 2011), Phoenix, AZ, USA, 11–14 December 2011; pp. 2374–2382.

50. Zhang, F.; Tian, C. Study on modeling and simulation of logistics sorting system based on flexsim. In Proceedings of the 2017 International Conference on Computer Network, Electronic and Automation (ICCNEA), Xi'an, China, 23–25 September 2017.
51. Kazerooni, A.; Chan, F.; Abhary, K. A fuzzy integrated decision-making support system for scheduling of FMS using simulation. *Comput. Integr. Manuf. Syst.* **1997**, *10*, 27–34. [CrossRef]
52. Tao, Y.; Chen, J.; Liu, M.; Liu, X.; Fu, Y. An estimate and simulation approach to determining the Automated Guided Vehicle fleet size in FMS. In Proceedings of the 2010 3rd IEEE International Conference on Computer Science and Information Technology (ICCSIT 2010), Chengdu, China, 9–11 July 2010; Volume 9, pp. 432–435.
53. Raj, T.; Attri, R.; Jain, V. Modelling the factors affecting flexibility in FMS. *Int. J. Ind. Syst. Eng.* **2012**, *11*, 350–374.
54. Jain, V.; Raj, T. Ranking of Flexibility in Flexible Manufacturing System by Using a Combined Multiple Attribute Decision Making Method. *Glob. J. Flex. Syst. Manag.* **2013**, *14*, 125–141. [CrossRef]
55. Rajput, H.S.; Jayaswal, P. A total productive maintenance (TPM) approach to improve overall equipment efficiency. *Int. J. Mod. Eng. Res.* **2012**, *2*, 4383–4386.
56. Rita, G.; Luca, G.; Francesco, L.; Bianca, R. On the analysis of effectiveness in a manufacturing cell: A critical implementation of existing approaches. *Procedia Manuf.* **2017**, *11*, 1882–1891. [CrossRef]
57. Nagarjuna, N.; Mahesh, O.; Rajagopal, K. A heuristic based on multi-stage programming approach for machine-loading problem in a flexible manufacturing system. *Robot. Comput. Manuf.* **2006**, *22*, 342–352. [CrossRef]
58. Chen, X.; Wu, J.; Zhou, T.; Li, Y. Summary of the Prediction Methods of Tool Remaining Life Based on Data Collection. *J. Phys. Conf. Ser.* **2021**, *1939*, 012055. [CrossRef]
59. Kulak, O. A decision support system for fuzzy multi-attribute selection of material handling equipments. *Expert Syst. Appl.* **2005**, *29*, 310–319. [CrossRef]
60. Nageswara, M.; Narayana, R.; Ranga, J. Integrated scheduling of machines and AGVS in FMS by using dispatching rules. *J. Prod. Eng.* **2017**, *20*, 75–84. [CrossRef]
61. Groover, M.P. *Automation, Production Systems, and Computer-Integrated Manufacturing*; Prentice Hall Press: New Delhi, India, 2006.
62. Choe, P.; Tew, J.D.; Tong, S. Effect of cognitive automation in a material handling system on manufacturing flexibility. *Int. J. Prod. Econ.* **2015**, *170*, 891–899. [CrossRef]
63. Oke, A. A framework for analysing manufacturing flexibility. *Int. J. Oper. Prod. Manag.* **2005**, *25*, 973–996. [CrossRef]
64. Singh, N.; Arora, N.; Rani, S. Performance Prediction of Flexible Manufacturing System using Queueing Networks. *Int. J. Syst. Softw. Eng.* **2016**, *4*.

designs

MDPI

Article

Implementation of Simulation Modeling of Single and High-Volume Machine-Building Productions

Nadezhda Savelyeva [1], Tatyana Nikonova [1,*], Gulnara Zhetessova [1], Khrustaleva Irina [2], Vassiliy Yurchenko [1], Olegas Černašėjus [3], Olga Zharkevich [1], Essim Dandybaev [1], Andrey Berg [1], Sergey Vassenkin [1] and Murat Baimuldin [1]

[1] Department of Technological Equipment, Abylkas Saginov Karaganda Technical University, Karaganda 100027, Kazakhstan; n.a.savelyeva@gmail.com (N.S.); zhetesova@mail.ru (G.Z.); juvv76@mail.ru (V.Y.); zharkevich82@mail.ru (O.Z.); 22526633@mail.ru (A.B.); sergey_nv_1979@bk.ru (S.V.); murat_owl@mail.ru (M.B.)

[2] Graduate School of Mechanical Engineering, Peter the Great St. Petersburg Polytechnic University, 195251 St. Petersburg, Russia; irina.khrustaleva@mail.ru

[3] Department of Mechanics and Materials Engineering, Vilnius Gediminas Technical University, LT-10223 Vilnius, Lithuania; olegas.cernasejus@vgtu.lt

* Correspondence: nitka82@list.ru; Tel.: +7-7057655599

Abstract: The authors of this article analyze the existing methods and models of technological preparation of machine-building industries. The structure of a three-level simulation model with network-centric control, the structures of individual elements of the simulation model, and the process of simulation modeling are described. The criteria for choosing a rational option for the processing technological route have been determined. During this research, a simulation program was implemented in C++. It allows you to select the optimal scenario for the operation of a production site based on two criteria: time and cost. The volume of implementation is about 2×10^3 lines of code. A diagram of the modeling algorithm for the implemented program and a description of the classes and their interactions are given in the article. The developed simulation model was tested at a machine-building enterprise using the example of the "Pusher" part, manufactured under single-unit production conditions. The technological equipment used for the manufacture of this part was formed in the form of input data of the simulation model. The results of simulation modeling for the selected part are described. For each variant of the technological processing route, the values of variable costs and the duration of the production cycle were determined.

Keywords: technological preparation of production; automation of production processes; small-scale production; single production; machine-building enterprises

1. Introduction

The prevailing trend in modern production organization involves shifting towards sophisticated digital, intelligent production technologies, and robotic systems. Embedded within this trend is the defining direction for the future of material goods and service production, which revolves around employing production networks featuring network-centric control [1]. Examples of modern production sites are complexes of multifunctional computer numerical control (CNC) machines, 3D printers, and robots, which are integrated into a network to create conditions for effective planning and optimal implementation of parallel technological processes [2–5].

One characteristic of technological processes is their ability to adapt to small-scale or custom production in various fields such as mechanical engineering, raw material processing, assembly of multi-component products, and so forth [6,7].

Technological preparation of production (hereinafter referred to as TPP) is one of the most important stages of the process for the production of parts. The Chamber of

Commerce and Industry provides options for solving problems in many areas: for instance, guaranteeing the feasibility of product design; planning technological processes and crafting technological documentation; designing and producing technological equipment; and the organization and management of the process of technological preparation of production [8]. Typically, production automation and manufacturing processes involve standardizing design elements and technological procedures [9], although in situations of individual or small-scale production, this strategy might not be efficient due to the significant preparatory expenses [10].

The objective of this endeavor is to construct a simulation model aimed at diminishing the duration and labor intensity involved in the technological preparation of individual and small-scale production, thereby enhancing the efficiency of production processes.

To build a new simulation model, it is necessary to use an analysis of various technological processing routes and the corresponding technological machinery used. One of the most effective methods for assessing the many options for technological processes and selecting the most suitable one is simulation modeling of production processes and multi-criteria analysis based on a selected set of criteria [11,12]. When speaking about the technological preparation of production, the authors will further refer to its model.

One of the factors for increasing the efficiency of an enterprise and its competitiveness is the automation of production processes [13]. Production processes also include technological preparation of production, the automation of which through digitalization ensures not only a reduction in production preparation time, but also optimizes the overall costs of manufacturing products. In addition, such automation ensures and allows you to adapt the technological process to changes in external conditions and quickly recalculate the technological process [14].

The main approaches of modern TPP models include the theory of computational complexity, analysis of design and technological elements, analysis of the similarity of design solutions, and a scheduling system.

Complexity theory, as the basis of TPP, allows one to calculate the complexity of a system based on the complexity of the elements included in the system [14–18]. The complexity of the system (S) can be represented as the sum of the products of the complexity of an element of a certain type (S_i) and the number of such typical elements (k_i):

$$S = \sum_{i=1}^{n} S_i \cdot k_i \quad (1)$$

Along with the complexity of individual elements of the system, an assessment is made of the complexity of the relationships between them and the load on the system with connections.

This theory was used as the basis for a number of techniques aimed at increasing the efficiency of the TPP. These methods allow for, based on the analysis of time indicators, the duration of the production cycle and the frequency of inter-operational pauses, while facilitating the design of numerous technological processing routes for a product batch do not enable the estimation of production costs for manufacturing the batch [19].

The theory, based on the analysis of design and technological elements (hereinafter referred to as DTE), makes it possible to increase the efficiency of TPP through various strategies and selection of equipment [20–22]. If we imagine the final product as a set of DTEs, i.e., an elementary surface or a set of elementary surfaces that have a common design purpose and are characterized by a common manufacturing route, then the DTE can be described by characteristic parameters. These include the diameter of the DTE, the ratio of the length of the DTE to the total axial length of the part, the location of the material for cylindrical surfaces (inside/outside), the shape of the generatrix for surfaces obtained by rotating the generatrix around the axis of rotation of the part or other features of the shape of the DTE, and the location of the DTE in detail. Assessing the degree of compliance of a cutting tool with a number of criteria allows one to set an algorithm for selecting the optimal set of tools [23].

Different strategies for using a selected set of tools allow you to construct a set of possible outcomes, that is, the order in which these tools are used. Strategies take into account machine processing time and tool change time, its cost, and other costs (for example, the cost of processing an elementary surface).

Using a mathematical apparatus, a multi-criteria optimization problem was formulated, setting the significance of the criteria described above by weighting coefficients (a_i) [24]:

$$J(x) = \sum_{i=1}^{m} a_i \cdot f_i(x); a > 0, \sum_{i=1}^{m} a_i = 1, \tag{2}$$

where $f_i(x)$—one of the criteria for the processing strategies under consideration.

This technique makes it possible to construct instrumental strategies for processing DTEs and determine the parameters of the tools used in them. However, this technique does not take into account an important parameter—the accuracy of DTE processing.

The methodology for analyzing the similarity of design solutions during the technological preparation of production is based on the formalization and comparison of design and technological solutions [25–27]. The basic concept of this technique includes a technological complex (hereinafter referred to as the T-complex), which is a set of various standard surfaces for which there is a trajectory in which these surfaces can be processed together. Such complexes correspond to certain technological methods, used separately or in combination, depending on the expected production conditions and workmanship. T-complexes are characterized by types of incoming surfaces; indicators of production and operational quality; and external attributes of the connection of this complex with the others. These connections can be represented in the form of a model—a graph, the nodes of which are the identifiers of the selected T-complexes, and the edges are the corresponding connections between them. The connection graph makes it possible to estimate constructive similarity in two ways, including by the composition of T-complex models and the structure of connections of T-complexes [28,29]:

$$S = \frac{2m}{b + c} \tag{3}$$

where S—the value of the selected assessment of the design similarity of the part models.

Depending on the selected method, different values are associated with the parameters.

The disadvantages of this approach include incomplete consideration of the design of the part and its technical parameters and the inability to evaluate many alternative pathways for technological processing. As a result, the length of time for completing the production cycle and the amount of production costs are not determined accurately.

The TPP scheduling system is based on the calculation of time intervals between already planned operations for a given set of parts planned for production in the time period under consideration [30–32]. In the process of scheduling, the complexity of performing each operation on specific equipment is assessed: processing time for individual technological operations, as well as the total processing time for a given technological route, and equipment downtime during the processing of technological routes. Thus, based on the results of successfully implemented technological routes, it is possible to determine the most suitable equipment that has the best performance for still unprocessed routes. The criterion of minimizing service time is a solution to the problem of optimizing the production process [33]:

$$k_{1(i)} = \min\left(\sum_{I \in I_S} W_{ij}\right) \tag{4}$$

where W_{ij}—time interval between the end of the $(j-1)$-th and the beginning of the j-th operation of the i-th part;

I_s—many order details S.

Planning is carried out taking into account the loading of technological equipment for processing the ith part.

Minimization of the time required to complete all work on a given set of parts is carried out using the following formula [20]:

$$k_3 = \min(\sum_{I \in I_S} R_{ik} - \sum_{I \epsilon I_S} F_i), \quad (5)$$

where R_{ik}—the amount of work that needs to be performed on the i-th order detail S.

With this theory, the planning of technological processes is conducted utilizing the foundation of available data on successful implementations of similar or identical technological processes. Single and small-scale production involves the production of complex products in a small series or single copy, which makes it impossible to use this model. The use of equipment with the highest productivity for all identical technological routes does not allow for optimizing the overall duration of the production cycle, as well as the loading of all available equipment. In addition, this model does not take into account production costs.

Thus, methods based on complexity theory make it possible to design many technological processing routes, but do not allow for estimating production costs for the manufacture of a given batch of products. The methodology for analyzing design and technological elements makes it possible to determine strategies for processing DTE and the necessary parameters and tools used in them. Unfortunately, the parameters used do not take into account the accuracy of DTE processing. Methods based on the method of similarity of design solutions do not allow taking into account the design features and technical parameters of the product. In addition, the methodology does not allow us to evaluate many options for technological processing routes, as well as determine the precise length of the production cycle. The scheduling method does not allow for the correction of the start time of already existing technological routes. This constraint results in a notable limitation on the overall number of potential production scenarios and prevents the selection of the most optimal production process variant.

Therefore, the scientific novelty of this work is the construction of a simulation model that describes the method for selecting the best option for the production process of technological preparation of single and small-scale production based on multi-criteria analysis. The purpose of the article is to enhance the effectiveness of preparing technology for single and small-scale production based on simulation modeling of technological processes. The objectives of the study are to identify the main criteria for choosing a rational variant of the technological process, implement a simulation program for individual and limited-scale production with the possibility of multi-criteria analysis based on selected criteria, and test the implemented program for simulating the manufacture of the "Pusher" part.

2. Materials and Methods

The main task of creating a model is to choose the most suitable scenario for the operation of a production site using the provided information.

It must be taken into account that, depending on the equipment used and processing methods, the same product may have many solutions [34,35].

As part of the work, a generator of implementations of technological processes for the production of a given range of products on a given equipment (hereinafter referred to as ITTP) was built. The main task of the ITTP is to generate the largest possible number of possible implementation options at both the structural and parametric levels.

The generation of various implementations occurs by searching through all possible priorities of processing operations for given technological processes. Each technological process, as well as equipment, is given priority in the form of positive integers. Moreover, as the number increases, the priority increases. The given preferences within the technological process influence the choice of application, while the preferences regarding the equipment determine the sequence in which equipment is selected for executing the technological route. If a request with higher priority arrives while servicing an application, the current application is not interrupted, and the incoming application awaits its turn.

Likewise, priorities establish the order for choosing applications from sources, representing parts associated with specific technological processes, and dictate the sequence for their processing on designated technological equipment.

Upon completing these procedures, the ITPP yields a comprehensive array of implementations, from which suitable options must be chosen to satisfy the relevant decision-maker.

Hence, we encounter a multi-criteria problem with m objective functions delineated on a finite set D [36].

Problems of multi-criteria optimization of technological processes arise at different stages of modeling. Within the framework of a given specific structural implementation of a technological process, "continuous" problems may arise related to the multi-criteria choice in the space of such continuous variables as the processing time of parts on various machines, processing costs, cost of machine downtime, etc. Similar problems can be solved for all possible (generated) structural implementations. At the next level of modeling, when all particular implementations have been optimized according to their controllable parameters, we have a finite set of implementations and each element of this set has a vector estimate—the duration required for executing the technological process, the cost of implementation, the downtime of machines, etc. The number of structural implementations generated can be quite significant.

Thus, the task arises of automating the decision-making process for choosing a rational option for implementing a technological process in accordance with the system of preferences of the decision maker. Such problems belong to "discrete" multi-criteria selection problems.

When designing appropriate decision support systems, one can focus on existing methods of multi-criteria selection [36–39].

2.1. Structure of the Simulation Model and Description of the Simulation Process

The research methodology and multi-criteria solution are based on the application of Pareto principles and the associated concepts of effective (Pareto optimal) and suboptimal solutions [40–42]. We examine a multi-objective optimization issue of the form:

$$f_i(x) \to \max,\ \ npu\ x \in D \tag{6}$$

$$f_i : D \to R,\ i = 1, \ldots,\ m; \tag{7}$$

$$D \subseteq R^m \tag{8}$$

Hence, m functions or functionals f_i are provided, which map the set of Dn-dimensional vectors $x = (x_1, \ldots,)$ to real numbers R. The selection of optimal values for the controlled parameters (x) is not performed across the entire n-dimensional space Rn, but solely within its subset D.

Equation (8) can be depicted as the optimal parameter selection problem $(x_1, \ldots, x_n)$ for a system, evaluated based on quality metrics $(f_1, \ldots, f_m)$. In this context, the constraint $x \in D$ signifies our technological and other constraints that determine the feasible values for x_i. In particular, the processing times for parts on machines and other equipment are obviously non-negative. Certain limitations can be established based on the existing information, which makes it possible to exclude obviously unsuccessful option x from consideration.

To address the outlined multi-criteria problem in this research, we will employ the linear convolution approach.

This technique of "scalarization" (convolution) of Equation (8) enables us to substitute the vector optimality criterion $f = (f_1, \ldots,)$ with a scalar criterion J: $D \to R$. It relies on the linear amalgamation of all partial objective functionals $f_1, \ldots,$ into one [43]:

$$J(x) = \sum_{i=1}^{m} \alpha_i f_i(x) \to \max;\ \text{at}\ x \in D;\ \alpha_i > 0,\ \sum_{i=1}^{m} \alpha_i \tag{9}$$

In this case, the weighting coefficients α_i can be viewed as indicators of the relative significance of individual criterion functionals f_i. Giving more weight to a criterion f_j implies it should contribute more to the overall sum, hence requiring a higher value for α_j. When dealing with significantly disparate criteria, determining the final set of coefficients αi is often challenging and typically relies on informal considerations, typically stemming from the outcomes of expert analysis.

The method outlined above for tackling the posed multi-criteria problem does not pinpoint a single optimal solution. Solutions associated with different sets of weighting coefficients are equally valid elements within sets of effective and suboptimal solutions. These solutions, as per the general framework of the decision problem, represent the cores of the respective binary relations (Pareto's principle and Slater's condition), thus fulfilling the requisite criteria. However, from a practical standpoint, such as in problems involving the selection of rational options for organizing technological processes, additional insights into the decision maker's preferences should be factored in. In this context, the Pareto principle merely serves to refine the pool of potential solution candidates, eliminating clearly inferior options from consideration.

Methods for selecting a single solution to a multi-criteria problem exist and are associated with the use of models and procedures designed to structure and quantitatively describe the subjective opinion of the decision maker (technologist). These methods are not discussed in this paper.

2.2. Description of the Methodology for the Simulation Modeling Process and Development of the Structure of the Simulation Model

The main task is to analyze the maximum possible number of options for processing technological routes for given products, in order to select the optimal work scenario according to two criteria: time and cost of processing.

A simulation model of technological preparation of production can be presented as a model structured into three tiers with network-centric management [43–45]:

Level 1: Management of technological macro-operations (hereinafter referred to as MO—these are sequences of messages exchanged between the built-in controllers of first-level objects and second-level computers that control the implementation of the technology) of machine tools and robots. Every component of the production system is supervised, involving evaluating the present condition of each system component, verifying the accurate implementation of ongoing tasks for each component, and transmitting data indicating the beginning or completion of any actions of elements of the production system, as well as errors in their operation. At the first level, control of each element of the production system is carried out.

Level 2: Management and control of technological processes, described in the form of sequences of macro-operations: the interaction of elements of the production system is monitored, as well as the transmission of data on the state of network objects and their environment to the third level. At the second level, the interaction of elements of the production system is monitored, as well as the transmission of data on the state of network objects and their environment to the third level.

Level 3: Hierarchical optimization and planning involving multiple criteria of production processes. At the third level, the "data lake" is analyzed, on the basis of which dynamic planning of the workshop is carried out. When planning, the execution of MO by production network objects is optimized, which takes into account the possibilities of parallel execution of MO, their synchronization, areas of acceptable values of state parameters, conditions for reliable execution of the plan, etc. Since specific success criteria are defined for each operating mode of each network object, the third level must solve the problems of multi-criteria planning, as well as deriving new or changing existing control rules or assessing the state parameters of network objects.

Within the framework of this study, the third level of the model is considered, at which the modeling and analysis of various options for the process of determining the technological processing paths and the corresponding technological equipment utilized for each batch of components is conducted. The model being discussed will be realized in the form of a queuing system, abbreviated as QS [46].

QS are models of systems into which applications (requirements) are received at random times from outside or inside. Each application received into the system must be serviced by the system. The service system is a set of maintenance equipment and personnel with the appropriate organization of the maintenance process. The basic concepts of QS include [47]:

1. A source that generates applications, and a set of sources create the input flow of applications into the system. As a rule, sources can be of two types, finite and infinite, which differ in the methods of generating requests.
2. Buffer memory (storage location of the request queue). As a rule, it is divided into two types: general and zone. The shared memory stores requests from various sources, and the order in which they are recorded is determined only by the buffering discipline. Zone memory is a buffer divided into zones, each of which records requests only from a specific source. Thus, the quantity of zones aligns with the number of sources.
3. Devices that service requests and create an output stream of requests after servicing.
4. Arrangement manager: sends a request for service or to buffer memory if there are no free devices and organizes the refusal or knocking out of an application from the buffer memory if there are no free places left in the buffer.
5. Selection manager: selects the device on which applications will be processed and selects a request from the memory buffer, if it exists there.

Figure 1 shows the standard structure of the QS:

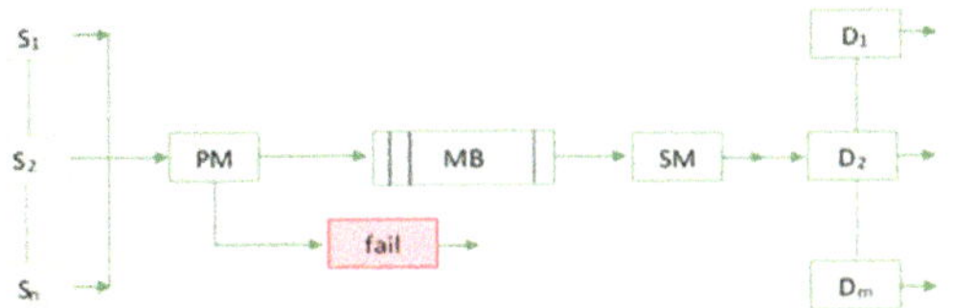

Figure 1. An example of the structure of a queuing system, where $И_i$ (i = 1 ... n)—sources, PM—production manager, MB—memory buffer, SM—selection manager, Π_i (i = 1 ... m)—device.

Unlike the standard structure, the following changes were made to the implemented model:

1. In the model, the sources are represented by technological processes (referred to as TP). TP denotes a precise sequence of tasks, starting from the delivery of raw materials and tools from the warehouse to the machines, and concluding with the storage of finished products of specific types at the warehouse. Moreover, the durations of all tasks are explicitly defined.
2. No buffer memory.
3. The system operates seamlessly as it cannot bypass any stages of the process.
4. The quantity of devices in the system is contingent upon the designated technological equipment required for executing the specified technological processes.
5. During the modeling process, applications are generated—a separate stage of the technological process. Simultaneously, the system handles one occurrence of each technological process. Consequently, the quantity of applications within the system does not surpass the overall count of technological processes.

3. Result

Algorithmization of Simulation Modeling

To analyze the widest range of potential technological processing routes for designated products and subsequently select the optimal operational scenario based on processing time and cost criteria, the simulation algorithm can be delineated into four primary phases: user input processing, preparatory phase, generation of implementation options, and selection of the optimal operational scenario (refer to Figures 2 and 3).

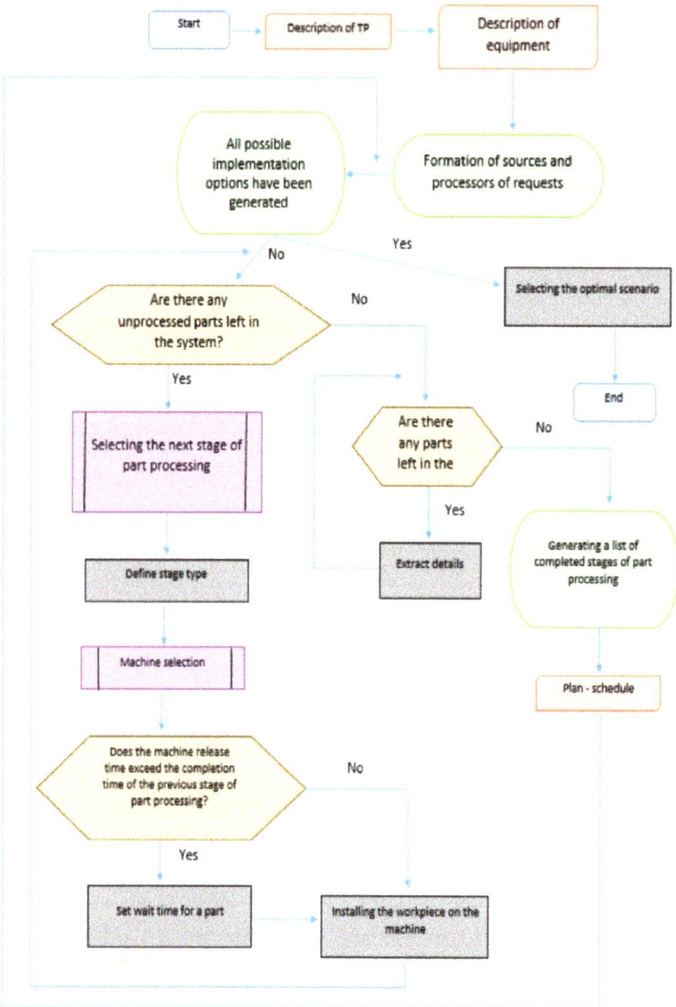

Figure 2. Simulation model algorithm diagram.

User Input: Input data are provided to the simulation model to ensure its proper functioning:

1. A compilation of TPs along with the number of implementations for each, comprising the operation type, time required, and implementation cost.
2. A collection of technological equipment—where one unit of a specified piece of equipment corresponds to one or more types of operations conducted.

Preparatory Stage: this involves the creation of processors and request sources, considering the user-provided information.

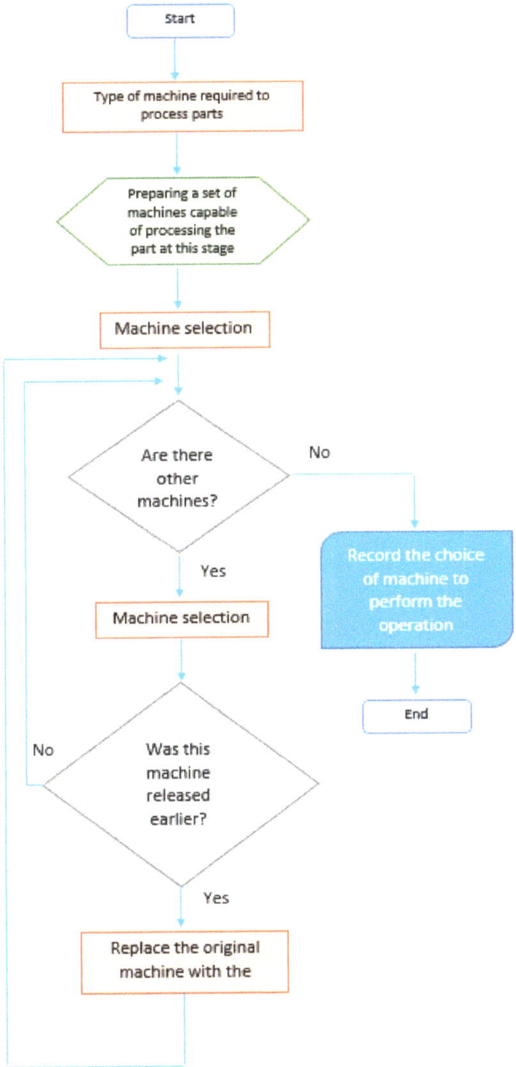

Figure 3. Scheme of the technology process description algorithm (TP).

Main Stage: This stage encompasses the operation of the ITTP. The process of generating implementation options allows you to generate the largest number of possible implementation scenarios by enumerating the priorities of operations for processing given data. Priorities allow you to set the sequence of selecting applications from sources, that is, parts corresponding to certain technological processes, as well as the sequence of their processing on specified equipment (machines) [48,49].

The optimal operational scenario is determined utilizing the linear convolution technique [50,51].

A simulation model consists of a set of simple elements, designed as basic abstractions. Based on these concepts, it is possible to formulate a basic algorithm for the implemented component of the simulation model.

To implement the TP, it is necessary to consistently perform all its stages. Each such stage can be represented in the form of a simulation model application. The application must contain the following information:

1. The TP's identification number implemented within the system (indicated by the id_ field);
2. The sequence number of the operation within this TP (id_);
3. The implementation number of the TP (id_);
4. The time when the application enters the system (birth_time_).

Fresh requests within the system await the availability of technological equipment capable of executing the relevant technological operation. Upon submitting a service request, the waiting duration (wait_time_) and the contractor's identifier (exec_id_) are logged. Consequently, for each implementation of a technological process, it becomes feasible to compute the delay time for executing technological operations. A comprehensive diagram of the Request class is depicted in Figure 4.

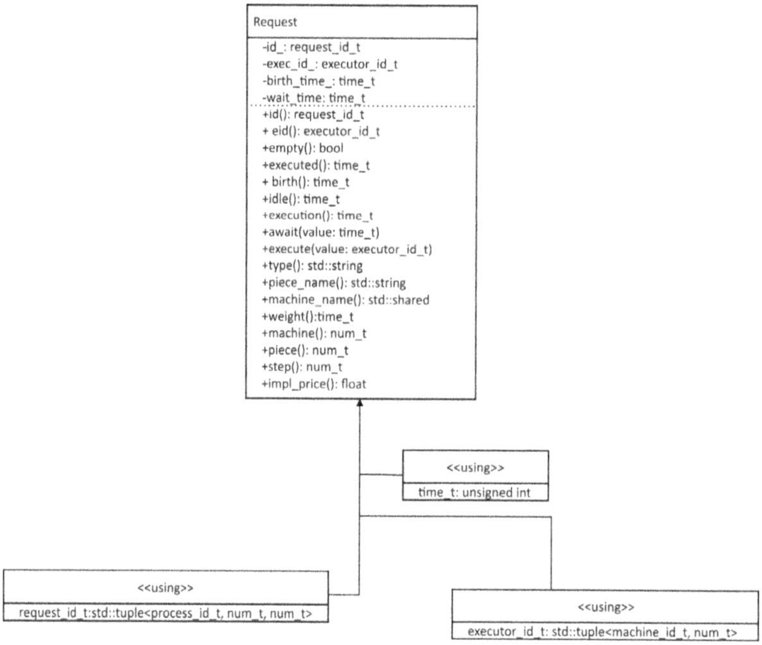

Figure 4. UML diagram of the Request class.

TPs are portrayed in the system as application sources, which monitor the number of implementations of a particular TP and generate a unique application identifier. The execution of the TP process is reiterated until the designated number of implementations is accomplished. As the TP stages unfold sequentially, each source furnishes only one application at a time. Consequently, the quantity of requests within the system corresponds to the number of concurrent processing operations. This pool of orders can be managed: for example, you can select orders of a certain type of operation.

The management of the application pool is overseen by the Device_union class. Order sources are depicted through the Device class, while TPs are portrayed by the Process class. Figure 5 presents a comprehensive UML diagram illustrating the detailed relationships among these classes and their interactions with the Request class.

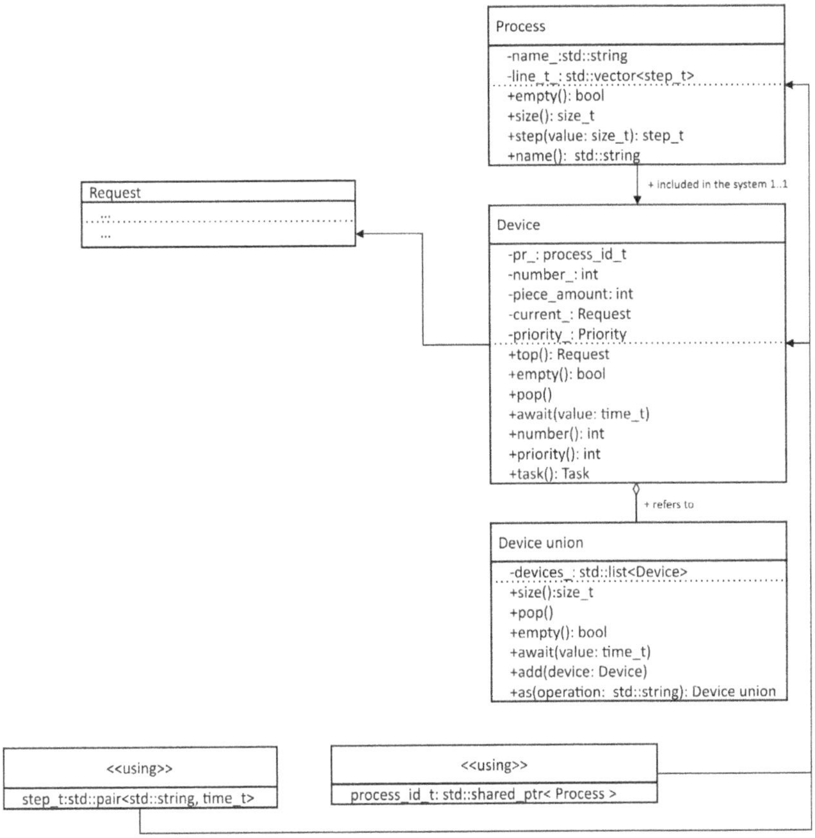

Figure 5. UML diagram of technological processes and sources of requests.

The technological equipment is embodied by the Machine class, encompassing the following details:

1. The designation of the process equipment (captured by the name_attribute);
2. The categories of operations it can execute (line_).

The resources associated with a specific piece of technological equipment are overseen by the Consumer class, which retains data regarding the most recent completed request. The information contained within the application is sufficient for determining the readiness of a particular piece of technological equipment to handle the subsequent application.

Throughout the modeling procedure, the Consumer_union class facilitates the allocation of the requisite unit of technological equipment based on the chosen request, considering the priorities and availability status of the units of technological equipment.

Below is a detailed UML diagram of technological equipment and application processors (Figure 6).

The complete class diagram is shown in Figure 7.

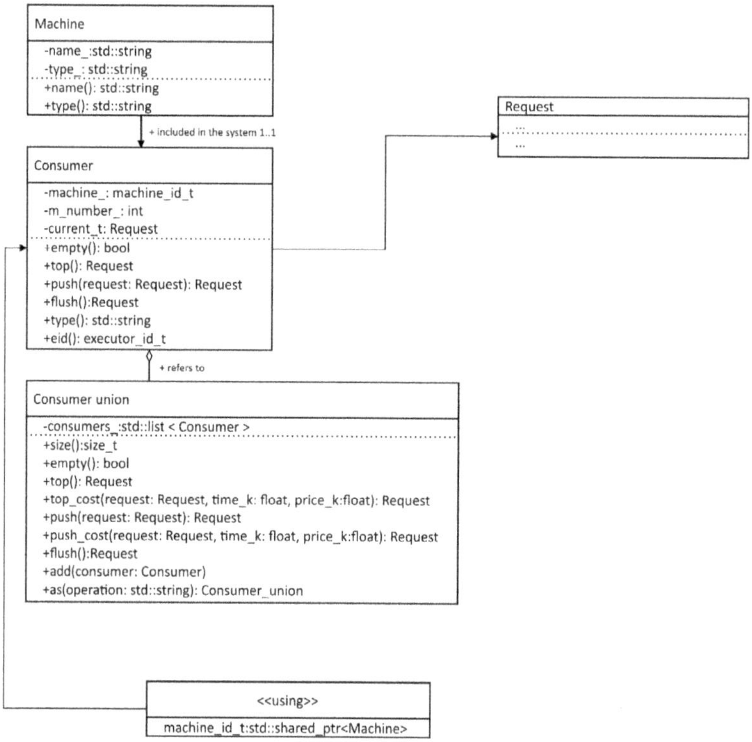

Figure 6. UML diagram of technological equipment and application processors.

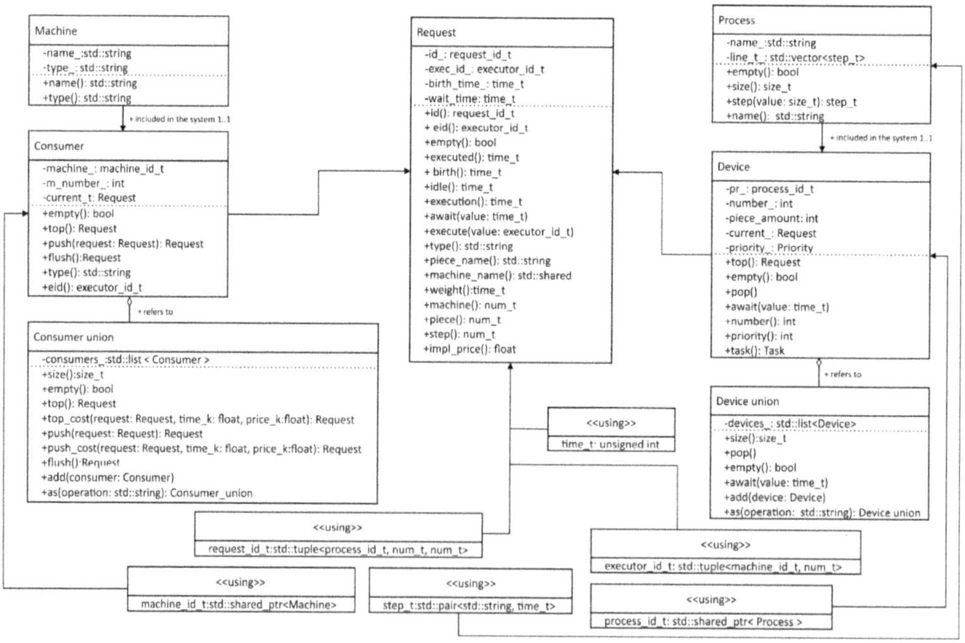

Figure 7. UML diagram of the complete simulation model.

4. Discussion

In existing models based on queuing systems, the production manager sends requests either directly for service or to buffer memory. If there are no free devices, the dispatcher organizes a refusal or the request from the buffer memory. If there are no free places left in the buffer the selection manager selects the device on which the requests will be processed, selects the request from the memory buffer if it is there. This whole process, in turn, makes it difficult to parameterize the state of the network according to specific success criteria for each operating mode of a certain model object. In the model proposed, the quantity of devices within the system is contingent upon the specified technological equipment essential for executing the designated technological processes. Consequently, in this model, there exists no buffer memory, thereby preventing the system from bypassing individual stages of the technological process and ensuring flawless operation. Throughout the modeling phase, applications are generated as a distinct step of the technological process. Concurrently, the system handles a singular instance of each technological process. Hence, the generation of applications within the system is managed in such a manner as to not exceed the overall count of technological processes. The developed simulation model was utilized in simulating the technological preparation of production for mimicking the production of the "Pusher" component.

The manufactured part "Pusher" (Figure 8) has the following technical parameters:

1. Dimensions: maximum diameter 24 mm, length 26 mm;
2. Exact dimensions:
   - outer diameter ϕ 24h6 with roughness Ra 0.8;
   - internal diameter ϕ 11H8 with roughness Ra 1.6;
   - internal threaded surface M8-7H, roughness Ra 3.2.
3. General tolerances for other dimensions: H14, h14, ±IT14/2.

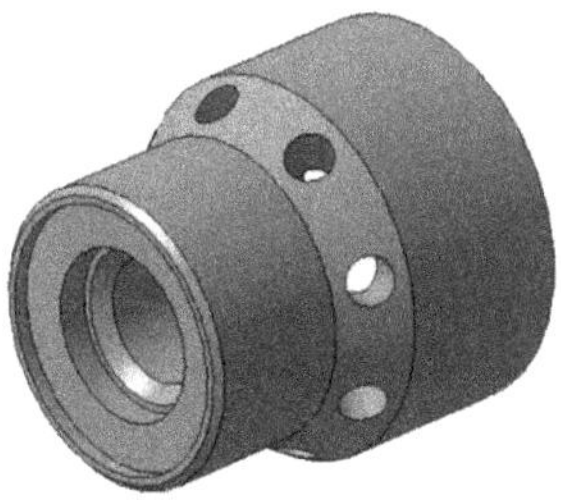

Figure 8. Solid model of the "Pusher" part.

Four types of machines are involved in the processing route:

- Thun—universal lathes;
- Tchpu—lathes with numerical control;
- TFNC—turning and milling machines with numerical control;
- KSh—cylindrical grinding machines.

Table 1 describes the types of machines and their method for processing technological units.

Table 1. Varieties of machine tools utilized and their machining techniques.

№	Machine Type	Type of Machinery Designation	Machining Technique	Designation of Processing Method
1	Tun	mach1	tapping threads	op1
			drilling	op2
			semi-finish turning	op3
			rough turning	op4
			finishing turning	op5

Table 1. *Cont.*

№	Machine Type	Type of Machinery Designation	Machining Technique	Designation of Processing Method
2	T_{CNC}	mach2	tapping threads	op1
			drilling	op2
			semi-finish turning	op3
			rough turning	op4
			finishing turning	op5
3	TF_{CNC}	mach3	tapping threads	op1
			drilling	op2
			semi-finish turning	op3
			rough turning	op4
			finishing turning	op5
			pre- grinding	op6
4	KS	mach4	tapping threads	op1

In accordance with Table 1, the input data for the simulation model for the given equipment were formulated:

4.1. Equipment

(:consumer 0:worker (:machine "mach1":line ("op1" "op2" "op3" "op4" "op5"):factor 1.0:):)
(:consumer 1:worker (:machine "mach2":line ("op1" "op2" "op3" "op4" "op5"):factor 1.2:):)
(:consumer 2:worker (:machine "mach3":line ("op1" "op2" "op3" "op4" "op5"):factor 1.3:):)
(:consumer 3:worker (:machine "mach4":line ("op6"):factor 1.0:):)
Explanation:

- Consumer ID—unique identifier for the designated equipment;
- Operator—details regarding the designated equipment;
- Machinery—designation of the specified equipment;
- Operation list—roster of operation types feasible on this equipment;
- Factor—cost factor (greater values correspond to higher part processing costs on this equipment).

For each block of surfaces, the following sequence of processing of the design and technological elements included in their composition was determined:

- MB(1/1): DTE(1H-1)1 → DTE(2H/2-1)2 → DTE(2H/1-1)2 → DTE(2H/2-1)1;
- MB(1/2): DTE(1H-1)2 → DTE(2H/1-4)1;
- MB(2/1): DTE(2B/1-2)1 → DTE(2B/1-1)2 → DTE(2B/2-1)1 → DTE(1B-1)1;
- MB(2/2): DTE(2B/1-2)3 → DTE(2B/1-3)4 → DTE(2B/2-2)2 → DTE(1B-1)2 → DTE(1B-2)3;
- MB(4/1): DTE(4B/1-2)1.

For the "Pusher" part, the following sequence of processing of surface blocks was defined: MB(1/2) → MB(2/1) → MB(1/1) → MB(2/2) → MB(4/1).

Technological elements are included in the design and technological elements are processed sequentially.

The given surface blocks were divided into final operations (Table 2):

Table 3 shows the relationship between processing methods and their cost and execution time:

In accordance with relation (7) and Tables 2 and 3, the TP input data for the simulation model were formulated, and the input data of the simulation model for a given technological process are as follows:

Table 2. Analysis of surface blocks for final operations.

№	Block of Surfaces		DTE	Processing Method	Machining Process Identification
1	MB(1/2)	1	DTE(1H-1)2	rough turning	op4
		2	DTE(2H/1-4)1	re-grinding	op6
				rough turning	op4
				semi-finish turning	op3
				finishing turning	op5
2	MB(2/1)	1	DTE(2B/1-2)1	drilling	op2
				semi-finish turning	op3
		2	DTE(2B/1-1)2	drilling	op2
		3	DTE(2B/2-1)1	drilling	op2
		4	DTE(1B-1)1	rough turning	op4
3	MB(1/1)	1	DTE(1H-1)1	rough turning	op4
		2	DTE(2H/2-1)2	rough turning	op4
		3	DTE(2H/1-1)2	rough turning	op4
		4	DTE(2H/2-1)1	rough turning	op4
4	MB(2/2)	1	DTE(2B/1-2)3	drilling	op2
				semi-finish turning	op3
		2	DTE(2B/1-3)4	drilling	op2
				semi-finish turning	op3
				rough turning	op5
		3	DTE(2B/2-2)2	drilling	op2
				semi-finish turning	op3
		4	DTE(1B-1)2	rough turning	op4
		5	DTE(1B-2)3	rough turning	op4
				semi-finish turning	op3
5	MB(4/1)	1	DTE(4B/1-2)1	drilling	op2
				thread cutting	op1

Table 3. Correlation of processing methods with cost and execution time.

№	Designation of Processing Method	Processing Method	Processing Time (Working Hours)	Amount of Costs (Conventional Units)
1	op1	tapping threads	32	4800
2	op2	drilling	18	5600
3	op3	semi-finish turning	12	8200
4	op4	rough turning	20	6800
5	op5	finishing turning	16	7400
6	op6	re-grinding	29	5800

4.2. Technological Process

(:device 0:task ((:process "TP-T":line (("op4" (:time 40:price 13,600:)) ("op3" (:time 12:price 8200:)) ("op5" (:time 16:price 7400:)) ("op6" (:time 29:price 5800:)) ("op2" (:time 18:price 5600:)) ("op3" (:time 12:price 8200:)) ("op2" (:time 36:price 11,200:)) ("op4" (:time 100:price 34,000:)) ("op2" (:time 18:price 5600:)) "op3" (:time 12:price 8200:)) ("op2" (:time 18:price 5600:)) ("op3" (:time 12:price 8200:)) ("op5" (:time 16:price 7400:)) ("op2" (:time 18:price 5600:)) ("op3" (:time 12:price 8200:)) ("op4" (:time 40:price 13,600:)) ("op3" (:time 12:price 8200:)) ("op2" (:time 18:price 5600:)) ("op1" (:time 32:price 4800:))):) 1):)

Explanation:

- Device №—identification number of the specified TP;
- Tsk—information about a given TP;
- Process—name of the specified TP;
- Line—list of operations that make up the given TP:
    (1) time—processing time of a given operation;
    (2) price—the cost of processing a given operation.
- No.—number of implementations of a given TP.

The output of the simulation model is presented in MRD (Machine-readable Data) format, where:

id: details concerning the ongoing application;
name: TP name;
num: current TP part number;
step: operation number in progress;
type: type of ongoing operation;
birth: timestamp of when the current request entered the system;
price: cost of the ongoing operation;
wait: start time of processing the current request;
executor: details regarding the equipment processing the application:
name: name of the equipment processing the current request;
num: equipment number processing the current request;
execution: completion time of application processing;
impl_price: cost after the application is processed.

A listing of the final schedule of the simulation model with the main set of cutting tools:

4.3. Part of the Simulation Listing Is Shown Below

(:id (:name TP-T:num 0:step 0:type op4:):birth 0:price 13,600:wait 0:executor (:name mach1:num 0:execution 40:impl_price 13,600:):)

(:id (:name TP-T:num 0:step 1:type op3:):birth 40:price 8200:wait 40:executor (:name mach1:num 0:execution 52:impl_price 8200:):)

(:id ():birth 0:price 0:wait 0:executor ():) (:id (:name TP-T:num 0:step 2:type op5:):birth 52:price 7400:wait 52:executor (:name mach1:num 0:execution 68:impl_price 7400:):)

(:id (:name TP-T:num 0:step 4:type op2:):birth 97:price 5600:wait 97:executor (:name mach1:num 0:execution 115:impl_price 5600:):)

(:id (:name TP-T:num 0:step 5:type op3:):birth 115:price 8200:wait 115:executor (:name mach1:num 0:execution 127:impl_price 8200:):)

(:id (:name TP-T:num 0:step 3:type op6:):birth 68:price 5800:wait 68:executor (:name mach4:num 3:execution 97:impl_price 5800:):)

Res_price:175,000 Res_time: 471

The full listing is provided in Appendix A.

As a result of simulating the production of the "Pusher" part, the optimal scenario for the operation of the production site was selected according to the two criteria of time and cost. The final duration of the production cycle was 471 working hours, and the final value of variable costs was 175,000 conventional units, which confirms that the created algorithm and simulation model are adequate under the conditions of a single production. In the future, it is planned to develop a simulation model with an alternative set of cutting tools. Typically, the choice of cutting tool affects the processing time of the part and its cost.

5. Conclusions

In this article, the authors solved the problem of automating the decision-making process of choosing a rational option for implementing a technological process in accordance with the system of preferences of the decision maker. Such problems belong to "discrete" multi-criteria selection problems.

1. The manuscript scrutinized technological production preparation models outlined in both Russian and international literature. In delving into the scientific literature, models focusing on complexity theory, scrutiny of individual design and technological facets, examination of design solution similarities, and scheduling were analyzed.

Through the examination of various methods and models aimed at enhancing the efficiency of production technological preparation, the following observations emerged:

Multi-criteria analysis is notably absent in the majority of production technological preparation models. The selection of optimal technological processing routes and processing strategies for individual part elements is not elaborated upon; rather, decisions are predominantly based on analyzing production cycle duration or production cost assessments.

2. Methods for estimating the value of inter-operational breaks, based on the method of mathematical statistics, do not allow an accurate assessment of the duration of the production cycle and a highly accurate prediction of the production time of the product.
3. Approaches relying on the similarity of design solutions during the design of a technological process fail to consider all the design intricacies of the component and its technical specifications. Moreover, they do not facilitate the assessment of numerous processing route options or the determination of production duration with high precision.
4. A number of the described methods use a production process planning method based on the analysis of identical operations that have already been implemented in the conditions of a particular enterprise. This method does not provide high accuracy in single and small-scale production types due to the wide variety of design and technological solutions.
5. In the model of technological production preparation, based on the scheduling method, when forming a production schedule, adjustments to the start time of already existing technological operations are not allowed, which significantly limits the number of simulated production scenarios. Consequently, this does not allow for choosing the most rational option for the production process.

To build a new simulation model, the authors used an analysis of various technological processing routes and the technological equipment used. One of the most effective methods for assessing multiple options for technological processes and selecting the most suitable one is simulation modeling of production processes and multi-criteria analysis based on a selected set of criteria.

Unlike the standard structure, the following changes were made to the implemented model:

1. In the model, the sources are represented by technological processes (referred to as TP). TP embodies a strict sequence of operations, encompassing the delivery of raw materials and tools from the warehouse to the machines, culminating in the retrieval of finished products of specified types at the warehouse. Additionally, the durations of all operations are precisely defined.
2. There is no provision for buffer memory.
3. The system operates flawlessly as it cannot bypass individual stages of the process.
4. The quantity of devices within the system is contingent upon the designated technological equipment essential for executing the specified technological processes.
5. Throughout the modeling process, applications are generated as a distinct step of the technological process. Concurrently, the system handles a singular instance of each technological process. Consequently, the number of applications within the system does not exceed the overall count of technological processes. Following the analysis, a simulation model algorithm was devised, and UML diagrams were crafted to delineate the system structure, classes, attributes, methods, and object relationships.

6. A simulation program was developed in C++ specifically tailored for single and small-scale production, with the objective of automating the technological processing process by automatically generating a plethora of work scenarios and subsequently selecting the optimal scenario for the production site based on two criteria: time and cost.
7. Upon simulating the production of the "Pusher" component, the resultant production cycle lasted 471 working hours, with variable costs totaling 175,000 conventional units.

In the future, based on the obtained structures and algorithms of the network-centered simulation model, and created UML diagrams, it is planned to develop a new software product that will combine the functions of production preparation systems (technical processes, equipment preparation, route development, preparation of operational maps) and MES (production control, schedule tracking, planning, and optimization). The peculiarity of this development will be that it will not be a direct competitor to large narrowly focused products that have a significant cost and in most cases are not available to small and medium-sized businesses, this software product offers a private solution to a number of mechanical engineering problems with low labor intensity of use.

In the following works, the purpose of rational cutting modes will be considered, based on the calculation of the total processing error and the quality loss function according to the Taguchi theory. The issues of assembly production will also be covered. It is planned to develop a methodology for selecting a rational option for the assembly technological route, and the variables that wield the most substantial influence on them, to implement the developed simulation model: mathematical dependencies for determining the production cycle, mathematical dependencies for determining the duration of the production cycle, and mathematical dependencies for determining the amount of costs.

Author Contributions: Acquisition of the financial support for the project leading to this publication; conducting the research and investigation process, specifically performing the experiments, or data/evidence collection, N.S.; application of statistical, mathematical, computational, or other formal techniques to analyze or synthesize study data; verification, whether as a part of the activity or separate, of the overall replication/reproducibility of results/experiments and other research outputs, T.N.; development or design of methodology; provision of study materials, G.Z.; scrub data and maintain research data (including software code, where it is necessary for interpreting the data itself) for initial use and later re-use; preparation, creation, and/or presentation of the published work, specifically visualization/data presentation, K.I.; oversight and leadership responsibility for the research activity planning and execution, including mentorship external to the core team and creation of models, V.Y.; conducting the research and investigation process, specifically performing the experiments, or data/evidence collection, O.Č.; management and coordination responsibility for the research activity planning and execution; preparation, creation, and/or presentation of the published work by those from the original research group, specifically critical review, commentary, or revision, including pre- or post-publication stages, O.Z.; programming and software development, E.D.; designing computer programs and implementation of the computer code and supporting algorithms, E.D.; designing computer programs; implementation of the computer code and supporting algorithms, A.B.; testing of existing code components, S.V.; preparation, creation and/or presentation of the published work, specifically writing the initial draft (including substantive translation), M.B. All authors have read and agreed to the published version of the manuscript.

Funding: This research was funded by the Science Committee of the Ministry of Science and Higher Education of the Republic of Kazakhstan (Grant No. AP14972804) "Development of Software for Technological Preparation of Production on the Basis of the Formalized Design Methodology" (Head N.A. Savelyeva).

Institutional Review Board Statement: Not applicable.

Informed Consent Statement: Not applicable.

Data Availability Statement: Data is contained within the article.

Conflicts of Interest: The authors declare no conflicts of interest. The funders had no role in the design of the study; in the collection, analyses, or interpretation of data; in the writing of the manuscript; or in the decision to publish the results.

Appendix A

(:id (:name TP-T:num 0:step 6:type op2:):birth 127:price 11,200:wait 127:executor (:name mach1:num 0:execution 163:impl_price 11,200:):)

(:id (:name TP-T:num 0:step 7:type op4:):birth 163:price 34,000:wait 163:executor (:name mach1:num 0:execution 263:impl_price 34,000:):)

(:id (:name TP-T:num 0:step 8:type op2:):birth 263:price 5600:wait 263:executor (:name mach1:num 0:execution 281:impl_price 5600:):)

(:id (:name TP-T:num 0:step 9:type op3:):birth 281:price 8200:wait 281:executor (:name mach1:num 0:execution 293:impl_price 8200:):)

(:id (:name TP-T:num 0:step 10:type op2:):birth 293:price 5600:wait 293:executor (:name mach1:num 0:execution 311:impl_price 5600:):)

(:id (:name TP-T:num 0:step 11:type op3:):birth 311:price 8200:wait 311:executor (:name mach1:num 0:execution 323:impl_price 8200:):)

(:id (:name TP-T:num 0:step 12:type op5:):birth 323:price 7400:wait 323:executor (:name mach1:num 0:execution 339:impl_price 7400:):)

(:id (:name TP-T:num 0:step 13:type op2:):birth 339:price 5600:wait 339:executor (:name mach1:num 0:execution 357:impl_price 5600:):)

(:id (:name TP-T:num 0:step 14:type op3:):birth 357:price 8200:wait 357:executor (:name mach1:num 0:execution 369:impl_price 8200:):)

(:id (:name TP-T:num 0:step 15:type op4:):birth 369:price 13,600:wait 369:executor (:name mach1:num 0:execution 409:impl_price 13,600:):)

(:id (:name TP-T:num 0:step 16:type op3:):birth 409:price 8200:wait 409:executor (:name mach1:num 0:execution 421:impl_price 8200:):)

(:id (:name TP-T:num 0:step 17:type op2:):birth 421:price 5600:wait 421:executor (:name mach1:num 0:execution 439:impl_price 5600:):)

(:id (:name TP-T:num 0:step 18:type op1:):birth 439:price 4800:wait 439:executor (:name mach1:num 0:execution 471:impl_price 4800:):)

(:id (:name TP-T:num 0:step 3:type op6:):birth 68:price 5800:wait 68:executor (:name mach4:num 3:execution 97:impl_price 5800:):)

(:id (:name TP-T:num 0:step 0:type op4:):birth 0:price 13,600:wait 0:executor (:name mach1:num 0:execution 40:impl_price 13,600:):)

(:id (:name TP-T:num 0:step 1:type op3:):birth 40:price 8200:wait 40:executor (:name mach1:num 0:execution 52:impl_price 8200:):)

(:id (:name TP-T:num 0:step 2:type op5:):birth 52:price 7400:wait 52:executor (:name mach1:num 0:execution 68:impl_price 7400:):)

(:id (:name TP-T:num 0:step 4:type op2:):birth 97:price 5600:wait 97:executor (:name mach1:num 0:execution 115:impl_price 5600:):)

(:id (:name TP-T:num 0:step 5:type op3:):birth 115:price 8200:wait 115:executor (:name mach1:num 0:execution 127:impl_price 8200:):)

(:id (:name TP-T:num 0:step 6:type op2:):birth 127:price 11,200:wait 127:executor (:name mach1:num 0:execution 163:impl_price 11,200:):)

(:id (:name TP-T:num 0:step 7:type op4:):birth 163:price 34,000:wait 163:executor (:name mach1:num 0:execution 263:impl_price 34,000:):)

(:id (:name TP-T:num 0:step 8:type op2:):birth 263:price 5600:wait 263:executor (:name mach1:num 0:execution 281:impl_price 5600:):)

(:id (:name TP-T:num 0:step 9:type op3:):birth 281:price 8200:wait 281:executor (:name mach1:num 0:execution 293:impl_price 8200:):)

(:id (:name TP-T:num 0:step 10:type op2:):birth 293:price 5600:wait 293:executor (:name mach1:num 0:execution 311:impl_price 5600:):)

(:id (:name TP-T:num 0:step 11:type op3:):birth 311:price 8200:wait 311:executor (:name mach1:num 0:execution 323:impl_price 8200:):)

(:id (:name TP-T:num 0:step 12:type op5:):birth 323:price 7400:wait 323:executor (:name mach1:num 0:execution 339:impl_price 7400:):)

(:id (:name TP-T:num 0:step 13:type op2:):birth 339:price 5600:wait 339:executor (:name mach1:num 0:execution 357:impl_price 5600:):)

(:id (:name TP-T:num 0:step 14:type op3:):birth 357:price 8200:wait 357:executor (:name mach1:num 0:execution 369:impl_price 8200:):)

(:id (:name TP-T:num 0:step 15:type op4:):birth 369:price 13,600:wait 369:executor (:name mach1:num 0:execution 409:impl_price 136,00:):)

(:id (:name TP-T:num 0:step 16:type op3:):birth 409:price 8200:wait 409:executor (:name mach1:num 0:execution 421:impl_price 8200:):)

(:id (:name TP-T:num 0:step 17:type op2:):birth 421:price 5600:wait 421:executor (:name mach1:num 0:execution 439:impl_price 5600:):)

(:id (:name TP-T:num 0:step 18:type op1:):birth 439:price 4800:wait 439:executor (:name mach1:num 0:execution 471:impl_price 4800:):)

(:id (:name TP-T:num 0:step 3:type op6:):birth 68:price 5800:wait 68:executor (:name mach4:num 3:execution 97:impl_price 5800:):)

Res_price:175,000 Res_time: 471

References

1. Chernorutsky, I.; Kotlyarov, V.; Shyamasundar, R.; Tolstoles, A.; Voinov, N. Implementation of reliable net-centric management of IoT industrial workshop for small-scale production. *IOP Conf. Ser. Mater. Sci. Eng.* **2018**, *497*, 012040. [CrossRef]
2. Manzei, C.; Schleupner, L.; Heinze, R. *Industrie 4.0 im Internationalen Kontext: Kernkonzepte, Ergebnisse, Trends*; VDE VERLAG GmbH: Berlin, Germany, 2016; p. 261.
3. Marcos, M.P.; Pitarch, J.L.; de Prada, C. Integrated Process Re-Design with Operation in the Digital Era: Illustration through an Industrial Case Study. *Processes* **2021**, *9*, 1203. [CrossRef]
4. Bako, B.; Božek, P. Trends in simulation and planning of manufacturing companies. *Procedia Eng.* **2016**, *149*, 571–575. [CrossRef]
5. Chlebus, E.; Krot, K. CAD 3D models decomposition in manufacturing processes. *Arch. Civ. Mech. Eng.* **2016**, *16*, 20–29. [CrossRef]
6. Adam, R.; Kotze, P.; Van der Merwe, A. Acceptance of enterprise resource planning systems by small manufacturing Enterprises. In Proceedings of the 13th International Conference on Enterprise Information Systems, Beijing, China, 8–11 June 2011; Volume 1. [CrossRef]
7. Browne, J.; O'Kelly, M.E.J.; Davies, B.J. Scheduling in a batch or job shop production environment. *Eng. Manag. Int.* **1982**, *1*, 173–184. [CrossRef]
8. Burdo, G.B. Improving the technological preparations for manufacturing production. *Russ. Eng. Res. Vol.* **2017**, *37*, 49–56. [CrossRef]
9. Siderska, J. Application of tecnomatix plant simulation for modeling production and logistics processes. *Bus. Manag. Educ.* **2016**, *14*, 64–73. [CrossRef]
10. Dong, S.; Medeiros, D.J. Minimising schedule cost via simulation optimization: An application in pipe manufacturing. *Int. J. Prod. Res.* **2012**, *50*, 831–841. [CrossRef]
11. San Cristobal, J.R. Critical Path Definition Using Multicriteria Decision Making: Promethee Method. *J. Constr. Eng. Manag.* **2013**, *29*, 158–163. [CrossRef]
12. Jato-Espino, D.; Castillo-Lopez, E.; Rodriguez-Hernandez, J.; Canteras-Jordana, J. A review of application of multi-criteria decision making methods in construction. *Autom. Constr.* **2014**, *45*, 151–162. [CrossRef]
13. Rosova, A.; Behun, M.; Khouri, S.; Cehlar, M.; Ferencz, V.; Sofranko, M. Case study: The simulation modeling to improve the efficiency and performance of production process. *Wirel. Netw.* **2022**, *28*, 863–872. [CrossRef]
14. Pompeev, K.P.; Timofeeva, O.S.; Yablochnikov, E.I.; Volosatova, E.E. Methods of Parts Digital Models Design for Problems Resolving in Technological Preparation of Production. In *Advances in Mechanical Engineering*; Lecture Notes in Mechanical Engineering; Evgrafov, A.N., Ed.; Springer: Berlin/Heidelberg, Germany, 2022; pp. 129–139. [CrossRef]
15. Torres-Jimenez, J.; Rangel-Valdez, N.; De-la-Torre, M.; Avila-George, H. An Approach to Aid Decision-Making by Solving Complex Optimization Problems Using SQL Queries. *Appl. Sci.* **2022**, *12*, 4569. [CrossRef]
16. Sipser, M. *Introduction to the Theory of Computation*, 2nd ed.; Thomson: Boston, MA, USA, 2006; p. 418. ISBN 0-534-95097-3.
17. Abramov, S.A. *Lectures on the Complexity of Algorithms*, 3rd ed.; MTsNMO: Moscow, Russia, 2020; p. 256. ISBN 978-5-443-91464-0.
18. Kalyakulin, S.Y.; Kuzmin, V.V.; Mitin, E.V.; Suldin, S.P.; Tyurbeeva, T.B. Designing the Structure of Technological Processes Based on Synthesis. *Mordovia Univ. Bull.* **2018**, *28*, 77–84. [CrossRef]
19. Khrustaleva, I.N.; Lyubomudrov, S.A.; Chernykh, L.G.; Stepanov, S.N.; Larionova, T.A. Automating production engineering for custom and small-batch production on the basis of simulation modeling. *J. Phys. Conf. Ser.* **2021**, *1753*, 012047. [CrossRef]

20. Ivanov, V.; Botko, F.; Kolos, V.; Pavlenko, I.; Hatala, M.; Antosz, K.; Trojanowska, J. Locating Chart Choice Based on the Decision-Making Approach. *Materials* **2022**, *15*, 3557. [CrossRef] [PubMed]
21. Trojanowska, J.; Kolinski, A.; Galusik, D.; Rocha Varela, L. A methodology of improvement of manufacturing productivity through increasing operational efficiency of the production process. In *Advances in Manufacturing*; Springer: Berlin/Heidelberg, Germany, 2018; pp. 23–32. [CrossRef]
22. Araujo, A.F.; Varela, M.L.; Gomes, M.S.; Barreto, R.C.; Trojanowska, J. Development of an intelligent and automated system for lean industrial production, adding maximum productivity and efficiency in the production process. In *Advances in Manufacturing*; Springer: Berlin/Heidelberg, Germany, 2018; pp. 131–140. [CrossRef]
23. Vukelic, D.; Zuperl, U.; Hodolic, J. Complex system for fixture selection, modification, and design. *Int. J. Adv. Manuf. Technol.* **2009**, *45*, 731–748. [CrossRef]
24. Abouel Nasr, E.; Al-Ahmari, A.; Khan, A.A.; Mian, S.H. Integrated system for automation of process, fixture and inspection planning. *J. Braz. Soc. Mech. Sci. Eng.* **2020**, *42*, 52. [CrossRef]
25. Liu, M.; Zhou, B.; Li, J.; Li, X.; Bao, J. A Knowledge Graph-Based Approach for Assembly Sequence Recommendations for Wind Turbines. *Machines* **2023**, *11*, 930. [CrossRef]
26. Chen, Z.; Bao, J.; Zheng, X.; Liu, T. Assembly information model based on knowledge graph. *J. Shanghai Jiaotong Univ.* **2020**, *25*, 578–588. [CrossRef]
27. Li, X.; Zhang, S.; Huang, R.; Huang, B.; Xu, C.; Kuang, B. Structured modeling of heterogeneous CAM model based on process knowledge graph. *Int. J. Adv. Manuf. Technol.* **2018**, *96*, 4173–4193. [CrossRef]
28. Xu, Z.; Liu, H.; Li, J.; Zhang, Q.; Tang, Y. CKGAT: Collaborative Knowledge-Aware Graph Attention Network for Top-N Recommendation. *Appl. Sci.* **2022**, *12*, 1669. [CrossRef]
29. Han, Z.; Mo, R.; Hao, L. Clustering and retrieval of mechanical CAD assembly models based on multi-source attributes information. *Robot. Comput. Integr. Manuf.* **2019**, *58*, 220–229. [CrossRef]
30. Stavropoulos, P.; Papacharalampopoulos, A.; Sabatakakis, K. Data Attributes in Quality Monitoring of Manufacturing Processes: The Welding Case. *Appl. Sci.* **2023**, *13*, 10580. [CrossRef]
31. Wang, X.; Liu, M.; Zhong, P.; Zhang, C.; Zhang, D. A Discrete Cooperative Control Method for Production Scheduling Problem of Assembly Manufacturing System. *Sustainability* **2023**, *15*, 13785. [CrossRef]
32. Xiong, H.; Shi, S.; Ren, D.; Hu, J. A survey of job shop scheduling problem: The types and models. *Comput. Oper. Res.* **2022**, *142*, 105731. [CrossRef]
33. Mokhtari, H.; Hasani, A. An energy-efficient multi-objective optimization for flexible job-shop scheduling problem. *Comput. Chem. Eng.* **2017**, *104*, 339–352. [CrossRef]
34. Ciurana, J.; Garcia-Romeu, M.L.; Ferrer, I.; Casadesus, M. A model for integrating process planning and production planning and control in machining processes. *Robot. Comput. Integr. Manuf.* **2008**, *24*, 532–544. [CrossRef]
35. García-Leon, A.A.; Dauzère-Pérès, S.; Mati, Y. An efficient Pareto approach for solving the multi-objective flexible job-shop scheduling problem with regular criteria. *Comput. Oper. Res.* **2019**, *108*, 187–200. [CrossRef]
36. Knopp, S.; Dauzère-Pérès, S.; Yugma, C. A batch-oblivious approach for Complex Job-Shop scheduling problems. *Eur. J. Oper. Res.* **2017**, *263*, 50–61. [CrossRef]
37. Azzouz, A.; Ennigrou, M.; Ben Said, L. A self-adaptive hybrid algorithm for solving flexible job-shop problem with sequence dependent setup time. *Procedia Comput. Sci.* **2017**, *112*, 457–466. [CrossRef]
38. Xing, L.N.; Chen, Y.W.; Yang, K.W. An efficient search method for multi-objective flexible job shop scheduling problems. *J. Intell. Manuf.* **2009**, *20*, 283–293. [CrossRef]
39. Shen, X.N.; Han, Y.; Fu, J.Z. Robustness measures and robust scheduling for multi-objective stochastic flexible job shop scheduling problems. *Soft Comput.* **2017**, *21*, 6531–6554. [CrossRef]
40. Fernández Pérez, M.A.; Raupp, F.M.P. A Newton-based heuristic algorithm for multi-objective flexible job-shop scheduling problem. *J. Intell. Manuf.* **2016**, *27*, 409–416. [CrossRef]
41. Moslehi, G.; Mahnam, M. A Pareto approach to multi-objective flexible job-shop scheduling problem using particle swarm optimization and local search. *Int. J. Prod. Econ.* **2011**, *129*, 14–22. [CrossRef]
42. Kacem, I.; Hammadi, S.; Borne, P. Pareto-optimality approach for flexible job-shop scheduling problems: Hybridization of evolutionary algorithms and fuzzy logic. *Math. Comput. Simul.* **2002**, *60*, 245–276. [CrossRef]
43. Gunantara, N. A review of multi-objective optimization: Methods and its applications. *Cogent Eng.* **2018**, *5*, 1502242. [CrossRef]
44. Ondov, M.; Rosova, A.; Sofranko, M.; Feher, J.; Cambal, J.; Feckova Skrabulakova, E. Redesigning the Production Process Using Simulation for Sustainable Development of the Enterprise. *Sustainability* **2022**, *14*, 1514. [CrossRef]
45. Iassinovski, S.; Artiba, A.; Fagnart, C. A generic production rules-based system for on-line simulation, decision making and discrete process control. *Int. J. Prod. Econ.* **2008**, *112*, 62–76. [CrossRef]
46. Bernard, A.; Perry, N. Fundamental concepts of product/technology/process informational integration for process modelling and process planning. *Int. J. Comput. Integr. Manuf.* **2003**, *16*, 557–565. [CrossRef]
47. Shah, D.; Shin, J. Randomized scheduling algorithm for queueing networks. *Ann. Appl. Probab* **2012**, *22*, 128–171. [CrossRef]
48. Bathaee, M.; Nozari, H.; Szmelter-Jarosz, A. Designing a New Location-Allocation and Routing Model with Simultaneous Pick-Up and Delivery in a Closed-Loop Supply Chain Network under Uncertainty. *Logistics* **2023**, *7*, 3. [CrossRef]
49. Malega, P.; Gazda, V.; Rudy, V. Optimization of production system in plant simulation. *Simulation* **2022**, *98*, 295–306. [CrossRef]

50. Kuznetsov, P.M.; Khoroshko, L.L. Digitalization of Multi-Object Technological Projecting in Terms of Small Batch Production. *Inventions* **2020**, *5*, 38. [CrossRef]
51. Simunovic, G.; Majdandzic, N.; Simunovic, K.; Lujic, R. Applying of the Typical Technological Operations in the Single Production. In *AMST'02 Advanced Manufacturing Systems and Technology*; International Centre for Mechanical Sciences; Kulianic, E., Ed.; Springer: Vienna, Austria, 2002; Volume 437, pp. 275–281. [CrossRef]

designs

MDPI

Article

Predicting Quality of Modified Product Attributes to Achieve Customer Satisfaction

Andrzej Pacana * and Dominika Siwiec

Faculty of Mechanical Engineering and Aeronautics, Rzeszow University of Technology, 35-959 Rzeszów, Poland; d.siwiec@prz.edu.pl
* Correspondence: app@prz.edu.pl

Abstract: In the era of the competitive environment, the improvement in current products is ensured through activities aimed at increasing a product's quality level and, consequently, reducing the amount of waste. The dynamically changing production environment and sudden changes in customer expectations force us to take precise and well-thought-out development steps. Furthermore, it is important to anticipate favourable product changes to prepare for market changes over time. This is still an open problem. The aim of this study was to develop a method to predict the quality of potential product prototypes resulting from the proposed modifications of the product features. This methodology takes into account current customer expectations. The method was created based on the principles of creating Quality Function Deployment (QFD) in the context of taking into account current and future customer expectations regarding product features. This is a new approach to analysing product quality within the principles of the traditional QFD method. The originality of the study is the technique used in the method to estimate the expected values of product features and their importance (weights), taking into account current customer expectations. Its originality is also manifested in drawing conclusions supporting the decision-making process of product improvement, because it involves ensuring the pro-quality modification of selected features of current products in the order that is most advantageous from the customer's point of view. The use of the proposed method allows for the analysis of the impact of modifying the current value of a product feature. The method is illustrated with an example of a vacuum cleaner for home use. However, the proposed method can be applied to the design of any product to predict products that will meet customer expectations.

Keywords: product design; predict; product quality; customer requirement; house of quality; mechanical engineering

1. Introduction

The improvement in product quality covers a product's customisation to expectations [1–3]. In this regard, it is necessary to have a skilful shape of the attributes of a product to achieve product quality that meets the level of expectation. This shaping includes acquiring the Voice of Customers (VoC) [4,5], and then involves its analysis and transformation in product attributes [6–8]. The product quality level mentioned is the level of compliance of a product with customers' expectations and refers only to specific product uses. On the other hand, customer expectations (i.e., their desires and assumptions in the context of the future) are customer satisfaction with what they received (current product) and what they expected (product they expected) [9–11]. Despite that, designing products is still a problem; therefore, a search for methods to solve this problem is needed.

After reviewing the literature, it was concluded that the House of Quality (HoQ) [12] i.e., Quality Function Deployment (QFD) [13], is the most popular and most-used method to design a satisfactory product [14–18]. This method is a correlation matrix. The method supports the process of design by collecting and verifying expectations, e.g., of customers and the producer, to achieve the required product quality level. Previous work as part of

HoQ concerns mainly the validity (weights) of customer requirements, e.g., mainly as part of the use of the AHP method (Analytic Hierarchy Process) [15,19–21]. Also, as part of HoQ, the customers' requirements were specified, mainly using the fuzzy Saaty scale and FAHP method (Fuzzy Analytic Hierarchy Process) [16,22,23]. For this purpose, the HoQ with the AHP method and the Kano model have been integrated relatively often [21,24,25], where the objective of this combination is to precisely determine the importance of customer expectations in the context of product satisfaction. This precision was achieved by using the 2-tuple linguistic [17], in which the customer's needs were characterised by description and number, which is a distance of the central value of the linguistic term. Despite this, the importance of customers' requirements in the HoQ was determined by implementing a mathematical model according to LGP (Linear Goal Programming), in which the ordinal scale was used [26]. Another example is a combination of the HoQ with the Yager algorithm [27] to determine customer requirement weights also in the case of indifferent requirements. Another example concerning the design of the expected product is the combination of the HoQ method with the TRIZ method (Theory of Innovative Problem Solving) [28–30]. This combination encompasses determining customers' requirements and their expectations as part of providing a satisfactory product.

Based on a literature review in the context of the popular HoQ method, it was shown that it is known how to design a product [18,23,27,31] and also how to verify the importance of product attributes [15,16,19,22,32]. As shown, the QFD method is one of the most popular methods for transforming customer expectations into technical features of the product. It has been modified many times, but its attributes have not yet been used as a framework for predicting the ranking of pro-quality modifications to existing product attributes. Therefore, there is no method or transformation method that would indicate how to modify the product to meet customer expectations. Hence, a method to predict the pro-quality modification of product attributes, i.e., the modification (changes) of important product attributes for customers, that allows achieving the expectation of customers of the product quality level is needed. This prediction is passing, guessing, or expecting a satisfactory product. Therefore, the lack of this method has been recognised as a gap.

It is possible to create the main question as follows: How can we predict the pro-quality modification of product attributes to achieve a product quality level that satisfies customers?

Hence, the purpose of this study is to propose a method to predict the quality of product attribute modifications considering the current customers' expectations. The proposed method is preferred to determine the order of product attribute changes based on customer information to achieve desired quality. Additionally, the method can provide the anticipation of beneficial activities of the manufacturer to achieve the product quality level expected by customers. The added value is the presentation of a method (according to the principles of the QFD method) according to which it is possible to predict which features of a product should be modified and how to achieve a competitive product. It is an original proposition to the application of the framework of the QFD method, which will support producers in making modifications to current product attributes ahead of the competition to design a product that is expected by customers. It is important to note that this method makes it easier to predict (in advance) what changes to make. Therefore, it provides time for appropriate preparatory activities for enterprises, and the achieved design results may be satisfactory to customers, which increases the accuracy of design decisions and, consequently, the quality level of the offered products.

The originality of the presentation of this method (according to the principles of the QFD method) regards its ability to predict which features of the product should be modified and how to achieve a competitive product.

The developed method is presented in the following parts of the article (Section 2). The method was characterised in ten main stages, taking into account the assumptions adopted for the method. The characteristics include a general methodology, so it can be used for any product. Section 3 tests and illustrates the method using the example of a home vacuum cleaner. The method test is presented stage by stage, as adopted in the general description of the method. Section 4 is the discussion, where the analysis is further deepened based on comparisons of the results obtained from the method test. The main advantages of the method, business implications, and limitations and directions for future research are presented. Section 5 is a summary covering the main findings of the study.

2. Materials and Methods

The method aims to predict the pro-quality modifications of product attributes considering current customers' expectations. The method was developed by transforming the House of Quality in the context of considering the current and future expectations of customers of the state of product attributes. A mathematical model was implemented to calculate the quality of attribute states and their importance. Additionally, the method allows one to make the product features dependent on their importance (weights). As a result, it is possible to rank the product attribute changes and also to predict expectations of the product quality level.

The structure of the research is as follows. The analysis of the product selected for the research is based on the customer requirements, which are obtained as part of the research of the survey. Customers rate the importance of product criteria and their satisfaction with these criteria in the current state and the proposed modified state (above and below the current state). These requirements are processed in subsequent stages of the method. Initially, the criteria important for the client are determined according to the Pareto–Lorenz principle. Pro-quality activities are planned for these important criteria. Then, the quality level is calculated for the current state of the product criteria and their modifications, and the impact of modifying the current values of the product attributes is estimated. A mathematical model developed for this purpose is used. As a result, a ranking of modifications of important product features is obtained in order to achieve the expected level of product quality. On this basis, a pro-quality modification of the product is planned, i.e., one that meets customer expectations.

The algorithm of the proposed method and the zones in a new House of Quality is shown in Figure 1.

The method in ten main stages was designed as follows:

Stage 1. Determining the purpose of the research.

Stage 2. Acquiring the customer's requirements.

Stage 3. Determining important product attributes based on customer expectations (II zone in HoQ).

Stage 4. Noting the current and modification values for important product attributes (I and V zones in HoQ).

Stage 5. Calculating the quality of product attributes on a point scale (III and VII zones in HoQ).

Stage 6. Calculating the quality of product attributes considering the weights of attributes (IV and VIII zones in HoQ).

Stage 7. Calculating the quality changes in the current product attribute values (IX zone in HoQ).

Stage 8. Calculating product quality for current important product attribute values (X zone in HoQ).

Stage 9. Calculating the impact of the attribute values of the modification of the current product (XI zone in HoQ).

Stage 10. Predicting the modification of the pro-quality product (XII zone in HoQ).

The short characteristic of the method stages is shown in the next part of the article.

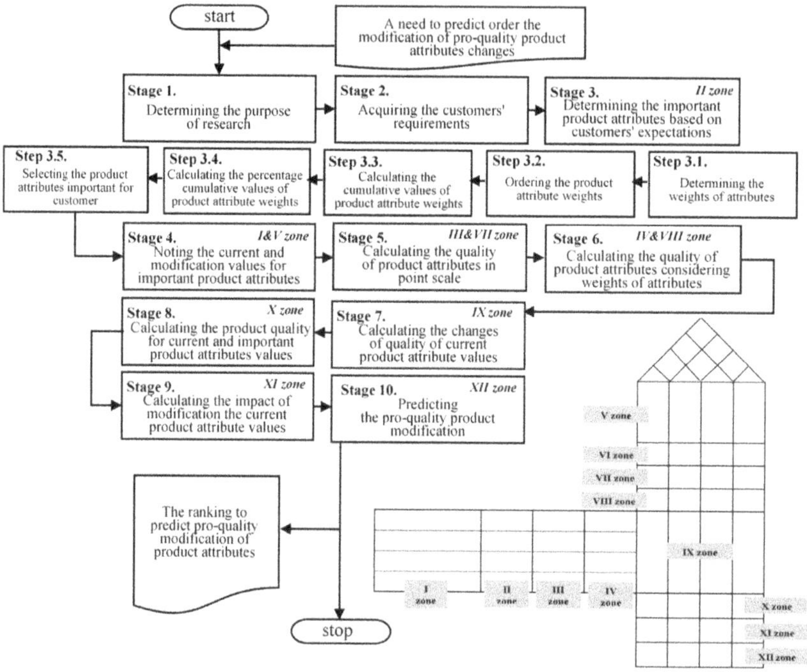

Figure 1. Algorithm to predict the pro-quality modification of product attributes considering the expectations of the current customers and the visualisation zones of the new House of Quality.

2.1. Stage 1: Determining the Purpose of the Research

As part of determining the purpose, it is necessary to use the SMART method (Specific, Measurable, Achievable, Relevant, Time-bound) [33]. In the proposed approach, the purpose is to predict a high-quality modification of product attributes.

2.2. Stage 2: Acquiring the Customers' Requirements

This purpose of this stage is to acquire customers' requirements, that is, Voice of Customers (VoC) [34,35] about product attributes. To achieve this, it is possible to use survey research, which is the most popular approach [7,36,37], with scoring on a Likert scale [26,38]. The sample size to predict expected product quality is preferably determined according to the method shown in [9]. Customers should be the users of the product. Furthermore, the entity that uses the proposed method should select the area of research depending on the market in which it will offer its modified product. As part of acquiring customers' requirements, determining the importance (weights) of product attributes, e.g., in the context of use of the product, should be allowed. Additionally, the customers' requirements have to refer to assessments of customer satisfaction of current product attributes and their modifications. For this reason, the characteristics of the product to be tested are initially defined. The selection of product attributes can be made based on the product catalogue (specification) by determining 14 to 25 attributes [19,39,40]. Next, by using a catalogue of a product, the current and modification values, e.g., values or range of values (parameters), must be noted for each attribute. The current value is the current state of the attribute. In turn, the modification value is the predicted state (future, not existing at present). From the obtained customers' requirements, the important attributes for customers will be selected, i.e., which attributes will have the greatest importance for product level quality.

2.3. Stage 3: Determining the Important Product Attributes Based on Customers' Expectations (II Zone in HoQ)

This purpose of this stage is to determine the attributes that are important (significant) to customers in the context of using the product. For these attributes, the pro-quality changes of the product features are predicted. The important features for customers are chosen based on the weights of product attributes according to the Pareto rule (20/80), i.e., even a small number of important attributes has a significant impact on product quality, where the impacts of other attributes are less [41]. This stage is shown in five main steps.

2.3.1. Step 3.1: Determining the Weights of Attributes

For each product attribute, the weights are calculated as an arithmetic average from assessments of the importance of the product attributes (1):

$$A_{w_i} = \frac{1}{n}\sum_{i=1}^{n} w_i \tag{1}$$

where A_w is the arithmetic average of the product attribute weights, w is the customer assessment of product attribute importance, n is the number of assessments, and i = 1, 2, 3, . . ., n.

The application of the arithmetic average results from the fact that it is an unencumbered estimator [42], and at the same time it has the highest reliability of the value of the expected random variable when the number of events is large enough (>100) or the distribution of the variable is normal. The median is better if a small number of observations is obtained or the distribution is not normal, e.g., there are outliers.

2.3.2. Step 3.2: Ordering the Product Attribute Weights

The first stage is to order the weights of the product attributes. For this purpose, it is necessary to sort in descending order the weights of the attributes of the product [41].

2.3.3. Step 3.3: Calculating Cumulative Values of Product Attribute Weights

Cumulative values are calculated on the basis of ordered values of product attribute weights. The first value is equal to the maximum value of the weight of the attribute of the product (2):

$$C_1 = A_{w_{max}} \tag{2}$$

In turn, other cumulative values are calculated with Equation (3):

$$C_i = C_{i-1} + A_{w_i} \tag{3}$$

where C_1 is the first value of the cumulative values, $A_{w_{max}}$ is the maximum product attribute weight, A_w is the product attribute weight, and i = 2, 3, 4, . . ., n.

The cumulative values of product attribute weights are calculated for all ordered values of product attribute weights.

2.3.4. Step 3.4: Calculating the Percentage Cumulative Values of Product Attribute Weights

The percentage cumulative values of the product attribute weights [43,44] are based on the cumulative values from Section 2.3.2. For this purpose, the following Equation (4) is used:

$$C_i^{\%} = \frac{C_i}{C_{max}} \times 100 \tag{4}$$

where $C^{\%}$ is the percentage cumulative value of the product attribute weight, C is the cumulative value of the product attribute weight, C_{max} is the maximum cumulative value of all product attribute weights, and i is 1, 2, 3, . . ., n.

The percentage cumulative values of the product attribute weights should be calculated for all cumulative values of the product attribute weights.

2.3.5. Step 3.5: Selecting the Product Attributes Important to Customers

The selection of features is made on the basis of the percentage values of cumulative product feature weights. For this, the Pareto rule (20/80) is used [43,44].

The percentage cumulative value of the weight of the product attribute close to 20% determines the group of important product attributes (GI) and the group of unimportant attributes for the customer (G_N) (5):

$$\begin{cases} C_i^{\%} \leq \sim 20\% \ \epsilon \ G_I \\ C_i^{\%} > \sim 20\% \ \epsilon \ G_N \end{cases} \quad (5)$$

where $C^{\%}$ is the percentage cumulative value of the product attribute weight, i = 1, 2, 3, . . .n, G_I is the group of important product attributes, and G_N is the group of unimportant attributes.

To determine important attributes, it is useful to visualise the results on the Pareto–Lorenz diagram, as shown, for example, in [43,44]. The weights of important product attributes for customers are noted in the II zone in the HoQ.

2.4. Stage 4: Noting the Current and Modification Values for Important Product Attributes (I and V Zones in HoQ)

The purpose of the fourth stage is to note the current and modification values for important product attributes. It refers to filling the I zone of the HoQ by current attribute values and filling the V zone of the HoQ by values of the modification of these attributes. These values were determined in the second stage of the method.

2.5. Stage 5: Calculating the Quality of Product Attributes on a Point Scale (III and VII Zones in HoQ)

The purpose of the fifth stage is to calculate the quality of the attributes of the product on a point scale. It refers to determining the sum of the points that the customers have given to important attributes of the product (in stage 2). The quality (current or modified) of the product attribute on a point scale is expressed by Equation (6):

$$S_{c_i} = \sum P_{c_i}, \quad S_{m_i} = \sum P_{m_i} \quad (6)$$

where P is the points given by customers, c is the current product attribute value, m is the modification product attribute value, and i = 1, 2, 3, . . ., n.

The quality of the current product attribute is noted in the III zone of the HoQ. In turn, the quality of the modified product attribute is noted in the VII zone of the HoQ.

2.6. Stage 6: Calculating the Quality of Product Attributes Considering Weights of Attributes (IV and VIII Zones in HoQ)

The purpose of the sixth stage is to calculate the quality of the attributes of the product considering the weights of the attributes. This stage refers to calculating the quality for each of the important attributes of the product. This quality is determined for current attributes (IV zone in HoQ) and for their modifications (VIII zone of HoQ).

The quality of the current product attribute is estimated as the quotient of the attribute weight and the sum of the customers' assessments, which refers to satisfaction with the current value of this attribute (7):

$$Q_{c_i} = A_{w c_i} \times S_{c_i} \quad (7)$$

where A_w is the weight of the product attribute, S is the quality of the product attribute, c is the current product attribute, and i = 1, 2, 3, . . ., n.

The quality of the modification product attribute is estimated as the quotient of the attribute weight and the sum of the customers' assessment, which refers to satisfaction with the modified value of this attribute (8):

$$Q_{m_i} = A_{wm_i} \times S_{m_i} \tag{8}$$

where A_W is the weight of the product attribute, S is the quality of the product attribute, m is the modification of the product attribute value, and i = 1, 2, 3, . . ., n.

The obtained quality values of important current and modified product attributes are noted, respectively, in the IV and VIII zones of the HoQ.

2.7. Stage 7: Calculating the Changes in Quality of Current Product Attribute Values (IX Zone in HoQ)

The purpose of the seventh stage is to calculate the quality changes in the current product attribute values (IX zone in HoQ). This refers to a comparison of the quality of the current and modified product considering the weights of these attributes, that is, (9):

$$J_i = \frac{Q_{c_i} + Q_{m_i}}{A_{wc_i} + A_{wm_i}} \tag{9}$$

where Q is the quality of the product attribute, A_W is the weight of the product attribute, c is the current value of the product attribute, m is the modified value of the product attribute, and i = 1, 2, 3, . . ., n.

The obtained values are noted in the IX zone of the HoQ.

2.8. Stage 8: Calculating the Product Quality for Current Important Product Attribute Values (X Zone in HoQ)

The purpose of the eighth stage is to calculate the quality of the product via the values of the current important attributes of the product. The quality of the product regarding the current values of the important product attributes is obtained from Equation (10):

$$I = \frac{Q_{c_i} + Q_{c_i} + \cdots + Q_{c_n}}{A_{wc_i} + A_{wc_i} + \cdots + A_{wc_n}} \tag{10}$$

where Q is the quality of the product attribute, A_W is the weight of the product attribute, c is the current value of the product attribute, and i—1, 2, 3. . ., n.

Based on the product quality for the current values of important product attributes, it is possible to estimate the changes (impact) in the modification of these attribute values. This is shown in the next stage of the method.

2.9. Stage 9: Calculating the Impact of Modification of Current Product Attribute Values (XI Zone in HoQ)

The purpose of stage nine is to calculate the impact of the modification of current product attribute values. This refers to determining the ratio of the change value of the current attribute quality to the level of quality of the modification. The purpose is to determine the relevance of changing the current product attribute value to the modified value, i.e., (11):

$$J_i' = \frac{J_i}{I} \tag{11}$$

where J is the quality of the attribute value, I is the quality of current attribute values, and i is 1, 2, 3. . ., n.

The higher the value of J_i', the greater (more beneficial) the impact of modifying the current value of the product feature from the customer's point of view.

2.10. Stage 10: Predicting the Pro-Quality Product Modification (XII Zone in HoQ)

The tenth stage predicts the modification of the product. The purpose of this stage is to create a ranking of the modification of important product attributes as part of achieving the expected level of product quality. It consists of arranging in descending order the value of the impact of modifying the current product attribute value (from the XI zone of the HoQ). The first position, the maximum value, is the priority modification, i.e., the most important in the view of the customers, which ensures a significant increase in product satisfaction.

Based on ranking, it is possible to predict which attributes should be primarily modified and also which attributes allow the achievement of a quality product. Because in the research important product attributes were included, it is assumed that each modification has a significant impact on the level of product quality. However, it is possible to schedule these modifications depending on the manufacturer's needs and the company's production capacity.

It should be remembered that the proposed method is a possible alternative for managers to support themselves in the decision-making process related to product improvement. Since the proposed method refers to potential prototypes of the same product, the results were not verified in comparison to those of other methods. Other methods operate according to different assumptions. Depending on the assumptions made, appropriate to the analysed product, the decision maker selects the appropriate method, and the selected method leads to suggestions regarding solutions that modify the product.

3. Results

The test research for the method was conducted using an example of a domestic vacuum cleaner. The vacuum cleaner test presented in this article should be treated as a pilot study. It shows the possibility of using the proposed methodology. The choice of the vacuum cleaner was based on its availability and universality, and it is a widely known and used product. As part of the first stage, the purpose of the research was determined, i.e., to predict the quality of the modification of vacuum cleaner attributes considering the expectations of current customers. According to the second stage, the customer expectations about product attributes were acquired. The survey was conducted electronically using MS Forms. The customers were people who used the vacuum cleaner, and therefore their opinions were considered relatively trustworthy. Research samples of 166 customers were obtained, and this sample size was a preliminary sample (for pilot studies). The sample size was calculated based on the method presented in [9]. Furthermore, following the authors of the studies [44], it was considered sufficient within the proposed method. A certain limitation was collecting information from respondents coming only from Poland and mainly from customers belonging to the Z formation. Customers' expectations were acquired as part of a survey using the five-point Likert scale. According to the assumptions obtained in the second stage of the method, the product was described by 20 attributes. The attributes were selected as part of brainstorming (BM) based on a product catalogue. Five employees familiar with the product participated in the brainstorming session. These were managers from the production plant whose knowledge and experience related to the product are the greatest. They know the real scope of product improvement possibilities but also know about the current, successful, and unsuccessful areas of improvement. However, in the future, the user of this proposed method will decide who will work in the brainstorming session because this is a very important decision. The research included the stage of satisfaction assessment with the current and modified values of the product features and the stage of determining the validity (weights) of the product attributes.

The current product attribute was determined by values and the metric unit according to the product specifications. In turn, the modifications of the product attributes were determined as values above and below current states. The customers' expectations were reliably obtained by comparing the current value with the modified value. Then, according to the third stage of the method, the important attributes of the product were determined. For these attributes, in the next part, the pro-quality changes were predicted. These attributes were selected using the Pareto (20/80) rule. Initially, for each of the product attributes, the weights were calculated as an arithmetic average of customers' assessments of the importance of product attributes. Subsequently, the attribute weights were ordered in descending order, and cumulative values and percentages of product feature weights were calculated. Then, according to the 20/80 rule, the important attributes and the unimportant attributes for the customers were selected. The results are shown in Table 1.

Table 1. Pareto–Lorenz analysis for product attribute weights.

No.	Product Attribute (A)	Attribute Weight (A_w)	Cumulative Value (C)	Percentage Cumulative Value ($C^{\%}$)	Group of Product Attributes
1	Vacuum in the suction pipe	3.66	0.06	6.04	Group of important product attributes (G_I)
2	Working range of the vacuum cleaner connected to the power cord	3.65	0.12	12.07	
3	Length of the power cord	3.62	0.18	18.06	
4	Vacuum cleaner motor power	3.61	0.24	24.03	
5	The noise level during operationof the vacuum cleaner	3.36	0.30	29.58	Group of unimportant attributes for customers (G_N)
6	Power cord winding system	3.33	0.35	35.09	
7	Vacuum tank capacity	3.16	24.40	40.31	
8	Thermal protection (against overheating)	3.13	27.53	45.49	
9	Type of vacuum cleaner dust filter	3.08	30.61	50.57	
10	Rubber protectors to protect furniture against knocking	3.04	33.65	55.60	
11	Number of accessories included with the vacuum cleaner (suction tubes and nozzles)	3.02	36.67	60.59	
12	Length of the suction pipe	2.84	39.51	65.28	
13	Vacuum bag type	2.84	42.35	69.97	
14	Weight of the vacuum cleaner	2.77	45.11	74.54	
15	Dimensions of the vacuum cleaner	2.73	47.85	79.06	
16	Possibility to control the vacuum cleaner vacuum in the working handle	2.69	50.54	83.51	
17	The type of material of the vacuum cleaner road wheels	2.66	53.20	87.90	
18	Electric brush socket	2.53	55.73	92.08	
19	Type of vacuum cleaner on/off	2.48	58.20	96.17	
20	Suction pipe diameter	2.32	60.52	100.00	

The important product attributes were the vacuum in the suction pipe, working range of the vacuum cleaner connected to the power cord, length of the power cord, and vacuum cleaner motor power. The weights of important attributes for customers were noted in the II zone of the HoQ, that is, (12):

$$\begin{cases} A_{w_1} = 3.66; & A_{w_3} = 3.62 \\ A_{w_2} = 3.65; & A_{w_4} = 3.61 \end{cases} \tag{12}$$

where A_w the weight of the product attribute, and 1, . . ., 4 is the number of attributes.

According to the fourth stage, the weights of important product attributes were described as current values (I zone of the HoQ) and modification of these values (V zone of the HoQ). The modification values of the product attributes were noted for the product state that was more favourably assessed by the customers (had a higher number of total points, which meant greater customer satisfaction). Subsequently, in the fifth stage, the quality of the product state was estimated on a point scale. For this purpose, the sum of points granted by customers for important product values was calculated (13):

$$\begin{cases} S_{c_1} = 527; & S_{m_1} = 658 \\ S_{c_2} = 557; & S_{m_2} = 670 \\ S_{c_3} = 550; & S_{m_3} = 661 \\ S_{c_4} = 538; & S_{m_4} = 689 \end{cases} \quad (13)$$

where S_c is the quality of the current product attribute on a point scale, S_m is the quality of the modified product attribute on a point scale, and 1, . . ., 4 is the number of attributes.

In the III zone of the HoQ, the number of points obtained for the current values was noted, and in the VII zone of the HoQ, the number of points obtained for the modification values was noted. According to the sixth stage, the quality of the attributes of the product was estimated considering the weights of the attributes (14):

$$\begin{cases} Q_{c_1} = 3.66 \times 527 = 1927.04; & Q_{m_1} = 3.66 \times 2406.06 = 2406.06 \\ Q_{c_2} = 3.65 \times 557 = 2033.39; & Q_{m_2} = 3.65 \times 2445.90 = 2445.90 \\ Q_{c_3} = 3.62 \times 550 = 1991.27; & Q_{m_3} = 3.62 \times 2393.14 = 2393.14 \\ Q_{c_4} = 3.61 \times 538 = 1944.58; & Q_{m_4} = 3.61 \times 2490.36 = 2490.36 \end{cases} \quad (14)$$

where Q_c is the quality of the current product attribute value, Q_m is the quality of the modification product attribute value, and 1, . . ., 4 is the number of attributes.

The results were observed in the IV and VIII zones of the HoQ. Then, as shown in the seventh stage, the changes in quality of the current product attribute values were estimated (15):

$$\begin{cases} J_1 = \frac{1927.04+2406.06}{3.66+3.66} = 592.5 \\ J_2 = \frac{2033.39+2445.90}{3.65+3.65} = 613.5 \\ J_3 = \frac{1991.27+2393.14}{3.62+3.62} = 605.5 \\ J_4 = \frac{1944.58+2490.36}{3.61+3.61} = 613.5 \end{cases} \quad (15)$$

where J is the change in quality of the current product attribute values and 1, . . ., 4 is the number of attributes.

The changes in the quality of the current product attribute values as part of converting them to modification product attribute values were observed in the IX zone of the HoQ. Next, as part of the eighth stage of the method, the quality of the product was estimated for the current important attribute values (16):

$$I = \frac{1927.04 + 2033.39 + 1991.27 + 1944.58}{3.66 + 3.65 + 3.62 + 3.61} = 542.99 \quad (16)$$

The estimated product quality for the current important attribute values was noted in the X zone of the HoQ. Then, as in the nine stages of the method, the influence of the modification of the current attribute values was estimated (17). The values obtained were noted in the XI zone of the HoQ:

$$\begin{cases} J'_1 = \frac{592.5}{542.99} = 1.091 \\ J'_2 = \frac{613.5}{542.99} = 1.130 \\ J'_3 = \frac{605.5}{542.99} = 1.115 \\ J'_4 = \frac{613.5}{542.99} = 1.130 \end{cases} \quad (17)$$

In the last stage of the method, the pro-quality modifications of the product were predicted. In this stage, the ranking (order) of the product attribute changes was predicted in the context of achieving the expected product quality level. It was considered that in the first order, the working range (above 19 m) and motor power (above 900 W) should be changed. It has been predicted that making changes to these attributes allows an expected increase in the product quality level. Then, it is necessary to consider changing the length of the power cord to above 15 m and the vacuum in the suction pipe to above 27,000 Pa. It is possible to consider the simultaneous modification of several product features, where this decision depends on the needs of the manufacturer and the company's production capabilities. The ranking of the predicted order of modifications of product attributes in the XII zone of the HoQ was observed. The new House of Quality designed to predict the pro-quality modification of the product attributes is shown in Figure 2.

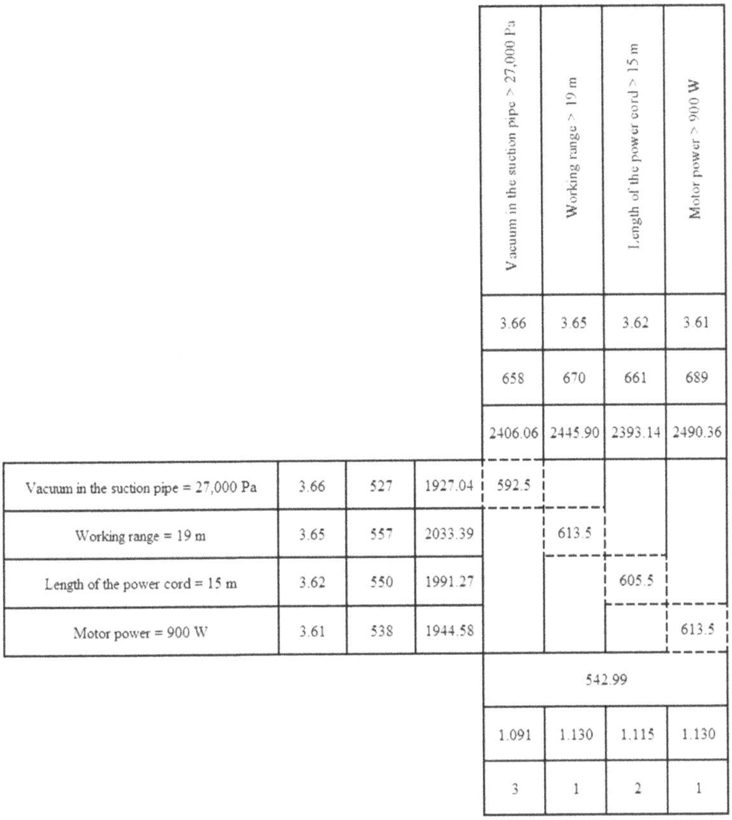

Figure 2. The new House of Quality to predict pro-quality modification of product attributes considering current customers' expectations.

Depending on the needs, it is possible to modify the presented House of Quality. For example, additional space in the HoQ can be filled with complaints or competitive products.

The test of the method shows that the proposed methodology is applicable to determining the ranking of potential product prototypes. This ranking would advise decision makers of which variants of the modified product would be most desired by customers and would be rated the highest. The proposed methodology can also be applied to intangible products, i.e., services. The proposed calculations are not complex and can be performed using a computer, significantly increasing the possibility of repeated use of the proposed method not only for one product but even for different products.

4. Discussion

As part of the test of the proposed method, the differences obtained between the estimated values were verified. It was shown that the values of the modification of the current product attributes (XI zone of the HoQ) and also the ranking of the predicted changes in the attributes (XII zone of HoQ) are characterised by slight differences.

It was concluded that relatively minor differences between the values of the predicted pro-quality modification of the product attributes resulted from minor differences between the values of the current product attribute levels, the values of the modification of the product attribute levels, and the product attribute weights. The differences between the values, which have an impact on the results of the method, are shown in Figure 3.

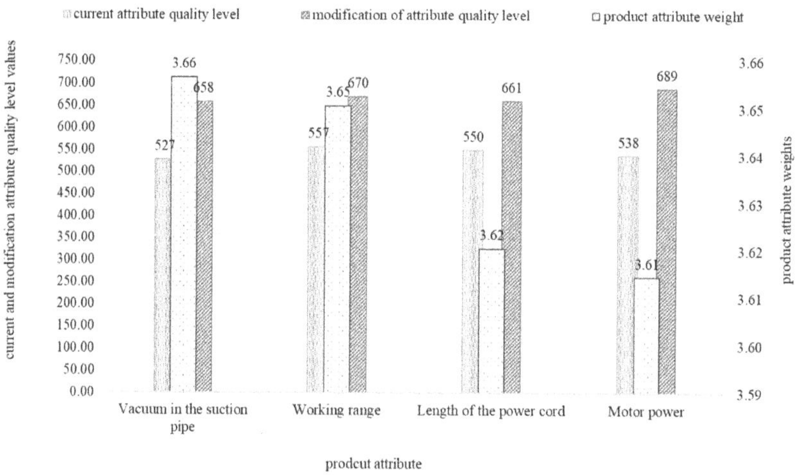

Figure 3. Difference between main values of the method.

The quality level of the attributes of the current product is relatively high. Despite that, the differences between the values of the current and modified attributes are important from a customer's point of view. The difference between the maximum current attribute value (557) and the maximum modification of the attribute value (689) is about 132 points. This is an increase in product quality of about 20%. Hence, the obtained values allow us to estimate the increase in the current product quality level after modification. The main advantages of the proposed method are as follows:

- Predicting pro-quality modifications of product attributes;
- Predicting product attribute changes based on current customers' expectations;
- Predicting the growth in the product quality level depending on product attribute modifications;
- Creating a ranking of product attribute modifications;
- Determining important product attributes and their modifications;
- Designing a product by including the importance of product attributes and the quality level of these attributes (current and modified);
- Supporting enterprise actions in the pro-quality design of a product.
- Additionally, the proposed method has the following business implications:
- Providing a low-cost method for designing important customer changes in an existing product;
- Predicting in advance a satisfactory product that will be competitive in the market;
- Making the right decisions about the need to modify product attributes and their order;
- Waste reduction, e.g., unnecessary changes to the product and time spent on strategic actions;
- An accurate definition of customer expectations.
- There are also disadvantages of the proposed method, as follows:
- The possibility of the occurrence of small differences (or a lack thereof) between predicted values, which generates a problem in creating ranking;

- A lack of a complex analysis of all product attribute values, which determines the lack of precision in determining the product quality level after modification;
- Creating voluminous matrices of HoQ, e.g., as a result of analysing a large number of product features.

A certain limitation of the method is the assumption that for all attributes, the modifications occur as values above and below the current state. This resulted from the assumptions regarding the uniformity of the method. The departure from this principle may constitute the basis for a new, perhaps even more utilitarian, attractive method. This would constitute an improvement of the currently proposed method and perhaps become the basis for further research. Improvements in the methodology should concern the analysis of a larger number of criteria, which is somewhat of a limitation of the current proposal. This is due to the fact that the number of criteria increases the number of possible modifications of these criteria to be considered, and thus the House of Quality becomes extensive, and the results obtained may be problematic in a comprehensive analysis. However, to solve this problem, we plan to develop several connected quality houses. As part of future research, we also plan to implement sustainable development criteria that will ensure prediction not only of product quality but also of the environmental impact or price changes depending on alternative product solutions. This is a challenge, but the proposed method, as well as the House of Quality itself, can be modified and extended, including by combining it with other techniques. At the same time, this is its advantage and versatility, not only in practice but also as an inspiration for other researchers. Also, as part of future research, we plant to extend of proposed House of Quality, e.g., by including subsequent modifications of product attributes, by including a transformation of assessments in the Saaty scale to reduce uncertainty in customers' assessments, or by implementing the method in software.

5. Conclusions

In a competitive environment, it is reasonable to look for various solutions to improving products. This mainly concerns shaping the current features of products as part of customising them to customers' expectations. However, designing a satisfactory product ahead of the competition is still a problem. This refers to the prediction of the quality level of the expected product, which can be designed based on the current customer expectations. For this purpose, a method was proposed to predict the pro-quality modification of product attributes considering current customers' expectations. The method was developed by transforming the House of Quality in the context of considering the values of the attributes of the current and expected product.

The proposed method was tested by using a domestic vacuum cleaner as an example. The purpose was to predict the quality of the modification of the attributes of the vacuum cleaner considering the expectations of current customers. By using a survey and the Likert scale, the requirements of 166 customers were obtained. In the research, 20 attributes of the vacuum cleaner were included, for which the customers assessed their satisfaction with the current and modified values of the features. In addition, customers determined the importance (weights) of the product attributes. According to the Pareto rule, the important attributes were selected. The important product attributes were the vacuum in the suction pipe, working range of the vacuum cleaner connected to the power cord, length of the power cord, and vacuum cleaner motor power. For these attributes, the current and favourable modification values were determined. As part of the calculation model, the quality of the product attributes was estimated. In addition, the importance and influence of the modification of the current product attribute values were estimated. As a result, the pro-quality product modifications were predicted. Therefore, a classification (order) of the changes of the product attributes in the context of achieving the expected level of product quality was predicted. It has been predicted that in the first order, the working range (above 19 m) and motor power (above 900 W) should be changed. The change in these attributes allows for an expected increase in the product quality level. The next anticipated changes were length of the power cord to above 15 m and then vacuum in the suction pipe to above 27,000 Pa. Herein, the manufacturer has the final decision as to which of the product attribute modifications will be implemented.

The test of the method allowed us to confirm that it is possible to design a new House of Quality to predict a ranking of the pro-quality modification of the current product attributes. Also, it is possible to predict which of the product attributes are most important and how to modify the design of the product to advance its satisfaction. Therefore, the proposed method can support manufacturers in designing competitive products that in the future will be satisfactory to customers.

Author Contributions: Conceptualization, D.S. and A.P.; methodology, A.P. and D.S.; software, D.S. and A.P.; validation, A.P. and D.S.; formal analysis, D.S. and A.P.; investigation, A.P. and D.S.; resources, D.S. and A.P.; data curation, A.P. and D.S.; writing—original draft preparation, D.S. and A.P.; writing—review and editing, D.S.; visualization, D.S. and A.P.; supervision, A.P. and D.S.; project administration, A.P. and D.S.; funding acquisition, A.P. All authors have read and agreed to the published version of the manuscript.

Funding: This research received no external funding.

Data Availability Statement: The data are included in the article or may be sent upon request to the corresponding author.

Conflicts of Interest: The authors declare no conflicts of interest.

References

1. Pacana, A.; Gazda, A.; Dušan, M.; Štefko, R. Study on Improving the Quality of Stretch Film by Shainin Method. *Przem. Chem.* **2014**, *93*, 243–245.
2. Pacana, A.; Siwiec, D. Method of Determining Sequence Actions of Products Improvement. *Materials* **2022**, *15*, 6321. [CrossRef]
3. Gawlik, R. Preliminary Criteria Reduction for the Application of Analytic Hierarchy Process Method. In *Evolution and Revolution in the Global Economy: Enhancing Innovation and Competitiveness Worldwide*; Fuxman, L., Delener, N., Lu, V., Rivera-Solis, L., Eds.; Global Business and Technology Association: New York, NY, USA, 2008.
4. Li, W.; Pomegbe, W.W.K.; Dogbe, C.S.K.; Novixoxo, J. Dela Employees' Customer Orientation and Customer Satisfaction in the Public Utility Sector. *Afr. J. Econ. Manag. Stud.* **2019**, *10*, 408–423. [CrossRef]
5. Siwiec, D.; Pacana, A. A New Model Supporting Stability Quality of Materials and Industrial Products. *Materials* **2022**, *15*, 4440. [CrossRef] [PubMed]
6. Realyvásquez-Vargas, A.; Arredondo-Soto, K.; Carrillo-Gutiérrez, T.; Ravelo, G. Applying the Plan-Do-Check-Act (PDCA) Cycle to Reduce the Defects in the Manufacturing Industry. A Case Study. *Appl. Sci.* **2018**, *8*, 2181. [CrossRef]
7. Ali, A.; Hafeez, Y.; Hussain, S.; Yang, S. Role of Requirement Prioritization Technique to Improve the Quality of Highly-Configurable Systems. *IEEE Access* **2020**, *8*, 27549–27573. [CrossRef]
8. Kwong, C.K.; Bai, H. A Fuzzy AHP Approach to the Determination of Importance Weights of Customer Requirements in Quality Function Deployment. *J. Intell. Manuf.* **2002**, *13*, 367–377. [CrossRef]
9. Kowalska, M.; Pazdzior, M.; Krzton-Maziopa, A. Implementation of QFD method in quality analysis of confectionery products. *J. Intell. Manuf.* **2018**, *29*, 439–447. [CrossRef]
10. Gajdzik, B.; Wolniak, R. Smart Production Workers in Terms of Creativity and Innovation: The Implication for Open Innovation. *J. Open Innov. Technol. Mark. Complex.* **2022**, *8*, 68. [CrossRef]
11. Ostasz, G.; Siwiec, D.; Pacana, A. Model to Determine the Best Modifications of Products with Consideration Customers' Expectations. *Energies* **2022**, *15*, 8102. [CrossRef]
12. Perezanovic, T.; Petrovic, M.; Bojkovic, N.; Pamucar, D. One approach to evaluate the influence of engineering characteristics in QFD method. *Eur. J. Ind. Eng.* **2019**, *13*, 299–331. [CrossRef]
13. Wolniak, R. The Use of QFD Method Advantages and Limitation. *Prod. Eng. Arch.* **2018**, *18*, 14–17. [CrossRef]
14. Ding, Y.; Niu, L.; Chen, Y.; Wang, M. Study on the Defect Structure of Carbon-Doped ZnO Materials. *Cryst. Res. Technol.* **2023**, *58*, 2300015. [CrossRef]
15. Ellman, A.; Wendrich, R.; Tiainen, T. Innovative Tool for Specifying Customer Requirements. In Proceedings of the Volume 1B: 34th Computers and Information in Engineering Conference, Buffalo, NY, USA, 17–20 August 2014; American Society of Mechanical Engineers: New York, NY, USA, 2014.
16. Geng, L.; Geng, L. Analyzing and Dealing with the Distortions in Customer Requirements Transmission Process of QFD. *Math. Probl. Eng.* **2018**, *2018*, 4615320. [CrossRef]
17. Li, M. The Extension of Quality Function Deployment Based on 2-Tuple Linguistic Representation Model for Product Design under Multigranularity Linguistic Environment. *Math. Probl. Eng.* **2012**, *2012*, 989284. [CrossRef]
18. Wang, H.; Wang, D.; Wang, P.; Fang, Z. A Novel Quality Risk Evaluation Framework for Complex Equipment Development Integrating PHFS-QFD and Grey Clustering. *Grey Syst. Theory Appl.* **2024**, *14*, 144–159. [CrossRef]

19. Huang, Y.M. On the General Evaluation of Customer Requirements During Conceptual Design. *J. Mech. Des.* **1999**, *121*, 92–97. [CrossRef]
20. El Badaoui, M.; Touzani, A. AHP QFD Methodology for a Recycled Solar Collector. *Prod. Eng. Arch.* **2022**, *28*, 30–39. [CrossRef]
21. Li, Y.; Tang, J.; Luo, X.; Xu, J. An Integrated Method of Rough Set, Kano's Model and AHP for Rating Customer Requirements' Final Importance. *Expert Syst. Appl.* **2009**, *36*, 7045–7053. [CrossRef]
22. Aoyama, K.; Matsuda, N.; Koga, T. A Design Method of Product Family for Unpredictable Customer Requirements Using Fuzzy Sets. In *New World Situation: New Directions in Concurrent Engineering: Proceedings of the 17th ISPE International Conference on Concurrent Engineering, Taipei, Taiwan, 6–10 September 2010*; Springer: London, UK, 2010; pp. 193–201.
23. Liu, Y.; Zhou, J.; Chen, Y. Using Fuzzy Non-Linear Regression to Identify the Degree of Compensation among Customer Requirements in QFD. *Neurocomputing* **2014**, *142*, 115–124. [CrossRef]
24. Sun, N.; Mei, X.; Zhang, Y. A Simplified Systematic Method of Acquiring Design Specifications From Customer Requirements. *J. Comput. Inf. Sci. Eng.* **2009**, *9*, 031004. [CrossRef]
25. YAMAGISHI, K.; SEKI, K.; NISHIMURA, H. Requirement Analysis Considering Uncertain Customer Preference for Kansei Quality of Product. *J. Adv. Mech. Des. Syst. Manuf.* **2018**, *12*, JAMDSM0034. [CrossRef]
26. Wang, Y.-M.; Chin, K.-S. A Linear Goal Programming Approach to Determining the Relative Importance Weights of Customer Requirements in Quality Function Deployment. *Inf. Sci.* **2011**, *181*, 5523–5533. [CrossRef]
27. Franceschini, F.; Maisano, D.; Mastrogiacomo, L. Customer Requirement Prioritization on QFD: A New Proposal Based on the Generalized Yager's Algorithm. *Res. Eng. Des.* **2015**, *26*, 171–187. [CrossRef]
28. Hameed, A.Z.; Kandasamy, J.; Aravind Raj, S.; Baghdadi, M.A.; Shahzad, M.A. Sustainable Product Development Using FMEA ECQFD TRIZ and Fuzzy TOPSIS. *Sustainability* **2022**, *14*, 14345. [CrossRef]
29. Melemez, K.; Di Gironimo, G.; Esposito, G.; Lanzotti, A. Concept Design in Virtual Reality of a Forestry Trailer Using a QFD-TRIZ Based Approach. *Turk. J. Agric. For.* **2013**, *37*, 789–801. [CrossRef]
30. Wang, C.; Zhao, W.; Wang, Z.; Zhang, K.; Li, X.; Guo, X. Innovative Design Strategy Based on Customer Requirements. *Open Mech. Eng. J.* **2014**, *8*, 930–935. [CrossRef]
31. Sun, S. The Availability Improvement of CNC Machine Tool Based on DEMATEL–ISM–QFD Integration Method. *Int. J. Interact. Des. Manuf. (IJIDeM)* **2023**, *17*, 69–77. [CrossRef]
32. Ginting, R.; Ishak, A. An Integrated of AHP-QFD methodology for Product Design: A review. *J. Ilm. Tek. Indsutri* **2020**, *8*, 69–78. [CrossRef]
33. Edwards, W.; Barron, F.H. SMARTS and SMARTER: Improved Simple Methods for Multiattribute Utility Measurement. *Organ. Behav. Hum. Decis. Process.* **1994**, *60*, 306–325. [CrossRef]
34. Ostasz, G.; Siwiec, D.; Pacana, A. Universal Model to Predict Expected Direction of Products Quality Improvement. *Energies* **2022**, *15*, 1751. [CrossRef]
35. Sakao, T. A QFD-centred design methodology for environmentally conscious product design. *Int. J. Prod. Res.* **2007**, *45*, 4143–4162. [CrossRef]
36. Chen, C.-H.; Khoo, L.P.; Yan, W. Evaluation of Multicultural Factors from Elicited Customer Requirements for New Product Development. *Res. Eng. Des.* **2003**, *14*, 119–130. [CrossRef]
37. Li, Z.; Tian, H. Research on Fuzzy Hierarchy Optimization Model of Product Family Parameters Based on Flexible Design of Clothing. *IOP Conf. Ser. Mater. Sci. Eng.* **2019**, *573*, 012002. [CrossRef]
38. van de Poel, I. Methodological problems in QFD and directions for future development. *Res. Eng. Des.* **2007**, *18*, 21–36. [CrossRef]
39. Hansen, E.; Bush, R.J. Understanding Customer Quality Requirements. *Ind. Mark. Manag.* **1999**, *28*, 119–130. [CrossRef]
40. Siwiec, D.; Pacana, A. Model of Choice Photovoltaic Panels Considering Customers' Expectations. *Energies* **2021**, *14*, 5977. [CrossRef]
41. Hoła, A.; Sawicki, M.; Szóstak, M. Methodology of Classifying the Causes of Occupational Accidents Involving Construction Scaffolding Using Pareto-Lorenz Analysis. *Appl. Sci.* **2018**, *8*, 48. [CrossRef]
42. Mishra, P.; Pandey, C.; Singh, U.; Keshri, A.; Sabaretnam, M. Selection of Appropriate Statistical Methods for Data Analysis. *Ann. Card. Anaesth.* **2019**, *22*, 297. [CrossRef]
43. Yazdani, M.; Hashemkhani, S.; Zavadskas, E. New integration of MCDM methods and QFD in the selection of green suppliers. *J. Bus. Econ. Manag.* **2016**, *17*, 1097–1113. [CrossRef]
44. Lestyánszka Škůrková, K.; Fidlerová, H.; Niciejewska, M.; Idzikowski, A. Quality Improvement of the Forging Process Using Pareto Analysis and 8D Methodology in Automotive Manufacturing: A Case Study. *Standards* **2023**, *3*, 84–94. [CrossRef]

designs

MDPI

Article

Scalable Compositional Digital Twin-Based Monitoring System for Production Management: Design and Development in an Experimental Open-Pit Mine

Nabil El Bazi [1,2,*], Oussama Laayati [2], Nouhaila Darkaoui [2,3], Adila El Maghraoui [2], Nasr Guennouni [2], Ahmed Chebak [2] and Mustapha Mabrouki [1]

[1] Laboratory of Industrial Engineering (LGIIS), Faculty of Science and Techniques (FST), University Sultan Moulay Slimane (USMS), Beni Mellal 23000, Morocco; m.mabrouki@usms.ma
[2] Green Tech Institute (GTI), Mohammed VI Polytechnic University (UM6P), Benguerir 43150, Morocco; oussama.laayati@um6p.ma (O.L.); nouhaila.derkaoui@esi.ac.ma (N.D.); adila.elmaghraoui@um6p.ma (A.E.M.); nasr.guennouni@um6p.ma (N.G.); ahmed.chebak@um6p.ma (A.C.)
[3] School of Information Sciences (ESI), Mohammed V University (UM5), Rabat 10100, Morocco
* Correspondence: nabil.elbazi@um6p.ma; Tel.: +212-666-343-887

Abstract: While digital twins (DTs) have recently gained prominence as a viable option for creating reliable asset representations, many existing frameworks and architectures in the literature involve the integration of different technologies and paradigms, including the Internet of Things (IoTs), data modeling, and machine learning (ML). This complexity requires the orchestration of these different technologies, often resulting in subsystems and composition frameworks that are difficult to seamlessly align. In this paper, we present a scalable compositional framework designed for the development of a DT-based production management system (PMS) with advanced production monitoring capabilities. The conducted approach used to design the compositional framework utilizes the Factory Design and Improvement (FDI) methodology. Furthermore, the validation of our proposed framework is illustrated through a case study conducted in a phosphate screening station within the context of the mining industry.

Keywords: digital twin; production management system; monitoring; industrial internet of things; framework; machine learning; forecasting; artificial intelligence; mining industry

1. Introduction

In the last few years, the idea of a DT has been evolving as a transformative agent in different industries. Physical objects, processes, and systems can now be virtually represented with the help of digital twins (DTs). This virtual model enables live tracking, exercising, and improving the functionality of system operations [1]. A DT makes use of past usage analysis, which is an important way to attain complete knowledge about the functionality of an IoTs device over its lifetime [2]. The application of this technology provides an immense boost to the operation and maintenance of such systems, especially in terms of hindering circumstances, namely mining [3]. The mining industry is renowned for being extremely tough; it needs all equipment to be highly available and reliable to ensure safety as well as to maximize productivity. The old ways of monitoring and preventive maintenance may be implemented, but they provide delayed responses that result in unexpected downtimes and reduced efficiency. But DT technology is now being introduced, which has virtual models made for real-world monitoring and simulation purposes. This technology has brought about a new ray of hope in terms of the productivity and efficiency of maintenance practices, where mining companies can delimit equipment production and decrease downtime.

In the competitive global market, deploying DT technology is pivotal for enhancing industrial efficiency. As a key component in the Industry 4.0 roadmap, DTs bridge physical and virtual realms, offering dynamic models that accurately reflect real-world entities. Within the context of the Fourth Industrial Revolution (Industry 4.0) [4], a scalable compositional DT framework emerges as a systematic solution for connecting the physical and virtual worlds [5].

This paper is dedicated to the elaboration of a scalable compositional framework, with a specific emphasis on the adoption of this framework in a real mining plant as a tool for production management improvement. A systemic approach, the framework incorporates a series of solutions intended to handle the complexity of modern industrial operations as well as streamline these operations to achieve smooth productivity and flexibility. Among others, this framework enables data entities to communicate with each other, thus spreading knowledge and linking information, creating a more intelligent contextually aware DT. Framework specifies a modular and integrated environment where users use DTs according to their needs and develop them.

The main contributions of this paper are as follows:

- The design of an architecture for a production management system tailored to the mining operations of the experimental pilot of our research, the experimental open-pit mine.
- The development and implementation of a scalable compositional framework for a DT, facilitating an efficient PMS.

This paper is organized as follows: Section 2 mainly presents the relevant background and related works on digital twin technology, monitoring systems, and production management systems. Section 3 contains the research methods and materials, which include a data modeling process and the FDI-based methods used to conduct the site survey at the industrial mining site to extract the value chain of the industrial mining site, as well as the design of the database architecture. Section 4 includes the digital twin infrastructure modular and scalable compositional monitoring framework. Section 5 presents the experimental data, the results obtained, and their analysis and interpretation. Section 6 concludes our study by highlighting the potentialities of the presented project and provides possible research avenues for further efforts.

2. Background and Related Works

Within this section, we intend to provide a comprehensive analysis of the key components of Industry 4.0 and automated machine systems that are created with the most recent technology and techniques, such as DT, monitoring systems, and PMS. These elements are what maintain the industrial machines' operation today, serving as the most important constituents in developing more efficient, precise, and adaptable production technologies. Furthermore, we will be tackling the foundational principles of these components, undressing the practicality of these ideas, and critiquing the recent research, as well as the practices in the field. By this means of disentangling and connecting these important points of divergence, a clear and well-researched understanding of the multidimensional and changing scenario of these issues, their relationships, and the field of science that guides the operational decisions in modern industry is the aim.

2.1. *Digital Twin*

2.1.1. Definitions and Associated Attributes

Digital twins (DTs), which originated as an aerospace concept, facilitate virtual versions of objects or systems. While it was originally used for aircraft analysis, its usage has spread to multiple areas. NASA uses a widely accepted definition that treats a simulation as a digital tool based on advanced models and real-time data, which can accurately resemble the original system. In simpler terms, DTs are akin to a pair of twins, who can either be tangible or digital [2,5]. Crucially, DTs encompass three key elements: a physical being, a virtual mimicry, and a network of paths for real-time synchronizing. DT can be presented in several ways, but the simplest explanation is that it provides scalable and secure connections between relevant data from the physical to the virtual world. Figure 1 illustrates the differences between a digital model, a digital shadow, and a digital twin, which are three different integration levels illustrated in the same figure. Generally, they are all associated with data exchange among physical, digital, and whatever else (model, shadow, twin) [6].

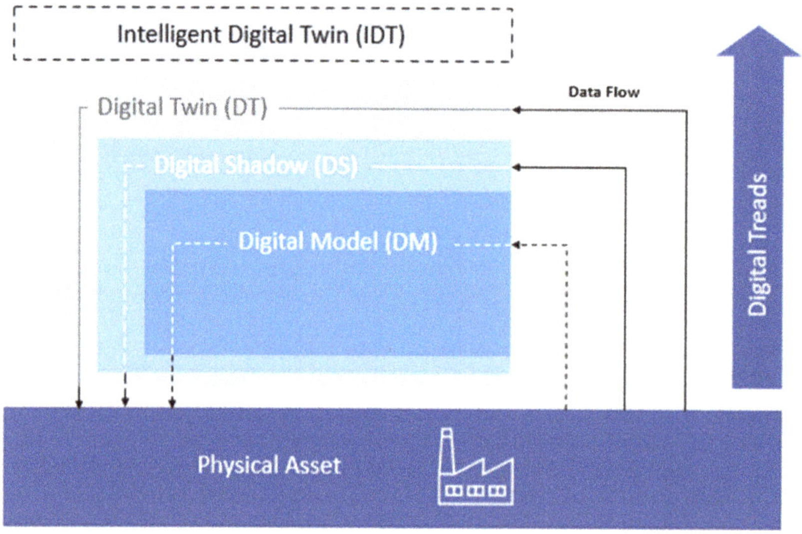

Figure 1. Integration level of digital twin.

The characteristics of DTs include real-time reflection; the physical and digital dimensions coexist, allowing for the synchronous real-time data of the current states to be made available in one place. A cycle of fusion and divergence between the real and virtual worlds is enabled based on the continuous integration of current and historical data. A DT has the potential to be self-updating, i.e., it dynamically updates its data simultaneously, allowing the virtual part to evolve by comparison with the real part [7].

By depicting the feature essential properties of DT for its incorporation, we have created a MoSCoW diagram, as represented in Figure 2, laden with their significance and impact. In the "Must-have" category, we have essential components like "Data Flow/Updates", which is the most livewire that keeps the DT algorithm in real-time modes, thereby giving it its relevance and accuracy. "Real-time Synchronous" operation, in turn, updates the DT with the immediate data and changing of conditions of the physical counterpart in a real-time manner [8], ensuring the quick response requirements. "Connectivity" is fundamental giving the possibility of working and exchanging ideas in a seamless setting with the sundry systems and "Simulatability" is thus used to simulate the behavior of the physical asset with the desired level of accuracy. The "Active Interaction Applicability" is a crucial contributor for the "Should Have" category, which enhances adaptability, "Data Fusion" increases the level of insights, and "Interpretability" helps with the generation of data which are actionable. In the "Could Have" realm, aspects like "Self-improvement", "autonomous" decision-making, and envisioning consequent roles for "Artificial Intelligence" provide some more capabilities, though they are not mandatory. Notably, otherwise, there is nothing described in the "Do not have" category, which means that the DT should accomplish this objective in order to obtain the best result in different industrial situations.

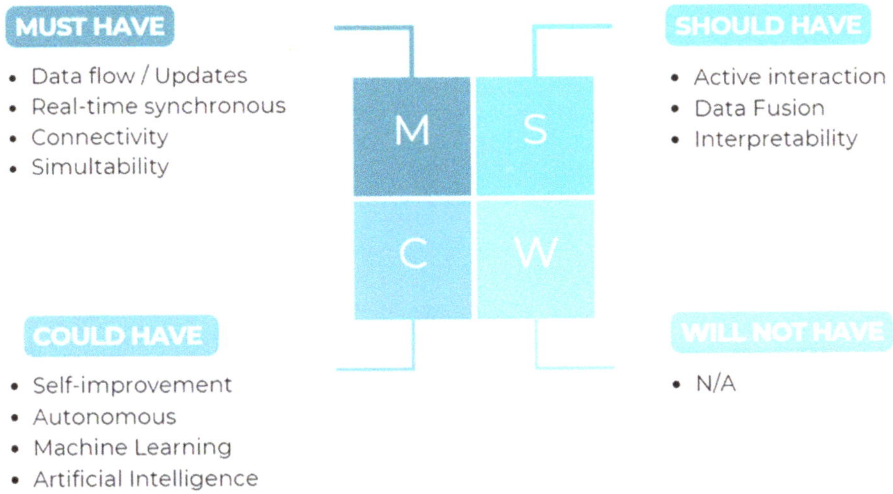

Figure 2. MoSCoW Diagram showing the attributes and features of digital twin.

2.1.2. Current State of Digital Twin Research

In the current state of DT research, worldwide efforts thrive, with numerous frameworks and architectures emerging. In the manufacturing context, the DT serves as a virtual representation mirroring the real-time operation of the physical manufacturing asset, encompassing machines, production lines, the shop floor, products, and workers. This digital counterpart empowers real-time monitoring and the prediction of future behavior, performance, and maintenance needs [1,5].

Notably, three primary application scenarios for DT exist: supervisory, involving real-time status provision for decision-making; interactive, where DT autonomously adjusts parameters upon disruptions; and predictive, where DT forecasts the asset's future state for a corrective action [9].

Within manufacturing, the DT rejuvenates multiple tasks:

- Equipment health management: DT enhances system and worker reliability, availability, and safety through seamless monitoring and informed maintenance decisions [10]. For example, it estimates the remaining useful life (RUL) of equipment components, enabling intelligent design and timely monitoring for predictive maintenance [11].
- Production control and optimization: Dynamic manufacturing environments require continuous monitoring and optimization [12,13]. DTs use real-time data to optimize throughput by adjusting controllable parameters [14]. They also react in real time to disturbances on the shop floor [15].
- Production scheduling: Traditional static production scheduling methods struggle with process uncertainty. DTs dynamically elaborate or verify schedules during disruptions. They even communicate with robots for optimal task scheduling.

The DT for manufacturing usually consists of a physical element, a virtual element, and a real-time information exchange between them, assisted by the IoTs, data collection and storage tools, big data analytics, and ML [15]. Making the privacy and security of data a top priority still stays. DT research has experienced rapid progress, especially in manufacturing, but defining the research areas remains a question because there is no unified definition [16]. To promote a base of knowledge shared among members and address implementation problems, further study is still required even though progress has been made in the realms of equipment health management, production control, optimization [17], and scheduling [18–20]. Nevertheless, there is still a noticeable gap, which is mostly defined by the lack of hands-on examples of real-live implementation. Although the technology is mostly theoretical, it has not been implemented in a clear manner, and its practical cases are limited [5,16].

These areas of research acknowledge the application of advanced analytics to disrupt conventional data analysis [21], such as equipment maintenance, simulations for the greater in-depth examination of processes and factories, and the utilization of emerging technologies like virtual and augmented reality for data accessibility improvement [22].

Although progress has been achieved, there are still gaps in research, e.g., the lack of practical implementation cases and suitable solutions for small- and medium-sized businesses. The adaptability and effectiveness of DT frameworks across different industries and scenarios also requires further study [23,24]. In the following sections, we examine the establishment of a DT-oriented integrated monitoring system that can enhance the effectiveness of a smaller enterprise to be in charge of its operation.

2.2. Monitoring System

In the realm of industry and manufacturing, monitoring systems have emerged as pivotal drivers of efficiency, safety, and productivity [25]. These systems, founded on the real-time tracking and analysis of processes, equipment, and operations, provide invaluable insights for decision-makers. Their multifaceted benefits encompass heightened operational control, predictive maintenance, quality assurance, and resource optimization, finding applications across diverse industrial domains [3]. However, challenges persist, including the integration of disparate data sources, ensuring data security, and scalability concerns.

It is the synergy between monitoring systems and technology of data that makes the position of research notable, creating a paradigm shift for the industry. The fusion of Electrical and Computer Engineering (ECE) and AI not only brings better control and predictive capabilities but also the capacity to simulate and optimize processes in a risk-free digital environment which is needed to conduct tests on "what if" scenarios [26]. It amplifies the perks of the monitoring systems and provides a great impact on industry applications, from observing mechanical processes on the factory floors to monitoring big industrial activities [24].

Nevertheless, the translation of the already-developed integrated systems from theory to practice appears to be a problem. Despite significant development, there is a lack of practical real-time prototypes existing currently that combine the specific systems and prove their applicability in different industries [24,25]. This offers a challenging demand for the development of practical, flexible, and cost-effective solutions that can effectively work in different industrial situations to be considered in future research [27]. In addition, we should take into account the security problem of the data, complexity of integration, and scalability problem [28] in order to provide the smooth implementation of the integrated monitoring systems empowered by DT technology [22,29].

2.3. Production Management System

PMSs have become recognized as a much more significant element in the industry and manufacturing sector of today. They carry out an essential function in performing the somewhat complicated tasks of production [4]. These elements are the main things to rely on for planning, scheduling, and sending out production signals. They ensure that resources are used effectively and nothing less than maximum productivity [27]. What is their behavioral formula is the data-driven approach, automation, and an analysis of the results. They support improvements in production that become more effective, speedy, and successful in today's turbulent markets. The main advantages of these systems are natural resource efficiency, cost savings, product excellence, and timely deliveries. Therefore, they are critical for modern manufacturing [7].

Nevertheless, the inclusion of such systems in the current networked world also poses serious challenges in regard to system integration, cybersecurity, and adaptation to different production zones [29]. A vital research viewpoint is the successive gaps between theoretical development and the actual realization of the scaled 'Pseudo-mobile work lifestyle' [30]. Nevertheless, significant advances have been made, yet the scarcity of complete real-life simulated cases of perfect integration of those transition tools, on the industrial level, remains.

The research gap reveals a very strong need: the development of solutions that are practical, flexible, and affordable and can be adjusted to comply with the specified requirements of any industry. In the quest to bridge this gap and fully harvest the capabilities of PMSs in manufacturing, there is a new and astonishing way to achieve this through the application of the DT concept.

DT framework implementation into PMSs is a turning point in which they can meet up. The integration of PMS dynamic capabilities and the real-time tracking, data integration, and analytics of DTs render a new horizon. DTs, which are known for their features such as real-time reflection, interaction, convergence, and self-evolution, can equip the PMSs with exceptional levels of accuracy and adaptability of operation [31].

Imagine a manufacturing environment where the DT of a production asset, be it a machine, production line, or even an entire factory, mirrors its real-world counterpart. This digital replica operates in tandem, offering real-time insights and predictions regarding the physical asset's behavior, state, performance, and maintenance needs [31,32]. It enables not only enhanced operational control but also predictive maintenance strategies, quality assurance, and optimization in a risk-free digital environment ideal for testing "what if" scenarios.

DTs could be the key to revolutionize PMS production control and optimization. The dynamic rapid changes and uncertainties in manufacturing environments require continuous monitoring. Since DTs are available, the PMS receives an inclusive perspective of the manufacturing asset in real-time when utilizing data [33]. This facilitates rapid and accurate corrections for overall throughput maximization [22].

Moreover, production scheduling, which has always had a hard time coping with the dynamic nature of production as a whole, might finally receive a boost from DTs. In reality, when disturbances happen on the shop floor, DTs responsible for the dynamic modeling and verification of schedules can do so efficiently and ensure resource allocation as well as prevent cascading effects due to unexpected events.

Meanwhile, it is undeniable that we also have several challenges to overcome. Data integration complications and the security and reliability of the integrated systems are the major challenges faced in the infusion of integrated systems into the healthcare industry [34]. To effectively make full use of the benefits of PMSs which have been delimited by DTs, more study and innovations are required [7]. This includes creating scalable, affordable, and flexible technical solutions that can take into consideration these challenges and also provide solutions based on the individual needs of industries [35]. This pillar of innovation challenges is calling us as we move to a future where the interfacing of PMSs and DTs transforms the manufacturing processes to unprecedented levels of efficiency, adaptability, and improvement [36].

3. Materials and Methods

The following section forms the cornerstone of our efforts to enhance mining operations. Here, we describe the following research methodology, as well as discuss the subsequent methods involved in developing the proposed digital twin-based production management system designed for the real-time monitoring and optimization of key performance indicators (KPIs). Data modeling is our starting point, and we use the modeling and improvement principles of a factory. Beginning with a comprehensive survey of the mine value chain, data points are collected, and a system architecture database is explored. To complete the picture, we introduce you to the ARIMA (Autoregressive Integrated Moving Average)-based production forecasting model, which is the key component in the context of our novel approach. The combination of these methods and materials forms the basis for an improved system adapted to increase the efficiency and accuracy of mining decision-making.

3.1. Research Methodology

The research methodology used in this study aims to develop a monitoring system based on a digital twin architecture for production management. This methodology—Figure 3—comprises eight sequential steps, each of which contribute to a comprehensive understanding and refinement of the proposed system. First, an exploratory literature review of the mapping type is carried out to gather evidence and identify existing knowledge gaps. This mapping review aims to provide a comprehensive overview or 'map' of the existing literature on a particular topic. This conducted literature review summarizes key characteristics, trends, and themes across a broad range of the literature. It helps to understand the scope of research on a topic and serves as a basis for further investigation or synthesis. Focus groups are then formed with heterogeneous stakeholders from various levels of the open-pit mine workforce, including managers, engineers, and technicians representing diverse disciplines within the craft body responsible for handling the mining operations. These focus groups are convened to discuss and evaluate the findings from the literature review. The adoption of a systems design approach based on the Factory Design and Improvement (FDI) technique is then validated. A survey of the mine value chain is then initiated to gather essential data for further analysis. Based on this baseline data, a system data model is constructed to provide a structured representation of the production environment. The subsequent development of a digital twin-based system framework integrates advanced technologies to enhance production monitoring capabilities. Following this framework's development, the case study of the proposed system is defined, with the screening station identified as an experimental pilot for evaluation purposes. Finally, the experiments and system integration are carried out by validating the proposed system with a real-time data flow and integrating it with the SCADA system data.

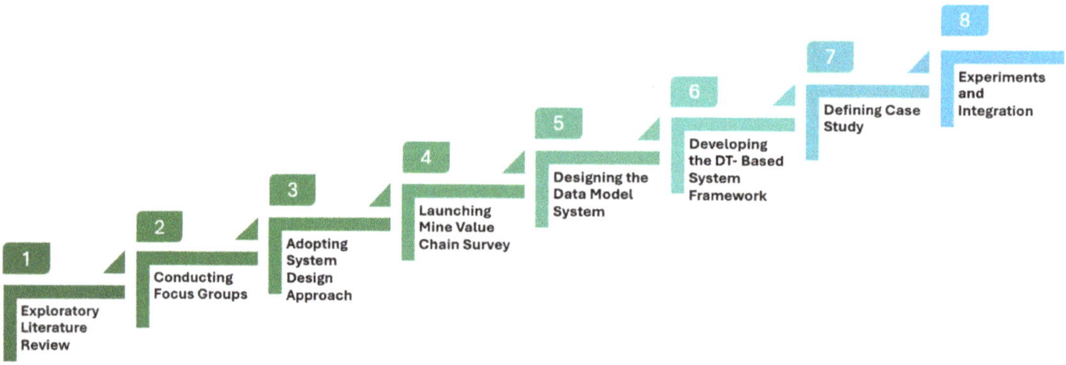

Figure 3. Followed research methodology.

3.2. Data Model

3.2.1. Factory Design and Improvement-Based Data Model

To achieve the aim of flexible integration and interoperability of our designed system, it is relevant to develop a data model assisting the connecting heterogeneity of the DT's components and the bidirectional data flows to ensure that there is a link with the physical part of the whole cyber–physical system (CPS). In this paper, our data model relies on the FDI Activity Model. Developed by the National Institute of Standards and Technology (NIST), the FDI serves as a foundation framework for our study, as unveiled in the presented diagram in Figure 4. The FDI formalizes essential [30] activities, functions of enterprise software, and crucial information for operational design and management tasks in the context of smart manufacturing systems [37]. It aligns with the standard work processes commonly observed in global manufacturing enterprises, encompassing aspects such as factory operations, manufacturing lines, processes, and equipment operations [30].

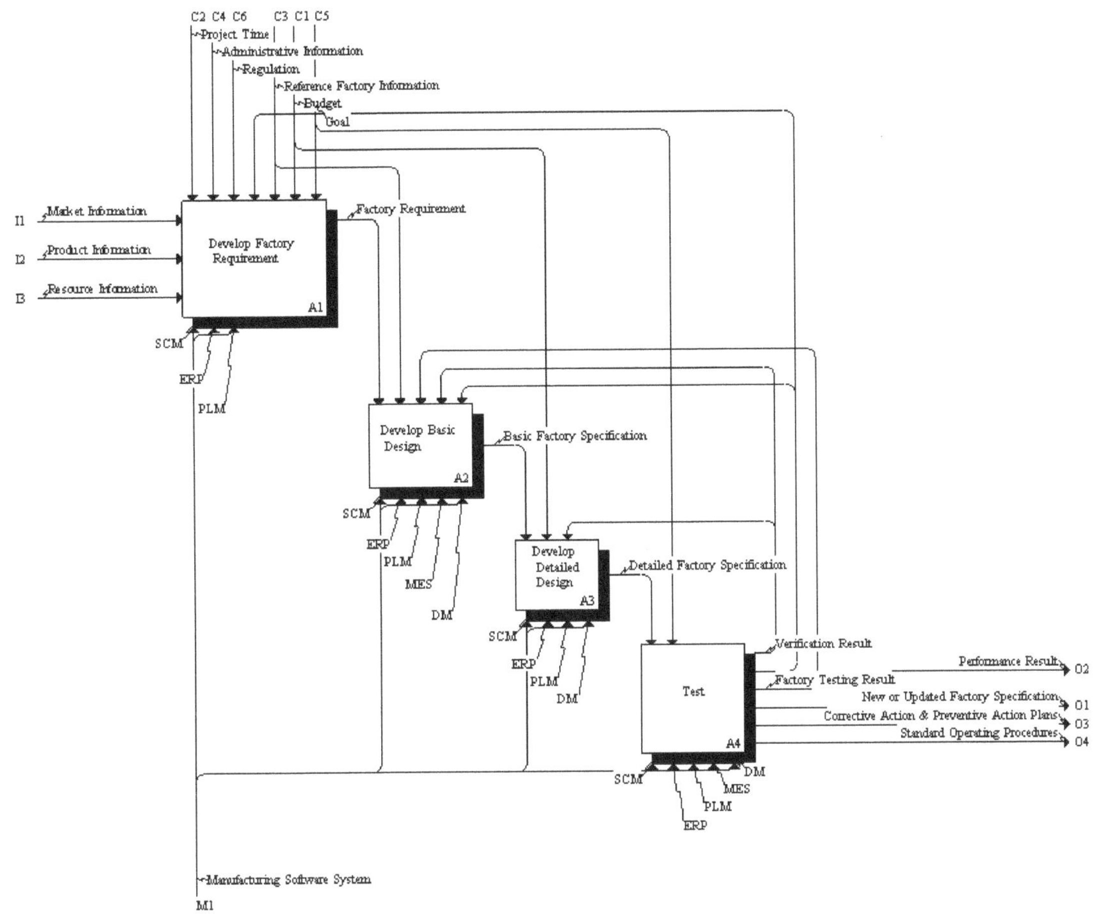

Figure 4. Factory Design and Improvement (FDI) reference activity model.

At its heart, the ICOM (input, control, output, mechanism) parameters, foundational to the smart manufacturing system (SMS), dictate its operational dynamics. Each component of the ICOM paradigm plays a distinct yet interlinked role in making decisions and performing actions regarding information [34].

3.2.2. Mine Value Chain Parameters Survey

By leveraging the FDI model, we gain a comprehensive overview of the operational processes within the open-pit mine's value chain [26], encompassing factory operations, manufacturing lines, processes, and equipment. This model empowers us to analyze performance measures, organizational structures, tools, systems, and associated data comprehensively. The FDI model divides the design process into four distinct phases, including development and design requirements, basic design tasks, detailed design tasks, and testing [30,34]. This structured breakdown of design activities has been demonstrated to expedite factory development projects significantly. By following this approach, we conduct a systematic inventory and mapping process to comprehensively document and analyze the facets within the mine value chain according to the ICOM (input, control, output, mechanism) parameters [30] listed in Table 1.

Table 1. FDI-based mine value chain parameters survey.

ICOM	Factors	Description
Input	Information	Product Information, Market Information, Resource Information, Production Schedule, Labor Information, Equipment Information,
Output	Key performance indicators (KPIs)	Cycle Time, Lead Time, Production Output, Work-In-Process (WIP), Return-On-Capital-Employed (ROCE),
Control	Work process	Product Lifecycle Management (PLM),
	Methodology	Operational Excellence (OpEx), PDCA
	People	Process Operators, Process Designer, Process Engineers, Process Managers
	Technology	Statistical method, stochastic method, simulation, co-simulation
Mechanism	Tools/system functions	PLM, Computer-Integrated Manufacturing (CIM) pyramid, SCADA, OEE

Though the focus of the production value chain survey includes three main stations, namely destoning, screening, and train loading, respectively, the input, control, output, and mechanism parameters within the FDI methodology offer a systematic approach to organizing and managing the operating processes logically. In the input stage, information is the paramount element, addressing the critical elements. This refers to Product Information giving out specifications of the products being manufactured, Market Information presenting insights into the target market for demand, competition, and trends, Resource Information furnishing the demand for tangible and intangible resources, Production Schedule detailing the sequencing and timing of tasks, Labor Information providing workforce details, and Equipment Information indicating the machinery and tools utilized in manufacturing. The output factor, key performance indicators (KPIs), serves as a guide for systems performance metrics such as Cycle Time, Lead Time, Production Output, Work in Process (WIP), and Return on Capital Employed (ROCE). Control mechanisms determine factors like Work Process, which focuses on Product Lifecycle Management (PLM) for systematic product supervision, methodology that highlights Operational Excellence (OpEx), PDCA, which stands for constant improvement, People, involving key personnel roles, and Technology, which emphasizes statistical methods, simulations, and co-simulations for precision. The Mechanism factor involves Equipment and System functionalities such as PLM, CIM, SCADA systems, and ERP, which drive process effectiveness and efficiency [38]. In general, the ICOM structure enables companies to have a roadmap in comprehending, evaluating, and refining complex operational processes that then emphasize the importance of accurate data inputs, performance metrics, control mechanisms, and the use of the latest technological tools in ensuring that operations efficiency and output optimization are achieved.

3.2.3. FDI Data Model Enabling Production Management System

The elucidated data model shown in Figure 5, is vividly evident in the intricate detailing of the overall mine's value chain. Central to this model is the *'Factory.Description'* element, anchoring the entire design by synchronizing the attributes and functionalities of the factory with the 3PR (Product, Process, Plant, Resources) approach, a holistic paradigm focusing on Product, Process, Plant, and Resources. The *'Product'* domain, with its attributes like weight, destination, quality, and profile, underscores the end goals and the quality benchmarks set by the production system. *'Resources'*, a pivotal segment, elucidates the tangible and intangible assets, accentuating the operational efficiencies, equipment reliability, and the strategic deployment of labor. The *'Process'* section, intertwined with operations such as destoning, screening, and train loading, exemplifies the step-by-step manufacturing mechanics, ensuring that no detail is overlooked. On the other hand, the *'Plant'* category offers a panoramic view of the facility's infrastructure, capturing aspects from layout designs to safety standards. With the inclusion of the 3PR approach, the model transcends traditional factory designs, harmonizing product quality, efficient resource allocation, streamlined processes, and optimal plant utilization. In essence, this data model, inspired by NIST's FDI methodology and enriched by the 3PR approach, presents a visionary overview for a future-ready, operational holistic customized database of our PMS.

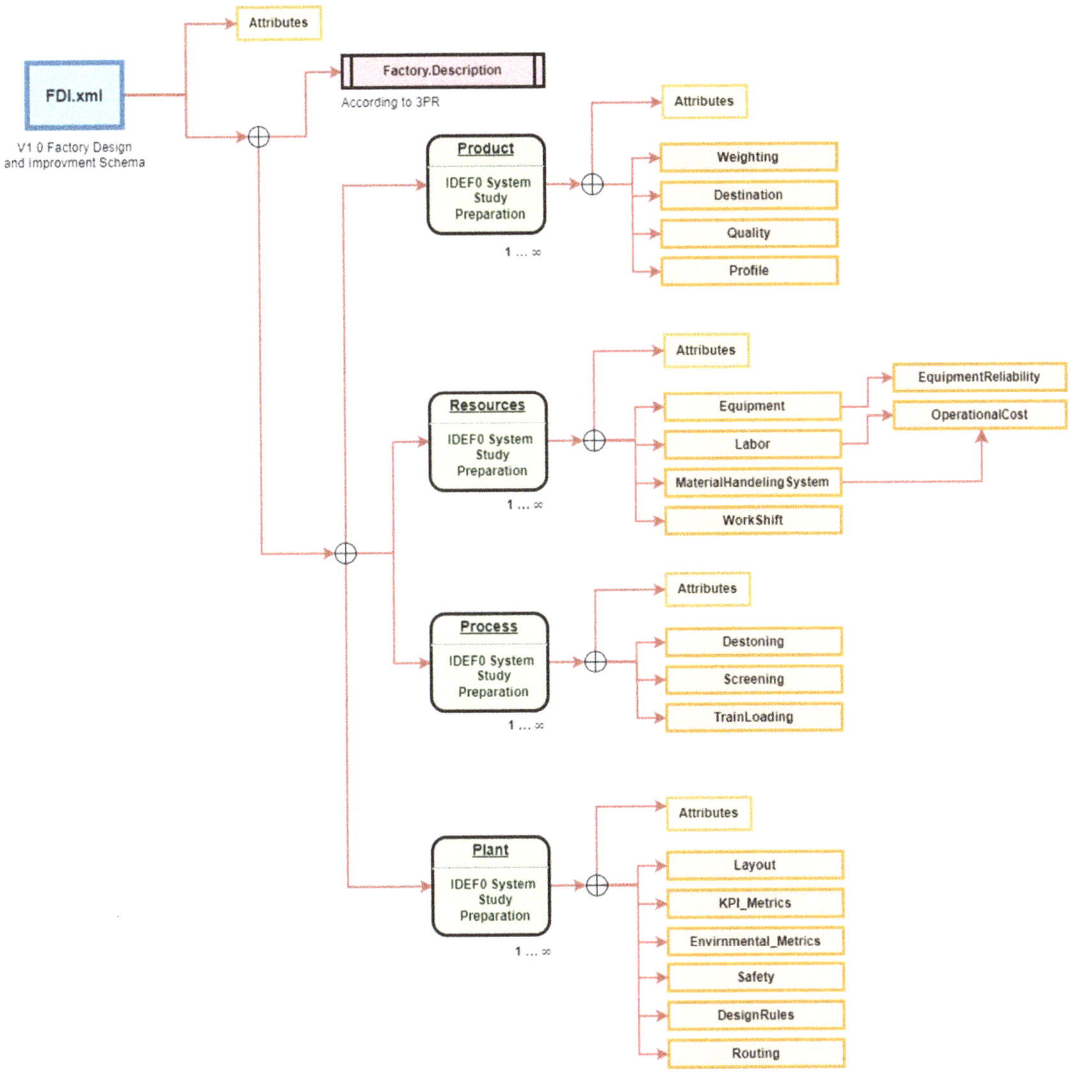

Figure 5. FDI-based data model enabling production management system.

3.3. *Autoregressive Integrated Moving Average Model*

3.3.1. Theoretical Background

The Autoregressive Integrated Moving Average (ARIMA) model stands as a robust statistical tool for the analysis and forecasting of time series data [39]. This model adeptly addresses various structures inherent in time series data, offering a straightforward yet powerful approach for making accurate forecasts [40].

The acronym ARIMA breaks down as follows:

AR (Autoregression): Emphasizing the dependent relationship between an observation and its lagged counterparts.

The Autoregressive (AR) model, which is among the earliest models employed in time series analysis, is a linear model. This model utilizes a combination of past values within a time window, to which an error is added, as illustrated by Equation (1):

$$X(t) = \sum_{i=0}^{i=p} a_i x(t-1) + e(t) \quad (1)$$

where $X(t)$ is the value of the series at time t, p is the order of the model, a_i denotes the autoregressive parameters, and $e(t)$ represents white noise, denoting the autoregressive model of order p as AR(p).

I (Integrated): Introducing differencing to achieve a stationary time series, mitigating trends and seasonality.

MA (Moving Average): Focusing on the relationship between an observation and the residual error from a moving average model based on lagged observations.

The moving average model, also known as 'Moving Average (MA)', is another linear model used for time series forecasting. Unlike the autoregressive model, it is based on the white noise of the series. This model is defined by Equation (2) and is referred to as a moving average of order q, denoted as MA(q):

$$X(t) = \sum_{i=0}^{i=p} a_i e^{t-i} + e^t \quad (2)$$

where a_i represents the moving average parameters, and e^n denotes the white noise of the series.

Each component is explicitly defined in the ARIMA model, denoted as ARIMA (p, d, q), where the parameters take integer values to specify the model type. The parameters are elucidated as follows:

p (Lag Order): Signifying the number of lag observations incorporated in the model.

d (Degree of Differencing): Representing the number of times raw observations undergo differencing to achieve stationarity.

q (Order of Moving Average): Indicating the size of the moving average window used in the model.

The ARIMA model is a combination of the two preceding linear models: AR and MA. It also includes an integration term (I) to account for the non-stationarity of time series. The equation for an ARIMA model is represented (3):

$$\Delta^d X_t = \sum_{i=1}^{p} a_i X_{t-i} + \sum_{i=1}^{q} \beta \varepsilon_{t-i} + \varepsilon_t \quad (3)$$

In constructing the linear regression model, the specified terms are integrated, and the data undergo differencing to attain stationarity, eliminating trends and seasonal structures that may negatively impact the model. It is noteworthy that any of these parameters can be set to 0, allowing the ARIMA model to emulate simpler models like ARMA, AR, I, or MA.

The adoption of an ARIMA model for time series analysis assumes that the underlying process generating the observations follows an ARIMA process. This assumption underscores the importance of validating the model's assumptions against both the raw observations and the residual errors of the forecasts [40,41].

To implement the ARIMA model in Python, the next step involves loading a simple univariate time series, initiating the practical application of the theoretical foundations discussed above.

The next diagram of Figure 6 illustrates the implementation workflow [42].

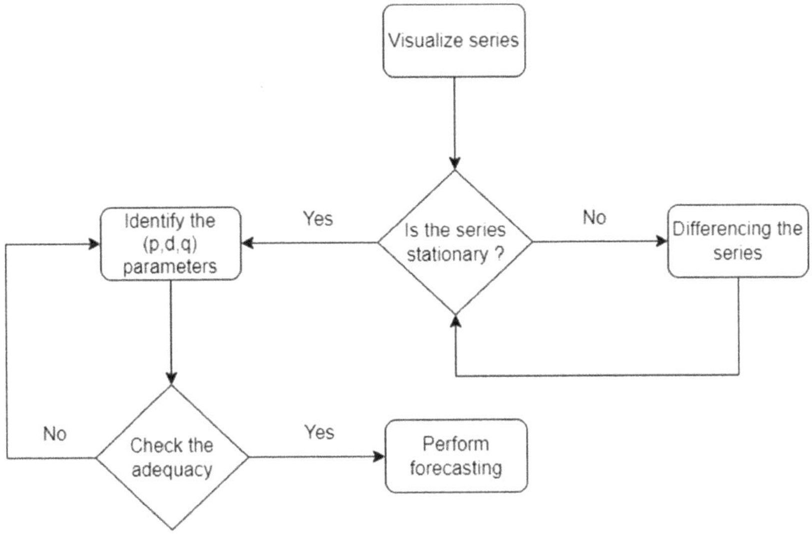

Figure 6. ARIMA model training flow.

3.3.2. Model Performance Metrics

We use different performance metrics to check how well our ML model performs on new, unseen data by comparing their predictions to the actual outcomes in a test dataset. We evaluate the suggested models using three key metrics [43]: mean absolute error (MAE), root mean squared error (RMSE), and mean absolute percentage error (MAPE). The specific formulas for these metrics are given by Equations (4)–(6).

- Mean Absolute Error

The mean absolute error (MAE) is a measure utilized to evaluate the discrepancies between paired observations that pertain to the same event. The computation of MAE involves the multiplication of the subsequent formula as depicted in Equation (4):

$$MAE = \frac{1}{n}\sum_{i=1}^{n}|y - y'| \quad (4)$$

The mean absolute error is calculated on the same scale as the data. However, since this accuracy metric is dependent on the scale, it is not suitable for comparing series with different scales. In the realm of time series analysis, the mean absolute error is commonly employed to gauge the forecast error.

- Root Mean Squared Error

The root mean square error (RMSE) is a common way to measure how much the predicted values from a model differ from the actual observed values. It is important to note that RMSE depends on the scale of the data, making it more suitable for evaluating forecasting errors within a specific dataset rather than comparing different datasets.

The equation for the RMSE is represented (5):

$$RMSE = \sqrt{\frac{1}{n}\sum_{i=1}^{n}(y - y')^2} \quad (5)$$

- Mean Absolute Percentage Error

The mean absolute percentage error (MAPE) is a metric used to assess the accuracy of a forecasting model by measuring the percentage difference between predicted and observed values, providing a relative measure of prediction accuracy.

The equation for the MAPE is represented (6):

$$MAPE = \frac{1}{n}\sum_{i=1}^{n}\left|\frac{y-y'}{y}\right| \times 100 \quad (6)$$

4. Scalable Compositional Digital Twin-Based Production Management System for Real-Time Monitoring and KPI Optimization in Mining Operations

The proposed Scalable DT-based PMS for real-time monitoring and KPI optimization in mining operations (Figure 7) introduces a groundbreaking approach to enhancing the efficiency and effectiveness of mining operations [44]. By seamlessly integrating edge computing, cloud computing, and artificial intelligence, the framework effectively gathers, analyzes, and visualizes data from various sources to create a comprehensive digital representation of the mining process. This virtual mirror, known as the DT, serves as a powerful tool for real-time monitoring, predictive production, and optimization of key performance indicators (KPIs).

The system efficiently manages the high volume of data generated by mining equipment through edge computing. The programmable logic controllers *Schneider Modicon 340* and local control panels also known as human machine interface *Schneider Magelis* are commonly positioned at the mining site to collect raw data from sensors, data loggers, and smart meters. These data are then transmitted to the multi-controller server *OPC Factory Server* (OFS) via *Modbus TCP* protocol between dispersed edge control devices (data logger, smart meters, relays, etc.).

Once collected, the data are logged onto the multi-controller Server Data Object File System (OFS). The SCADA system then utilizes these data for monitoring and SCADA functionalities, e.g., real-time visualization and monitoring [44]. Subsequently, the data are stored in the backbone of our system, the *PostgreSQL-pgAdmin* database, which serves as a centralized repository for both historical and current data. This centralized repository facilitates efficient data retrieval and analysis for various purposes.

To prepare the data for further analysis, the framework lies on the bloc named Data Preprocessing, which performs multiple Python libraries, such as *Numpy*, *Pandas*, *Seaborn*, and *Matplot*. This involves cleaning and formatting the data, removing outliers, missing values and inconsistencies, and ensuring data compatibility with the DT and predictive models. Furthermore, the preprocessed data come to the prediction stage, within the milestone component that revolves around leveraging advanced ML functionalities, particularly those embedded in the Auto Regression Integrated Moving Average (ARIMA) algorithm, where this block delves into the intricate mechanisms of the ML model that relies on ARIMA, utilizing historical data collected from diverse mining fields. This integration embodies a forward-looking approach that aligns with the industry's drive toward efficient production prediction. The advantages of the ARIMA model are presented in Section 3.3.

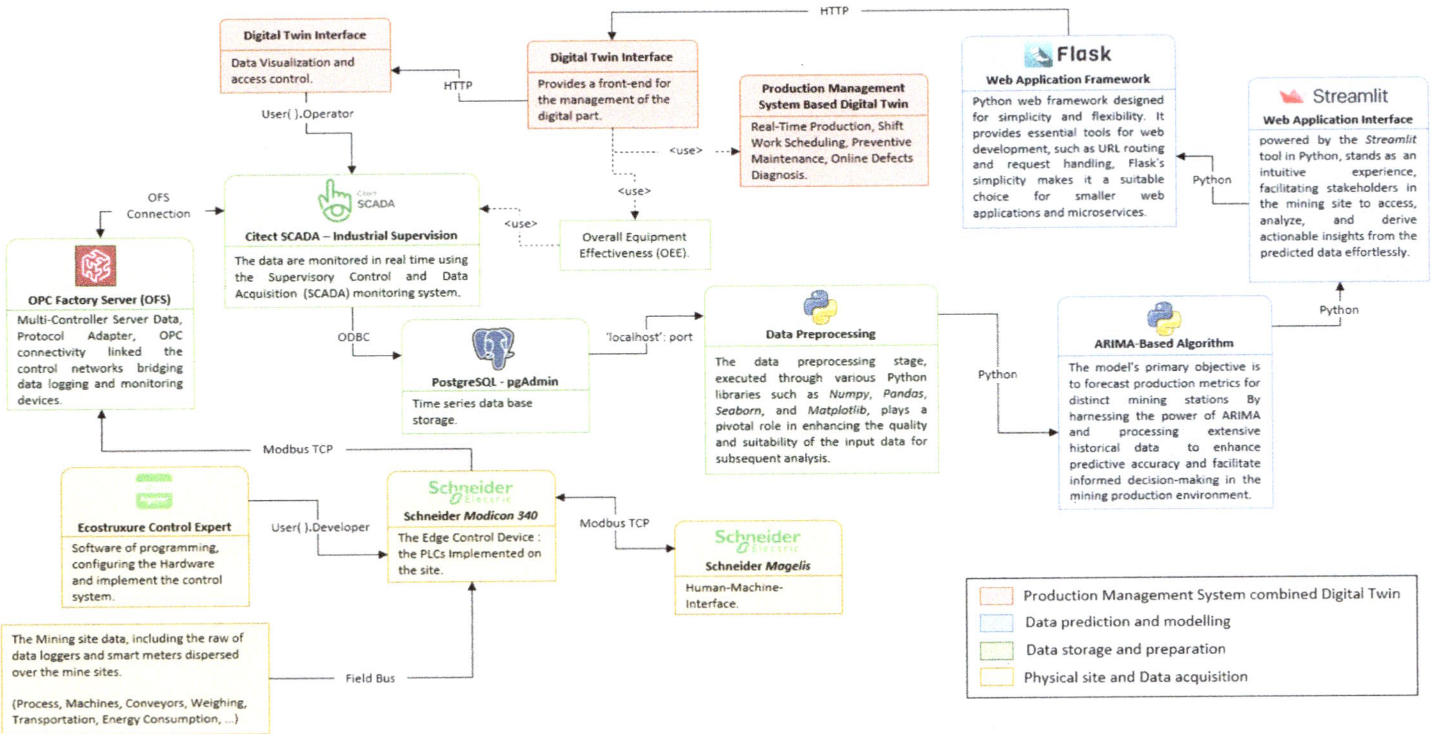

Figure 7. Scalable compositional digital twin-based production management system framework.

5. Experiment and Results

5.1. Data Description

The dataset under consideration originates from real-time measurements collected at the phosphate screening station over the course of 20 months, starting from 1 January 2021 to 31 August 2022. Specifically, the data pertain to the tonnage of phosphate that has been screened and is in a wet state upon exiting the screening station. The information is sourced directly from the mining site, with measurements taken daily and in real-time.

The flow of the operational process of the screen station is inextricably connected to the data collection process. A built-in weighing scale, consisting of the conveyor belt system, registers the tonnage of the phosphate that goes through the screen. These measurements are taken in intervals to specifically correspond with the end of each working shift. This total will provide an overall daily metric that is produced by the summation of the tonnage from all three shifts.

Figure 8 shows that the dataset includes the daily tonnage of wet phosphate as it is screened and monitored over 20 months. The temporal format of the dataset corresponds with the changes and operations cycles, providing a useful tool for analyzing and forecasting the trends of phosphorite production.

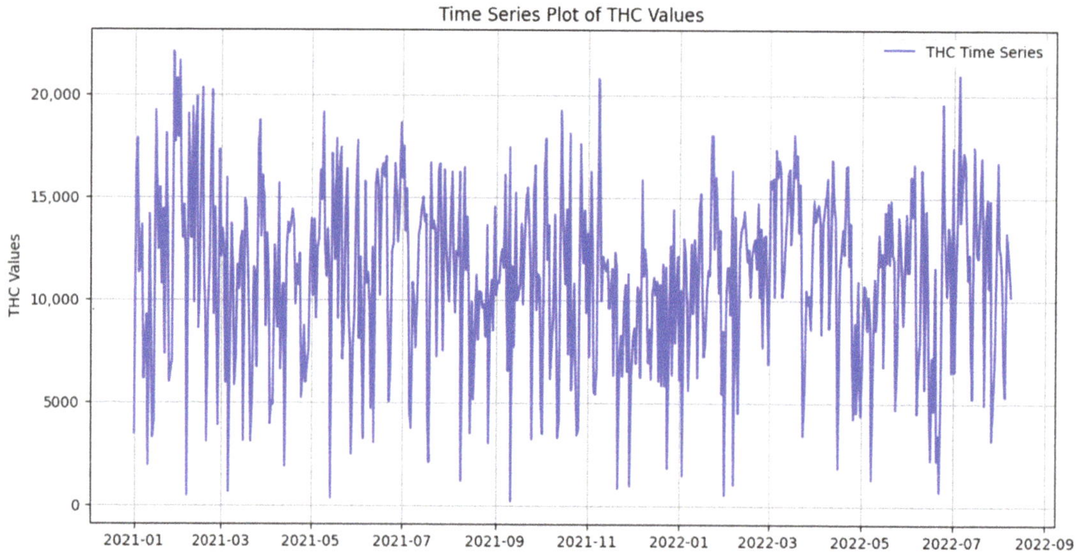

Figure 8. Screened tonnage of phosphates time series.

5.2. ARIMA Model Training

5.2.1. Time Series Stationarity

During this stage, a thorough analysis of the time series behavior is conducted to extract essential information for the model's creation. According to the model training workflow depicted in Figure 6, the process initiates with visualizing the series, followed by an examination of its stationarity and the identification of the parameters [41].

After that, we visualized the time series plot (Figure 8), where we employed the augmented Dickey–Fuller (ADF) test. The result in Table 2 reveals an ADF metric that dictates the stationarity of the series with an important low p-value of 0; all the other ADF metrics are shown in Table 2.

Table 2. Augmented Dickey–Fuller (ADF) test metrics.

ADF-Statistic		−19.273189906	
***p*-value**		0.0	
Critical values	1%	5%	10%
	−3.4415777	−2.8664932	−2.569407

5.2.2. ARIMA Parameters' Identification

In the subsequent phase, our attention turns to the crucial task of identifying the parameters p, d, and q for the ARIMA model. To achieve this, our initial step involves plotting the Autocorrelation Function (ACF) and Partial Autocorrelation Function (PACF). As depicted in Figure 9, a visual analysis of these plots guides us in estimating the values for p and q. Specifically, the ACF aids in estimating the Moving Average (q) part, while the PACF assists in estimating the Autoregressive (p) part.

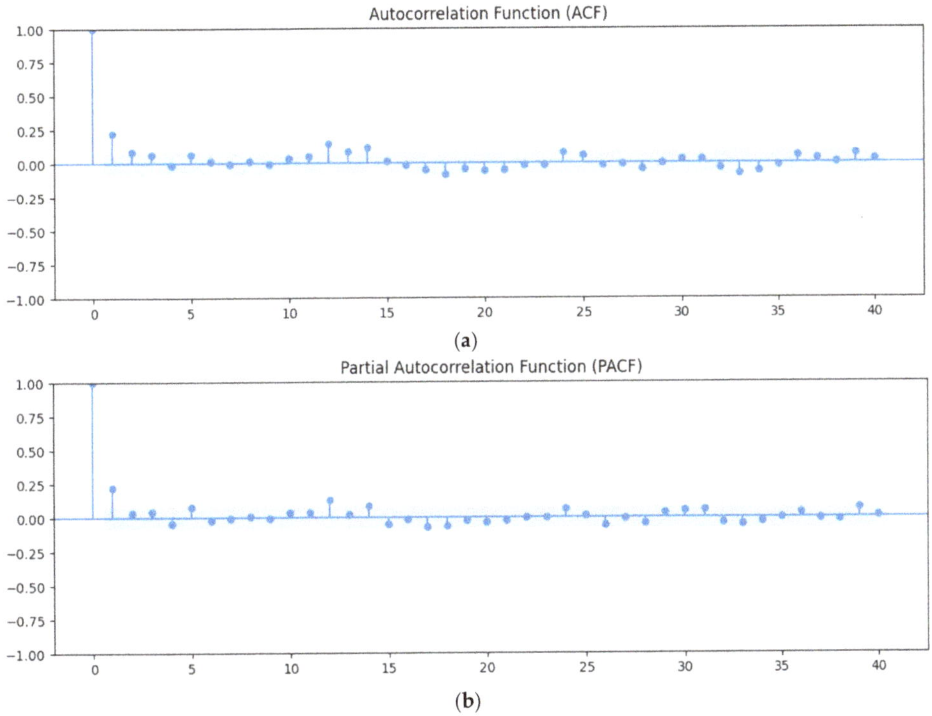

Figure 9. Plots of Autocorrelation Function and Partial Autocorrelation Function: (**a**) Autocorrelation Function; (**b**) Partial Autocorrelation Function.

Upon careful examination of Figure 9, a potential ARIMA candidate emerges as (1,0,1). This determination is rooted in the PACF plot, where the upper confidence interval is crossed at the first lag (p = 1), and the ACF plot similarly exhibits a crossing at the first lag (q = 1). With our time series confirmed as stationary (d = 0), this configuration presents itself as promising. In light of the stationarity, we initially set d = 0. Subsequent exploration involves testing alternative integer values in proximity to zero, encompassing 1 and 2, to identify the optimal differencing parameter.

Following this, we find that the ARIMA model (1,2,1) provides the most suitable results through the comparison of different combinations, (1,0,1), (1,1,1), and (1,2,1), which have the lowest errors. This procedure once more emphasizes the role of parametric studies, which in turn stresses the importance of choosing the right model in improving the model's performance.

While the visual analysis carries some subjectivity and more than one ARIMA parametric combination may be plausible, further diagnostic tools are used to improve our choice of the forecast model. That is when the Akaike Information Criterion (AIC) and Bayesian Information Criterion (BIC) become important instruments to use. These criteria help in the quantitative assessment of various parameter settings, and this aids in optimization.

- Akaike Information Criterion (AIC)

The AIC assesses both the appropriateness of a model to the data and its overall complexity. Typically, the selected model is the one with the lowest AIC value, indicating a balance between model fit and complexity.

$$AIC = -\log L\ + 2k \tag{7}$$

where L is the likelihood function, and k is the number of model parameters.

- Bayesian Information Criterion (BIC) (8)

$$BIC = -\log L + KlnT \tag{8}$$

where L is the likelihood function, k is the number of parameters, and T is the number of observations. The AIC and BIC scores, as defined above, are minimized. The model and parameters p, d, and q chosen are those that minimize these criteria.

In order to calculate the AIC and BIC metrics, we follow the algorithm that is presented below—Algorithm 1. We first set the range of possible values for p, d, and q, these being integers, as the starting point. The subsequent algorithm applies an iterative search over all possible combinations contained within the predefined range. For each running of the algorithm, the AIC and BIC criterion values are calculated, and the one with the lowest score is selected. This detailed strategy ensures the quality of the parameter combinations optimization and fitting, and finally the identification of the ideal combination that satisfies both the AIC and BIC criteria.

Algorithm 1: AIC and BIC Calculation Algorithm

```
1 → Load the dataset
2 → Define a range for p, d and q values
        p = range (0, 3)
        d = range (0, 3)
        q = range (0, 3)
3 → Initialize minimum AIC and minimum BIC as infinity
4 → Initialize best parameters as (0, 0, 0)
5 → for every combination of p, d, and q values Do:
        Fit the ARIMA model with the current combination
        Calculate AIC for the model
        Calculate BIC for the model
        if current AIC value is lower then
        Update the minimum AIC for best parameters.
      end if
        if current BIC value is lower then
        Update the minimum BIC for best parameters.
      end if
  end for
6 → Print the best parameters identified based on both AIC and BIC
```

The computation performed with the Python code that encloses the supplied Algorithm 1 returns the optimal parameters based on the Akaike Information Criterion (AIC) are (1,2,2), with a criterion value of 11,436.483. Further, for BIC (Bayesian Information Criterion), the optimal parameters are (1,1,1), with the associated criterion value being 11,451.199.

During the choosing of our parameters, we put our visual assortment, derived from visual analysis, to severe checkup. This entails evaluating both theoretically proposed combinations as well as the ones found through visual observation. The integrated approach combines both theoretical and empirical aspects to generate a robust and comprehensive model development protocol.

The Table 3 presents an overall comparison of a plethora of options that our ARIMA model has to offer. The framework of our comparative analysis presents both empirical and theoretical considerations. Specifically, the table encompasses three distinct approaches: the first is the visual analysis, the second is via AIC, and the third is the BIC. The comparison is organized based on four critical criteria: mean absolute error (MAE), mean absolute percentage error (MAPE), root mean squared error (RMSE), and criterion value that comes from the combination of them. On the other hand, the table also indicates that the smallest MAPE, 0.35, is for the first combination, (1,2,1). It must be pointed out that the values of MAE and RMSE are virtually the same for the first and the third combination, suggesting that an ideal lower value is achieved. Thus, we select configuring our model with the parameter swap combinations (1,2,1) which appear to be the best as they result in the most effective ML-based model.

Table 3. Comparative analysis of (p, d, q) optimal combinations.

	Analysis Based on Visual Observations	Akaike Information Criterion (AIC)	Bayesian Information Criterion (BIC)
(p, d, q)	**(1, 2, 1)**	(1, 2, 2)	(1, 1, 1)
Related criterion value	-	11,436.483	11,451.199
Mean absolute error (MAE)	**3553.32**	3628.66	3547.81
Mean absolute percentage error (MAPE)	**0.35**	0.59	0.60
Root mean squared error (RMSE)	**4386.69**	4386.69	4383.29

5.2.3. Dataset Splitting

To build our predictive model, we need to split our dataset into two main parts: the training set, which is 80% of the dataset and is used for teaching the model, and the test set, which is 20% of the dataset and is held back to evaluate the model. To create a high-performance model, we specify six different splits from our dataset. This will ensure complete training and foolproof testing of the model, thus giving a good insight into how it works in different scenarios. These six splits -shown in Table 4- constitute what we know as a time series cross-validator, which is a model that increases effectiveness by combining different training and test datasets into consideration. This reveals that the analysis of the model will be comprehensive and insightful.

Table 4. Dataset splits forming the training and testing sets.

Percentage of Dataset in Use per Each Split	Iteration (Split #)	Time Series Cross-Validator Combination	
		Training Set (# of Used Raw)	Testing Set (# of Used Raw)
29.18%	1st	88	83
43.34%	2nd	171	83
57.51%	3rd	254	83
71.67%	4th	337	83
85.84%	5th	420	83
100%	6th	503	83

5.3. Model Evaluation

5.3.1. Performance Metrics for ARIMA Model Predictions

To examine the efficiency of the ARIMA model, we utilize a set of performance metrics for a comparison of how good or relevant our predictions are. Through this method of evaluating the model, the predictions from the model are compared to the actual results from the test dataset, thus providing a comparative analysis that is used in determining real-world performance. These three basic metrics are the ones that matter the most—MAE, RMSE, and MAPE.

As depicted in Table 3, our ARIMA model's performance is summarized with key metrics: MAE (3553.32), MAPE (0.35), and RMSE (4386.69). These results remain promising and intriguing when compared to the statistical metrics of our dataset, specifically the maximum, minimum, mean, and standard deviation values (Table 5).

Table 5. Dataset statistical values.

Maximum Value	Minimum Value	Mean Value	Standard Deviation Value
22,092	220	11,178.58362	4314.32453

For a clearer understanding of the model's performance, we provide a graphical representation (Figure 10) that compares the predicted values from the ARIMA model with the actual values in the test dataset. This visual depiction not only evaluates accuracy but also offers insights into the model's ability to capture trends and patterns in the time series.

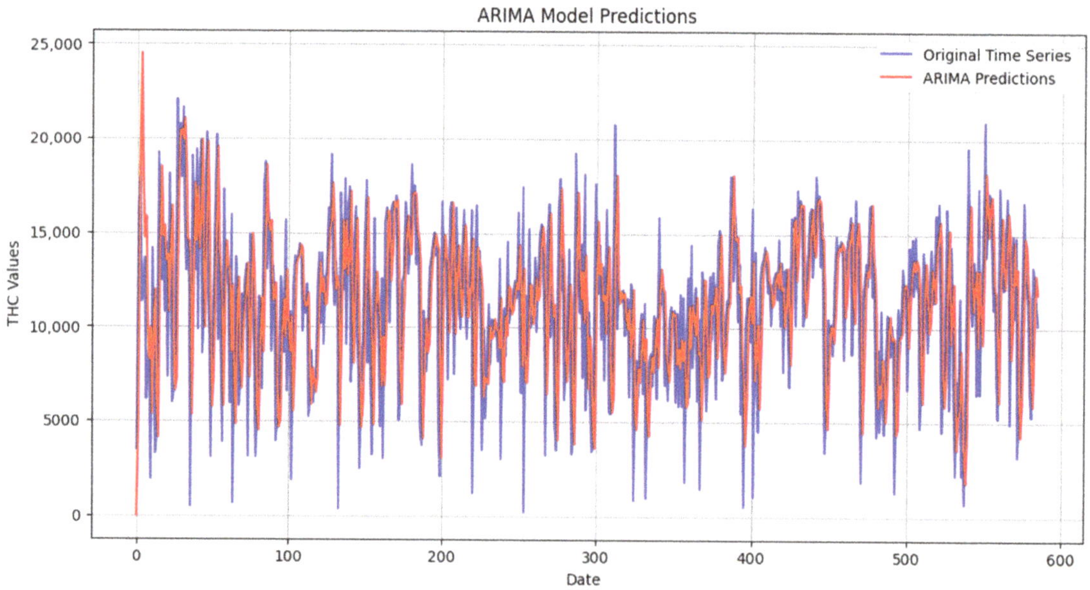

Figure 10. Original and predicted times series of screened wet phosphate.

5.3.2. Inference and Model Validation with External Time Series

As we go into the stage of monitoring how our ARIMA model works and if it can be processed better, we reach the point of introducing a time series that comes from an external source and that is not limited to the time parameter of our original dataset. The following critical procedure involves using a dataset with a three-month time series, from 1 January to 31 March 2023. Also, the other data anchor emerges from the same screening station as the initial wet phosphate samples. We will test the model, therefore, by carrying out inference passing, a rigorous evaluation of its accuracy in a separate time frame. This attempt seeks to make an overall evaluation, identifying its potential to adjust to the changing environments and the ability to predict under temporal conditions.

Based on Figure 11, it can be seen that our ARIMA model very well encompasses the temporal patterns and well captures the trends within the time series data for the verification part. The alignment of model performance and reliability herein points to the efficiency and trustworthiness of our built ARIMA-based model. Additionally, our building ARIMA model has a satisfactory performance in the evaluation metrics, with MAE, RMSE, and MAPE values of 4185.95, 5092.9, and 0,25. These parameters overall provide evidence of the model's performance in terms of precision and closeness of the forecasts to the observed values.

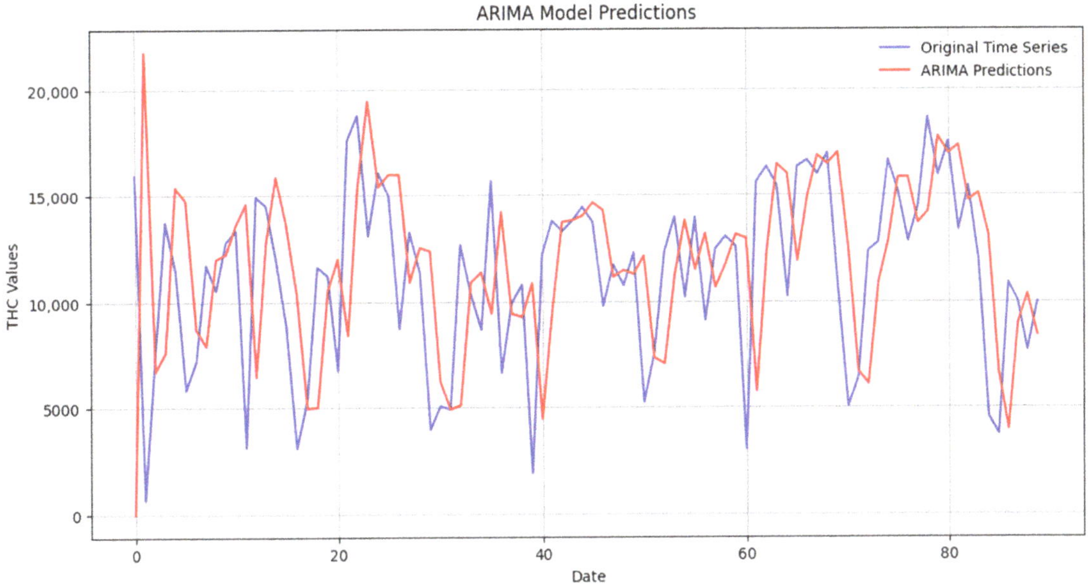

Figure 11. ARIMA model-based validation within external time series data.

5.4. Proposed Production Management System-Based Digital Twin Implementation

5.4.1. Implementation Setup

We begin our dedicated deployment procedure by gathering time series data from our primary database. Through a real-time link, the SCADA monitoring system updates these data on a minute basis. By filling in the missing values and combining the daily volume of screened phosphates from three work shifts, we systematically preprocess these data. Secondly, we use an ML-based model to perform the prediction. An important cornerstone of our system is *Flask* [45], a flexible web framework which is a very important element in supporting the hosting of our PMS application. Acting as a bridge, *Flask* provides advanced functionalities for processing *HTTP* (*hypertext transfer protocol*) requests, routes, and views. It enables users to access our application through a web browser with ease [46]. Moreover, *Flask* interfaces effortlessly with our graphical interfaces, which are based on *Streamlit*. This comprehensive deployment ensures visualization and interactivity within our DT application. Thereafter, we link the developed interfaces of our PMS into the SCADA supervision main views displayed in the control room associated with the screening station, as illustrated in Figure 12. This combined configuration will develop a next-generation monitoring system that enables SCADA functionalities, such as ensuring the availability and continuous supervision of performance and quality, known as Overall Equipment Effectiveness (OEE). Additionally, it provides an outlook of the expected production, including the tonnage of screened phosphate output from our screening station. Furthermore, it empowers operators to reconcile estimated/expected production with realized performance.

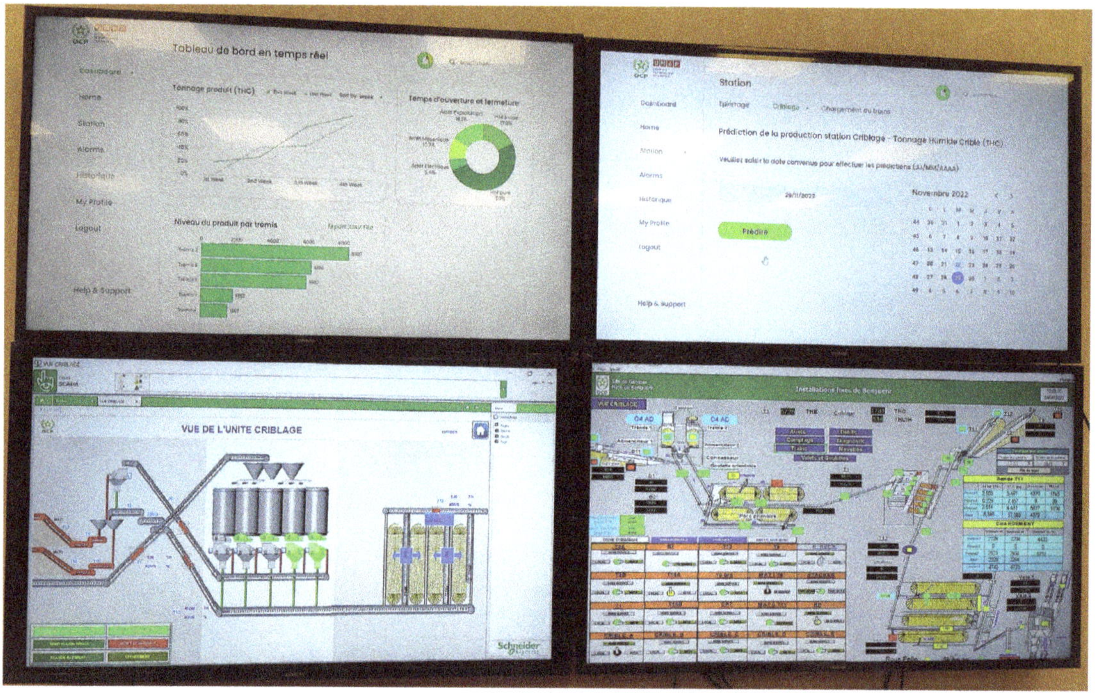

Figure 12. Integration of production management system (PMS) in the control room combined with SCADA monitoring of screening station.

5.4.2. Production Management System Digital Dashboard

Our PMS has an integrated digital dashboard, which is the most crucial element contacting the system framework-based DT. Its main emphasis lies in the control of the monitoring activities of the screening station. The digital dashboard contains a suite of functionality such as real-time monitoring [47,48], Overall Equipment Effectiveness (OEE) [49], and graphical analytics that depict the historical trends and patterns of the screening station. The entire package of these features allows for a complete picture of the station functioning, as well as more effective decision-making. Moreover, a forecasting function, for the screening of wet phosphate production, is incorporated to increase its capabilities. The new system will be based on an ML model that uses its history data from the station [50]. The forecasting function gives a chance to view the estimated output and device operation effectiveness and hence enable our production management [47,48,51].

Figure 13 shows various operational aspects of the proposed system. Figure 13a presents the homepage, revealing the useful aspects of production in real-time. The visual elements include the graphics representing the tonnage of the product and the phosphates produced from each hopper of the screening station. Secondly, the pie chart section exhibits the uptime hours and downtime hours as a dedicated segment, presenting a comprehensive categorization of the anomalies (mechanical issues, electrical issues). This figure allows one to have a complete illustration of the screening station's performance level. As seen in Figure 13b, the milestone in our prototype PMS comes with a forecasting function that offers us a real-time monitoring capability. This is a function that reconciles the values in production. This is achieved by comparing the estimated and realized values. In essence, it enables the supervision of operators in the decision-making process by means of a gap and difference measurement. The integrity of this function is vital for the performance of operators to monitor and react accordingly to the various production scenarios, thus allowing for more informed and strategic solutions.

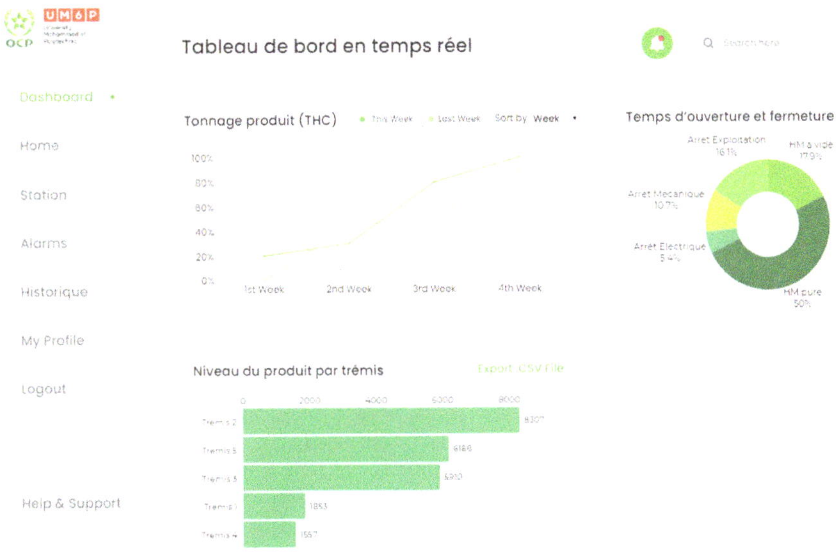

(a)

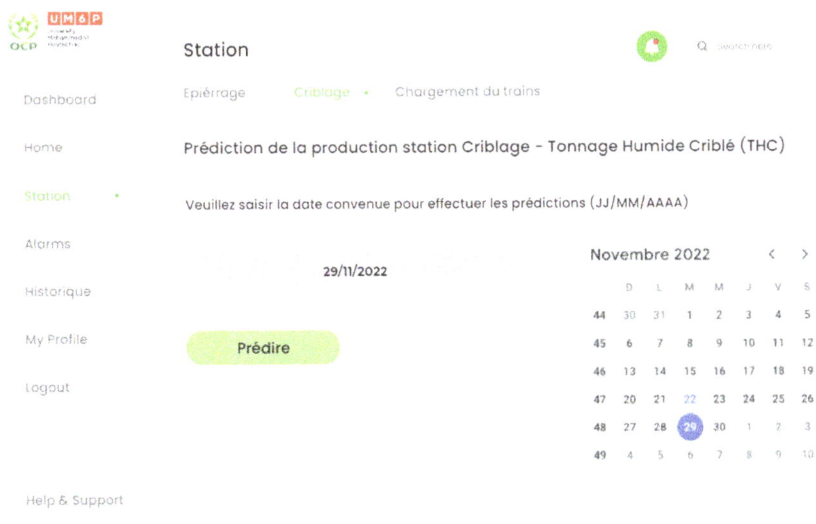

(b)

Figure 13. Production management system application digital dashboard: (**a**) graphical user interface—dashboard homepage; (**b**) graphical user interface—production forecasting page.

6. Conclusions and Outlook

DT technology is emerging as an invaluable tool used for gaining a comprehensive understanding of and proactively anticipating potential scenarios that physical assets may encounter. Despite numerous platforms outlined in the literature for DT development, the emphasis has primarily been on specific vertical contexts, such as multi-layered structures and architectures [52,53]. A significant evolution is imperative to achieve efficient and compositional DTs. Addressing the current needs in this domain, there is a demand for preferably open scalable platforms for the development and composition of DTs. This paper contributes to the existing literature by introducing a compositional DT framework designed to replicate the actual tonnage produced by the phosphate screening station. This framework enhances real-time monitoring capabilities through a systematic PMS, facilitating the forecasting of production. The proposed system provides some valuable information and suggestions for the operators; thus, the operators can operate the system smoothly and manage the production at screening stations well. The DT provides a digital dashboard designed to enable operators to have a complete status of machines, production time, Overall Equipment Effectiveness (OEE), and order scheduling. We propose an integrable framework for composing DT suits for different situations. DT is self-contained and works with AI-based models, particularly through ARIMA in ML. *Streamlit* makes this possible for DT through analytics and trends tracking. It is interactive and visualizes the IoTs monitoring data in a real-time manner, which are simple and complex. Besides this, the forecasting tool offers information that supports decision-making and helps in the process of reconciling actual and forecasted values of the stations.

To show the platform's usefulness, we designed a case for the production line process in the mining industry. This stage is aimed at screening phosphate and monitoring/predicting its tonnage at the station's output. DT shows the status of the station to the end user in a direct way by using the SCADA composite in the framework. Further, they are offered manufacturing KPIs which are easy to view on the related digital dashboard interface of our DT.

In our future growth plans for our proposed composition framework, we aim to expand its scope to include the entire mine value chain, including destoning and train loading stations, within the production management system (PMS). This expansion will enable a comprehensive approach to mine operations' management. We are also committed to enhancing the capabilities of the ML-based framework by increasing the role of ML algorithms in the integration processes. This evolution will go beyond predictive maintenance and shift scheduling optimization, enriching the functionality of the system. Our overall goal is to establish a robust and fully functional PMS that comprehensively addresses all essential parameters related to production efficiency. By achieving this goal, we aim to create a robust structure that optimizes operations throughout the mine value chain. To ensure a clear and comprehensive understanding of our objectives, strategies, and expected benefits, we recognize the need to improve the description of the content. We are committed to refining our documentation to provide stakeholders with a more transparent insight into the intended outcomes and benefits of implementing these new capabilities.

Author Contributions: Conceptualization, N.E.B.; methodology, N.E.B., N.G. and M.M.; software, N.E.B.; validation, M.M. and A.C.; formal analysis, O.L. and A.E.M.; investigation, N.D.; resources, N.E.B. and N.D.; data curation, N.E.B., N.D. and O.L.; writing—original draft preparation, N.E.B.; writing—review and editing, N.G. and A.E.M.; visualization, N.E.B.; supervision, O.L.; project administration, M.M.; funding acquisition, A.C. All authors have read and agreed to the published version of the manuscript.

Funding: This research was funded by the Green Tech Institute of UM6P and the Experimental Mine program (MineEx), a research and innovation program led by Mohammed VI Polytechnic University (UM6P) and OCP Benguerir.

Data Availability Statement: Data are contained within the article.

Acknowledgments: During the preparation of this manuscript, the author utilized ChatGPT to actively improve the grammar, coherence, and readability of its sections.

Conflicts of Interest: The authors declare no conflicts of interest.

References

1. Wang, K.-J.; Lee, Y.-H.; Angelica, S. Digital Twin Design for Real-Time Monitoring—A Case Study of Die Cutting Machine. *Int. J. Prod. Res.* **2021**, *59*, 6471–6485. [CrossRef]
2. Shangguan, D.; Chen, L.; Ding, J. A Digital Twin-Based Approach for the Fault Diagnosis and Health Monitoring of a Complex Satellite System. *Symmetry* **2020**, *12*, 1307. [CrossRef]
3. Elbazi, N.; Tigami, A.; Laayati, O.; Maghraoui, A.E.; Chebak, A.; Mabrouki, M. Digital Twin-Enabled Monitoring of Mining Haul Trucks with Expert System Integration: A Case Study in an Experimental Open-Pit Mine. In Proceedings of the 2023 5th Global Power, Energy and Communication Conference (GPECOM), Cappadocia, Turkiye, 14 June 2023; IEEE: Nevsehir, Turkiye; pp. 168–174.
4. Slob, N.; Hurst, W. Digital Twins and Industry 4.0 Technologies for Agricultural Greenhouses. *Smart Cities* **2022**, *5*, 1179–1192. [CrossRef]
5. El Bazi, N.; Mabrouki, M.; Laayati, O.; Ouhabi, N.; El Hadraoui, H.; Hammouch, F.-E.; Chebak, A. Generic Multi-Layered Digital-Twin-Framework-Enabled Asset Lifecycle Management for the Sustainable Mining Industry. *Sustainability* **2023**, *15*, 3470. [CrossRef]
6. Elbazi, N.; Mabrouki, M.; Chebak, A.; Hammouch, F. Digital Twin Architecture for Mining Industry: Case Study of a Stacker Machine in an Experimental Open-Pit Mine. In Proceedings of the 2022 4th Global Power, Energy and Communication Conference (GPECOM), Cappadocia, Turkiye, 14 June 2022; IEEE: Nevsehir, Turkiye; pp. 232–237.
7. Popescu, D.; Dragomir, M.; Popescu, S.; Dragomir, D. Building Better Digital Twins for Production Systems by Incorporating Environmental Related Functions—Literature Analysis and Determining Alternatives. *Appl. Sci.* **2022**, *12*, 8657. [CrossRef]
8. Vodyaho, A.I.; Zhukova, N.A.; Shichkina, Y.A.; Anaam, F.; Abbas, S. About One Approach to Using Dynamic Models to Build Digital Twins. *Designs* **2022**, *6*, 25. [CrossRef]
9. De Benedictis, A.; Flammini, F.; Mazzocca, N.; Somma, A.; Vitale, F. Digital Twins for Anomaly Detection in the Industrial Internet of Things: Conceptual Architecture and Proof-of-Concept. *IEEE Trans. Ind. Inform.* **2023**, *19*, 11553–11563. [CrossRef]
10. Torzoni, M.; Tezzele, M.; Mariani, S.; Manzoni, A.; Willcox, K.E. A Digital Twin Framework for Civil Engineering Structures. *Comput. Methods Appl. Mech. Eng.* **2024**, *418*, 116584. [CrossRef]
11. Kibira, D.; Shao, G.; Venketesh, R. Building A Digital Twin of AN Automated Robot Workcell. In Proceedings of the 2023 Annual Modeling and Simulation Conference (ANNSIM), Hamilton, ON, Canada, 23–26 May 2023; IEEE: New York, NY, USA, 2023; pp. 196–207.
12. Vilarinho, S.; Lopes, I.; Sousa, S. Design Procedure to Develop Dashboards Aimed at Improving the Performance of Productive Equipment and Processes. *Procedia Manuf.* **2017**, *11*, 1634–1641. [CrossRef]
13. Fathy, Y.; Jaber, M.; Nadeem, Z. Digital Twin-Driven Decision Making and Planning for Energy Consumption. *J. Sens. Actuator Netw.* **2021**, *10*, 37. [CrossRef]
14. Piras, G.; Agostinelli, S.; Muzi, F. Digital Twin Framework for Built Environment: A Review of Key Enablers. *Energies* **2024**, *17*, 436. [CrossRef]
15. Kandavalli, S.R.; Khan, A.M.; Iqbal, A.; Jamil, M.; Abbas, S.; Laghari, R.A.; Cheok, Q. Application of Sophisticated Sensors to Advance the Monitoring of Machining Processes: Analysis and Holistic Review. *Int. J. Adv. Manuf. Technol.* **2023**, *125*, 989–1014. [CrossRef]
16. Balogh, M.; Földvári, A.; Varga, P. Digital Twins in Industry 5.0: Challenges in Modeling and Communication. In Proceedings of the NOMS 2023-2023 IEEE/IFIP Network Operations and Management Symposium, Miami, FL, USA, 8 May 2023; IEEE: Miami, FL, USA; pp. 1–6.
17. Zhao, Y.; Yan, L.; Wu, J.; Song, X. Design and Implementation of a Digital Twin System for Log Rotary Cutting Optimization. *Future Internet* **2024**, *16*, 7. [CrossRef]
18. Peng, A.; Ma, Y.; Huang, K.; Wang, L. Digital Twin-Driven Framework for Fatigue Life Prediction of Welded Structures Considering Residual Stress. *Int. J. Fatigue* **2024**, *181*, 108144. [CrossRef]
19. Sifat, M.M.H.; Choudhury, S.M.; Das, S.K.; Ahamed, M.H.; Muyeen, S.M.; Hasan, M.M.; Ali, M.F.; Tasneem, Z.; Islam, M.M.; Islam, M.R.; et al. Towards Electric Digital Twin Grid: Technology and Framework Review. *Energy AI* **2023**, *11*, 100213. [CrossRef]
20. Singh, S.K.; Kumar, M.; Tanwar, S.; Park, J.H. GRU-Based Digital Twin Framework for Data Allocation and Storage in IoT-Enabled Smart Home Networks. *Future Gener. Comput. Syst.* **2024**, *153*, 391–402. [CrossRef]
21. El Hadraoui, H.; Laayati, O.; Guennouni, N.; Chebak, A.; Zegrari, M. A Data-Driven Model for Fault Diagnosis of Induction Motor for Electric Powertrain. In Proceedings of the 2022 IEEE 21st Mediterranean Electrotechnical Conference (MELECON), Palermo, Italy, 14–16 June 2022; pp. 336–341.
22. Onaji, I.; Tiwari, D.; Soulatiantork, P.; Song, B.; Tiwari, A. Digital Twin in Manufacturing: Conceptual Framework and Case Studies. *Int. J. Comput. Integr. Manuf.* **2022**, *35*, 831–858. [CrossRef]

23. Allam, Z.; Sharifi, A.; Bibri, S.E.; Jones, D.S.; Krogstie, J. The Metaverse as a Virtual Form of Smart Cities: Opportunities and Challenges for Environmental, Economic, and Social Sustainability in Urban Futures. *Smart Cities* **2022**, *5*, 771–801. [CrossRef]
24. Asad, U.; Khan, M.; Khalid, A.; Lughmani, W.A. Human-Centric Digital Twins in Industry: A Comprehensive Review of Enabling Technologies and Implementation Strategies. *Sensors* **2023**, *23*, 3938. [CrossRef]
25. Zhu, Z.; Liu, C.; Xu, X. Visualisation of the Digital Twin Data in Manufacturing by Using Augmented Reality. *Procedia CIRP* **2019**, *81*, 898–903. [CrossRef]
26. Elbazi, N.; Hadraoui, H.E.; Laayati, O.; Maghraoui, A.E.; Chebak, A.; Mabrouki, M. Digital Twin in Mining Industry: A Study on Automation Commissioning Efficiency and Safety Implementation of a Stacker Machine in an Open-Pit Mine. In Proceedings of the 2023 5th Global Power, Energy and Communication Conference (GPECOM), Cappadocia, Turkiye, 14 June 2023; pp. 548–553.
27. Mohammed, K.; Abdelhafid, M.; Kamal, K.; Ismail, N.; Ilias, A. Intelligent Driver Monitoring System: An Internet of Things-Based System for Tracking and Identifying the Driving Behavior. *Comput. Stand. Interfaces* **2023**, *84*, 103704. [CrossRef]
28. Choi, S.; Woo, J.; Kim, J.; Lee, J.Y. Digital Twin-Based Integrated Monitoring System: Korean Application Cases. *Sensors* **2022**, *22*, 5450. [CrossRef] [PubMed]
29. Bendaouia, A.; Abdelwahed, E.H.; Qassimi, S.; Boussetta, A.; Benzakour, I.; Amar, O.; Hasidi, O. Artificial Intelligence for Enhanced Flotation Monitoring in the Mining Industry: A ConvLSTM-Based Approach. *Comput. Chem. Eng.* **2024**, *180*, 108476. [CrossRef]
30. Choi, S.; Kang, G.; Jung, K.; Kulvatunyou, B.; Morris, K. Applications of the Factory Design and Improvement Reference Activity Model. In *Advances in Production Management Systems. Initiatives for a Sustainable World*; Nääs, I., Vendrametto, O., Mendes Reis, J., Gonçalves, R.F., Silva, M.T., Von Cieminski, G., Kiritsis, D., Eds.; IFIP Advances in Information and Communication Technology; Springer International Publishing: Cham, Switzerland, 2016; Volume 488, pp. 697–704. ISBN 978-3-319-51132-0.
31. Saihi, A.; Awad, M.; Ben-Daya, M. Quality 4.0: Leveraging Industry 4.0 Technologies to Improve Quality Management Practices—A Systematic Review. *Int. J. Qual. Reliab. Manag.* **2021**, *40*, 628–650. [CrossRef]
32. Alexopoulos, K.; Tsoukaladelis, T.; Dimitrakopoulou, C.; Nikolakis, N.; Eytan, A. An Approach towards Zero Defect Manufacturing by Combining IIoT Data with Industrial Social Networking. *Procedia Comput. Sci.* **2023**, *217*, 403–412. [CrossRef]
33. Foit, K. Agent-Based Modelling of Manufacturing Systems in the Context of "Industry 4.0.". *J. Phys. Conf. Ser.* **2022**, *2198*, 012064. [CrossRef]
34. Jung, K.; Choi, S.; Kulvatunyou, B.; Cho, H.; Morris, K.C. A Reference Activity Model for Smart Factory Design and Improvement. *Prod. Plan. Control* **2017**, *28*, 108–122. [CrossRef]
35. Zayed, S.M.; Attiya, G.M.; El-Sayed, A.; Hemdan, E.E.-D. A Review Study on Digital Twins with Artificial Intelligence and Internet of Things: Concepts, Opportunities, Challenges, Tools and Future Scope. *Multimed. Tools Appl.* **2023**, *82*, 47081–47107. [CrossRef]
36. Choi, S.; Jun, C.; Zhao, W.B.; Do Noh, S. Digital Manufacturing in Smart Manufacturing Systems: Contribution, Barriers, and Future Directions. In *Advances in Production Management Systems: Innovative Production Management Towards Sustainable Growth*; Umeda, S., Nakano, M., Mizuyama, H., Hibino, H., Kiritsis, D., Von Cieminski, G., Eds.; IFIP Advances in Information and Communication Technology; Springer International Publishing: Cham, Switzerland, 2015; Volume 460, pp. 21–29. ISBN 978-3-319-22758-0.
37. Choi, S.; Kim, B.H.; Do Noh, S. A Diagnosis and Evaluation Method for Strategic Planning and Systematic Design of a Virtual Factory in Smart Manufacturing Systems. *Int. J. Precis. Eng. Manuf.* **2015**, *16*, 1107–1115. [CrossRef]
38. Liu, J.; Ji, Q.; Zhang, X.; Chen, Y.; Zhang, Y.; Liu, X.; Tang, M. Digital Twin Model-Driven Capacity Evaluation and Scheduling Optimization for Ship Welding Production Line. *J. Intell. Manuf.* **2023**, *34*. [CrossRef]
39. Yadav, R.S.; Mehta, V.; Tiwari, A. An Application of Time Series ARIMA Forecasting Model for Predicting Nutri Cereals Area in India. 2022. Available online: https://www.thepharmajournal.com/archives/2022/vol11issue3S/PartQ/S-11-3-85-221.pdf (accessed on 27 January 2024).
40. Ning, Y.; Kazemi, H.; Tahmasebi, P. A Comparative Machine Learning Study for Time Series Oil Production Forecasting: ARIMA, LSTM, and Prophet. *Comput. Geosci.* **2022**, *164*, 105126. [CrossRef]
41. Fan, D.; Sun, H.; Yao, J.; Zhang, K.; Yan, X.; Sun, Z. Well Production Forecasting Based on ARIMA-LSTM Model Considering Manual Operations. *Energy* **2021**, *220*, 119708. [CrossRef]
42. Implementation of Time Series Forecasting with Box Jenkins ARIMA Method on Wood Production of Indonesian Forests | AIP Conference Proceedings | AIP Publishing. Available online: https://pubs.aip.org/aip/acp/article-abstract/2738/1/060004/2894351/Implementation-of-time-series-forecasting-with-Box?redirectedFrom=fulltext (accessed on 29 January 2024).
43. El Maghraoui, A.; Ledmaoui, Y.; Laayati, O.; El Hadraoui, H.; Chebak, A. Smart Energy Management: A Comparative Study of Energy Consumption Forecasting Algorithms for an Experimental Open-Pit Mine. *Energies* **2022**, *15*, 4569. [CrossRef]
44. Pajpach, M.; Pribiš, R.; Drahoš, P.; Kučera, E.; Haffner, O. Design of an Educational-Development Platform for Digital Twins Using the Interoperability of the OPC UA Standard and Industry 4.0 Components. In Proceedings of the 2023 3rd International Conference on Electrical, Computer, Communications and Mechatronics Engineering (ICECCME), Tenerife, Spain, 19 July 2023; IEEE: Tenerife, Spain; pp. 1–6.
45. Mufid, M.R.; Basofi, A.; Al Rasyid, M.U.H.; Rochimansyah, I.F.; Rokhim, A. Design an MVC Model Using Python for Flask Framework Development. In Proceedings of the 2019 International Electronics Symposium (IES), Surabaya, Indonesia, 27–28 September 2019; pp. 214–219.

46. Srikanth, G.; Reddy, M.S.K.; Sharma, S.; Sindhu, S.; Reddy, R. Designing a Flask Web Application for Academic Forum and Faculty Rating Using Sentiment Analysis. *AIP Conf. Proc.* **2023**, *2477*, 030035. [CrossRef]
47. Schulze, A.; Brand, F.; Geppert, J.; Böl, G.-F. Digital Dashboards Visualizing Public Health Data: A Systematic Review. *Front. Public Health* **2023**, *11*, 999958. [CrossRef] [PubMed]
48. Gonçalves, C.T.; Gonçalves, M.J.A.; Campante, M.I. Developing Integrated Performance Dashboards Visualisations Using Power BI as a Platform. *Information* **2023**, *14*, 614. [CrossRef]
49. Laayati, O.; El Hadraoui, H.; El Magharaoui, A.; El-Bazi, N.; Bouzi, M.; Chebak, A.; Guerrero, J.M. An AI-Layered with Multi-Agent Systems Architecture for Prognostics Health Management of Smart Transformers: A Novel Approach for Smart Grid-Ready Energy Management Systems. *Energies* **2022**, *15*, 7217. [CrossRef]
50. Islam, M.A.; Sufian, M.A. Employing AI and ML for Data Analytics on Key Indicators: Enhancing Smart City Urban Services and Dashboard-Driven Leadership and Decision-Making. In *Technology and Talent Strategies for Sustainable Smart Cities*; Singh Dadwal, S., Jahankhani, H., Bowen, G., Yasir Nawaz, I., Eds.; Emerald Publishing Limited: Leeds, UK, 2023; pp. 275–325; ISBN 978-1-83753-023-6.
51. Jwo, J.-S.; Lin, C.-S.; Lee, C.-H. An Interactive Dashboard Using a Virtual Assistant for Visualizing Smart Manufacturing. *Mob. Inf. Syst.* **2021**, *2021*, e5578239. [CrossRef]
52. Honghong, S.; Gang, Y.; Haijiang, L.; Tian, Z.; Annan, J. Digital Twin Enhanced BIM to Shape Full Life Cycle Digital Transformation for Bridge Engineering. *Autom. Constr.* **2023**, *147*, 104736. [CrossRef]
53. Li, W.; Li, Y.; Garg, A.; Gao, L. Enhancing Real-Time Degradation Prediction of Lithium-Ion Battery: A Digital Twin Framework with CNN-LSTM-Attention Model. *Energy* **2024**, *286*, 129681. [CrossRef]

designs

MDPI

Article

An Approach for Predicting the Lifetime of Lead-Free Soldered Electronic Components: Hitachi Rail STS Case Study

Paolo Renna [1,*], Michele Ambrico [2], Vito Romaniello [2] and Thomas Russino [1]

1 School of Engineering, University of Basilicata, 85100 Potenza, Italy; thomas.russino@outlook.it
2 Hitachi Rail STS, Via Argine 425, 80147 Napoli, Italy; michele.ambrico@hitachirail.com (M.A.); vito.romaniello@hitachirail.com (V.R.)
* Correspondence: paolo.renna@unibas.it

Abstract: Throughout much of the 20th century, Sn–Pb solder dominated electronics. However, environmental and health concerns have driven the adoption of lead-free alternatives. Since 2006, legislation such as the European Union's RoHS Directive has mandated lead-free solder in most electronic devices, prompting extensive research into high-performance substitutes. Lead-free solders offer advantages such as reduced environmental impact and improved reliability but replacing Sn–Pb presents challenges in areas like melting point and wetting ability. This transition is primarily motivated by a focus on protecting environmental and human health, while ensuring equal or even improved reliability. Research has explored lead-free solder's mechanical properties, microstructure, wettability, and reliability. However, there is a notable lack of studies on its long-term performance and lifetime influence. To address this gap, mathematical models are used to predict intermetallic bond evolution from process conditions, validated with experimental tests. This study contributes by extending these models to predict bond evolution under typical operating conditions of devices and comparing the predictions with actual intermetallic thickness values found through metallographic sections.

Keywords: electronic board; load-free solder; estimated lifetime; experimental study

1. Introduction

For much of the 20th century, Sn–Pb solder was the industry standard for electronics. However, growing concerns about lead's toxicity has led to a major swift towards lead-free alternatives in recent decades [1–3]. Since 2006, European legislation like the Restriction of Hazardous Substances Directive (RoHS) has banned lead in most electrical and electronic devices. This shift away from traditional Sn–Pb solder reflects a critical prioritization of public health and environmental protection. While Sn–Pb offered some performance benefits, its drawbacks ultimately made its continued use unacceptable. Lead exposure poses serious health risks, including neurological damage and reproductive problems [4]. Additionally, lead pollution harms ecosystems through air and water contamination. Concerns about lead's health and environmental impacts led to legislation like the European Union's Restriction of Hazardous Substances Directive (RoHS). Since 2006, this law has effectively banned lead-based solder (Sn–Pb) in most electronics. When Directive 2011/65/EU (RoHS 2) replaced Directive 2002/95/EC (RoHS 1), existing exemptions were transferred to Annex III of RoHS 2. The maximum validity for exemptions increased from four to five years for most categories of electrical and electronic equipment (EEE), except for categories 8 and 9, which have a seven-year validity. According to Article 5(2), exemptions listed in Annex III as of 21 July 2011 follow this maximum validity unless specified otherwise. Consequently, over 30 exemptions would have expired on 21 July 2016 if they were not extended by the industry. The European Commission contracted the Oeko-Institut and Fraunhofer IZM to review these exemption applications from June 2015 to July 2016 [5].

This triggered a surge in the research and development of high-performance, lead-free alternatives. Ideally, these new solders would be just as good or even better than Sn–Pb, but without the toxic risks. Lead-free solders offer some advantages. They allow for the use of less harmful "no-clean" or water-soluble fluxes, reducing heat stress during soldering. Additionally, they can be more resistant to electrical wear, oxidation, and corrosion, improving the overall reliability of the solder joint. However, replacing a well-understood material like Sn–Pb solder comes with challenges. Researchers are constantly working to improve the performance and reliability of lead-free solders, addressing issues like higher melting points, weaker wetting ability, and long-term stability. Overall, the shift to lead-free solders in the electronics industry reflects a critical focus on protecting human health and the environment. While challenges remain, ongoing research is making lead-free alternatives ever more effective, paving the way for a more sustainable future in electronics manufacturing. Several previous studies have explored different aspects of lead-free solder [6], including its mechanical properties [7], microstructure and intermetallic layers [8], wettability [9], and overall reliability [10]. While these studies provide valuable insights, none have specifically investigated the impact of lead-free solder on the lifespan of solder joints. This paper uses mathematical models to predict how intermetallic bonds change over time based on soldering conditions. These models are validated by testing boards made by Hitachi in collaboration with independent labs. Previously, researchers have proposed various formulas to estimate intermetallic thickness growth. These formulas, based on experiments and data from the literature, can provide somewhat accurate predictions. Our models focus on lead-free alloys, especially SAC (tin–silver–copper). However, they can be adapted to work with outdated lead-based alloys as well. These models predict both the initial intermetallic thickness after soldering and its growth during stress tests. These tests include high-temperature exposure, thermal shocks, and thermal cycles simulating real-world conditions (−40 °C to 125 °C). Our original contribution is extending these models to predict how bonds change under normal operating conditions. These predictions, conducted only for boards with SAC solder, were then compared to the actual intermetallic thickness measured in metallographic sections. The rest of this manuscript is organized as follows: Section 2 provides a literature review on lead-free solders. Section 3 describes the proposed model, while Section 4 presents the test trials and numerical results. Section 5 discusses the metallographic sections in comparison to the model estimations. Finally, conclusions and future developments are presented in Section 6.

2. Literature Review

Lead-free solders are becoming increasingly important in the electronics industry due to the environmental and health concerns associated with lead-based solders. Cheng et al. [6] reviewed the recent research on lead-free solders, focusing on their properties and performance. Lead-free solders have been shown to be reliable in a variety of applications. However, there are some challenges that need to be addressed, such as higher melting temperatures and reduced ductility.

Jung and Jung [9] focused on the importance of solder wettability for reliable microelectronic packaging and explored how the wetting balance test is used to assess it. They highlighted the wetting balance test (also known as Meniscograph) as the most common method for quantitatively measuring the wettability between molten solder and a substrate.

Fan et al. [11] investigated how the surface finish of electronic component pads (where the solder connects) affects the microstructure and intermetallic compound (IMC) growth in tin–bismuth (Sn–Bi) and tin–bismuth–silver (Sn–Bi–Ag) solders. They highlighted the importance of pad surface finish in influencing IMC growth in Sn–Bi and Sn–Bi–Ag solders. Their work suggests that Cu–OSP, or the addition of silver to the solder, is a possible strategy to mitigate excessive IMC growth and potentially improve the reliability of solder joints.

Dale et al. [12] studied the fatigue life (durability under repeated stress) of a specific lead-free solder alloy (Sn3.0Ag0.5Cu) under combined loading conditions. They underlined the importance of considering combined loading conditions when evaluating the reliability of solder joints in electronic devices. Their study suggests that solder joints may be more susceptible to fatigue failure in applications where they experience both shear and tensile stress.

Wang et al. [13] proposed a new analytical model to predict the growth of intermetallic compounds (IMCs) in Cu–Sn–Cu sandwich structures. This model offers a simpler yet effective tool for researchers and engineers to design reliable Cu–Sn–Cu structures for electronic applications.

Peng et al. [14] explored Cu–Sn intermetallic compounds (IMCs) growth kinetics for high-temperature resistant packaging in semiconductor power devices. They established a 3D kinetics model and equations for Cu_6Sn_5 and Cu_3Sn growth. Experimental validation shows temperature and Cu particle size affect growth kinetics, while Sn content impacts solidification time.

Ramli et al. [15] presented a review that summarizes factors impacting intermetallic compounds (IMCs) formation and growth, which are crucial for joint structure. Their key findings were: (1) Minor alloying elements significantly affect primary and interfacial IMCs, altering microstructure and IMC growth rates; (2) Surface finishes material heavily influences IMC layer thickness and composition via dissolution during soldering; (3) Higher aging temperatures and longer times increase IMC thickness; (4) Smaller solder volumes lead to thicker IMC layers due to faster copper concentration growth in smaller solder balls.

Ismail et al. [16] conducted a comprehensive review of 171 research articles and papers from Scopus, Google Scholar, and Dimensions databases to examine the impact of surface roughness on wettability and intermetallic compound (IMC) layer formation in lead-free solder joints. Their analysis revealed a predominant focus on soldering processes and materials, with limited attention to surface roughness effects. Future research could explore post-treatment methods, such as additives or reinforcement, to improve lead-free solder material properties. Understanding the correlation between soldering processes and surface roughness in lead-free solder materials is crucial for advancing knowledge in this field and optimizing solder joint performance.

The focus of the models is on lead-free alloys, particularly SAC (tin–silver–copper), but they can be adapted to leaded alloys, which are now obsolete. The predictions address both the intermetallic thickness achievable after the soldering process and its growth following testing through accelerated exposure to heat, thermal shocks, and thermal cycles simulating operating conditions from −40 °C to 125 °C. The original contribution of this work was to extend the use of these mathematical relationships to predict bond evolution under normal device operating conditions. These predictions, made exclusively for boards with SAC alloy soldering, were then compared with the intermetallic thickness values observed through metallographic sections.

3. Proposed Model

Solder joints in electronic components typically undergo two types of fatigue failure: ductile rupture, often due to creep, which involves localized fracture at the interface between the intermetallic layer and the solder, or brittle rupture with localized fracture within the intermetallic layer or along the surface finish. Ductile rupture for these solders is due to thermomechanical fatigue, as the operating temperatures of electronic assemblies range from a minimum of −40 °C to a maximum of 125 °C; ductile rupture is the result of progressive damage due to thermomechanical cycles that cause the formation of a crack. Tin-based solders such as SAC operate at relatively high temperatures compared to the melting temperature of the alloy, which means that yielding occurs at relatively low applied stress values, while creep and diffusion processes are quite rapid. The solder itself can exhibit strain rate-independent as well as plastic and strain rate-dependent behavior; therefore, the solder layer can be simplified as a combination of a spring and a viscous damper, while the intermetallic layer exhibits elastic behavior and is modeled as a spring. Brittle rupture also depends on the thickness of the intermetallic layer; in fact, it seems that the strain rate value at which the transition from ductile rupture to brittle rupture occurs becomes lower and lower as the intermetallic layer grows, until only brittle ruptures are observed as the life of the board increases. The intermetallic layer grows over time, especially with exposure to heat during the board's working cycles.

At this point, it is clear that the thickness of the intermetallic layer is important for the reliability of the soldered joint: if it is an intrinsic quality of a solder joint made on electronic boards, and its presence in an adequate quantity denotes good quality of the solder joint under consideration, it is also important that there are no excessive thicknesses downstream of the soldering process, nor that their growth is excessive during operation, in order to avoid premature and sudden failures. It should be remembered that not only the thickness of the intermetallic layer affects the reliability of the soldered joint, but also its morphology; the microstructure of a soldered joint changes over time, with the grains of the solder growing larger and coarser. Furthermore, if the intermetallic layer becomes thicker, the intermetallic compound grains dispersed in the solder also become larger, thus losing the strengthening effect mentioned above.

It should be clarified that the initial intermetallic thickness after the soldering process is an unknown parameter; it has not been experimentally measured because metallographic sections were performed on tested electronic boards (which have already undergone accelerated aging cycles). However, as mentioned earlier, considering the soldering parameters, an initial intermetallic thickness X_0 (after the process) has been calculated, to which the thickness variation ΔX predicted through the test parameters will be added. The sum obtained will give the total predicted intermetallic thickness X_{tot} after the test, which will then be compared with the intermetallic thickness observed through metallographic sections of the tested boards. It should be noted that, since the intermetallic thickness of the secondary Cu_3Sn compound is not clearly visible in the metallographic sections and is negligible in size, it can be assumed that the calculated total intermetallic thickness is that of the primary Cu_6Sn_5 compound. In summary, once the process parameters, such as temperature and time, are defined, and once the modes, times, and temperatures of the accelerated tests are determined, all the necessary elements are available to use the formulas.

Regarding lead-free alloys used to solder component terminals to the copper base (or pad) with possible nickel surface finish, the following Formula (1) is used to calculate the intermetallic thickness in a generic manner [17].

$$X(t,T) = kt^n \tag{1}$$

This formula is expressed as a function of time and temperature and is a power law where the intermetallic thickness depends on the growth rate parameter k (which varies with temperature and is measured in μm/s) and the dimensionless growth exponent n. These two parameters vary based on temperature and the type of surface finish used (nickel or copper). Generally, the value of n is approximately 1/3 if the diffusion of the reactive species forming the intermetallic compounds is limited to grain boundaries, whereas the value of n is approximately 1/2 if the diffusion of the metallic species is volumetric, in which case the intermetallic thickness growth is faster [18]. In turn, k can be expanded as a further expression, which is referred to as Formula (2).

$$k = k_0 e^{-\frac{Q}{RT}} \tag{2}$$

In this expression, k_0 represents the unit growth constant, Q is the apparent activation energy, R is the Boltzmann constant, and T is the absolute temperature. The value of k expressed in this extended form will be subsequently used in the intermetallic thickness prediction formula during testing. By combining Formulas (1) and (2), an extended expression for calculating the intermetallic thickness is obtained, as shown below in Formula (3).

$$X(t,T) = \left(k_0 e^{-\frac{Q}{RT}}\right) t^n \tag{3}$$

In the case of the initial thickness calculation, the literature provides values of k and n, eliminating the need for using the extended formula for k. Tables 1 and 2 below show the literature values of k and n for SAC alloys at various temperatures and surface finishes.

Table 1. Values of k and n for Sn–Ag–Cu alloy on copper surface finish [19].

Temperature (°C)	k (μm/sn)	n
225	0.37	0.20
235	0.31	0.27
245	0.46	0.27
260	0.65	0.29
280	0.66	0.28

Table 2. Values of k and n for Sn–Ag–Cu alloy on copper coated with nickel [19].

Temperature (°C)	k (μm/sn)	n
225	0.09	0.40
235	0.15	0.51
245	0.21	0.50
260	0.37	0.50

The process parameters are then defined for Surface Mount Devices (SMD) and Through-Hole Technology (THT) components (see Figure 1), with the former soldered in the oven using reflow soldering and the latter using wave soldering. Regarding the soldering of the analyzed components, an SMD component has copper terminals, with the base (or pad) it is soldered on being copper but presenting a nickel surface finish. The reflow soldering process occurs with a dwell time above the liquidus temperature around 65–70 s, while the peak temperature above the liquidus is approximately 235–245 °C. For Through-Hole Technology (THT) components, the terminal is again made of copper, while the through-hole is copper and is also coated with a nickel finish. The peak temperature reached by the hole in contact for 4–5 s with the molten alloy wave is approximately 250 °C.

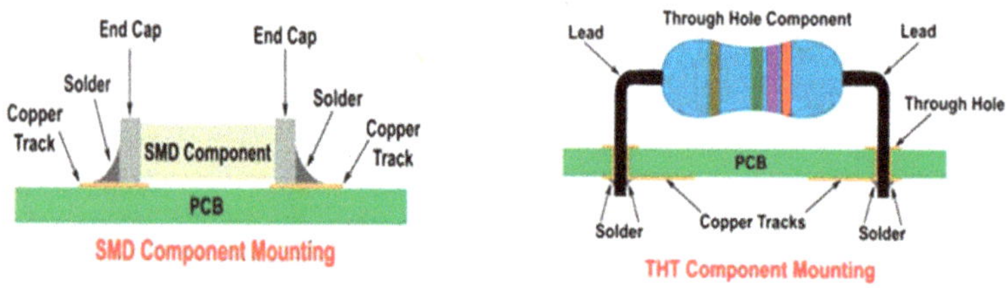

Figure 1. Surface Mount Devices and Through-Hole Technology components.

With these parameters defined, a prediction of the intermetallic thickness X_0 downstream of the soldering process can be outlined. By then replacing the tabulated values of k and n with the process temperatures and substituting the time variable in Formula (1) with the available data, values of thickness X_0 in μm are obtained, as shown in Tables 3 and 4, from which the larger one will be taken to evaluate the worst-case scenario.

Table 3. Prediction of intermetallic layer thickness X_0 for SMD components after the soldering process.

	Surface Mount Device (SMD)							
	Terminal in Cu				Pad in Cu/Ni			
Temperature (°C)	235		245		235		245	
Time (s)	65	70	65	70	65	70	65	70
X_0 (μm)	0.96	0.98	1.42	1.45	1.26	1.31	1.69	1.76

- X_0 maximum of 1.45 μm for the copper terminal.
- X_0 maximum of 1.76 μm for copper pads with a nickel finish.

Table 4. Prediction of intermetallic layer thickness X_0 for THT components after the soldering process.

THT (Through-Hole Component)			
Terminal (Cu)		**Hole (Cu/Ni)**	
X_0	4.6 s	X_0	4.6 s
245 °C	0.69 μm	245 °C	0.45 μm
260 °C	1.00 μm	260 °C	0.80 μm

- X_0 maximum of 1.45 μm for the copper terminal.
- X_0 maximum of 0.80 μm for the copper hole with a nickel finish.

4. Estimation of the Growth Delta of the Intermetallic Thickness Following the Test Trials

Once the thickness X_0 downstream of the process is estimated, the growth ΔX of the intermetallic thickness can be evaluated using Formula (3). When using Formula (3) for ΔX, some considerations must be made regarding the constants used, which pertain to SAC-type alloys. The analyzed solder joints are composed of these alloys. For SAC alloys, values found in the literature will be used for n and k_0 [18], specifically, $n = 0.52$ and $k_0 = 0.0178\ m/s^{0.52}$ (value to be converted to $\mu m/s^{0.52}$).

The other constants include the Boltzmann constant R (equal to 8.314 J/K mol) and the apparent activation energy Q. The apparent activation energy for diffusion and intermetallic thickness growth phenomena is not uniquely determined and depends on many factors, even at small scales. It is experimentally obtained by researchers in reverse, starting from the observed intermetallic thickness value, through graphs that plot ln(k) against $1/T$ [18]. It is also reported in the literature that the apparent activation energy can be calculated very approximately by multiplying the absolute melting temperature value of the intermetallic compound by 85. For example, considering the primary compound that melts at 415 °C, it results in a value of 58.5 kJ/mol [19]; another value reported in the literature is 57.7 kJ/mol [18]. However, since obtaining a precise value for each case is impossible, based on the many values found in the literature, the apparent activation energy generally ranges around 50–60 kJ/mol for the primary intermetallic compound Cu_6Sn_5 and more generally for the total intermetallic thickness, with extreme values that can exceed 80 kJ/mol [20] predictions. Given the exponential nature of Formula (3), the maximum ΔX that can be calculated is the one resulting from the minimum observed activation energy, which will therefore be set at 50 kJ/mol to assess the worst-case scenario. Regarding the test phases, after the soldering process, the boards undergo three stages of exposure to high and low temperatures:

(1) The first phase involves constant temperature exposure in an oven at 125 °C for 12 h.
(2) The second phase consists of thermal shock for 50 cycles, with rapid transitions from −40 °C to 85 °C at a heating/cooling rate of 41.66 °C/min (the transition between temperatures occurs in less than 3 min), with a constant stay of 15 min at each of the two temperatures, totaling 36 min per cycle.

(3) The third phase spans 55 h in total, repeating a thermal cycle with a constant stay at −40 °C and 85 °C for 20 min at each of the two temperatures, with transitions between temperatures at a rate of 5 °C/min, which is gentler than that of the thermal shocks. A complete cycle lasts 90 min, so the total number of cycles in 55 h will be 36.666.

By calculating the total exposure times to the respective temperatures for each phase of the cycle, neglecting the transition phases between temperatures, we obtain:

- Total stay at 125 °C for 12 h (43,200 s).
- Thermal shock with a total stay of 12 h and 30 min (45,000 s) for each of the two temperatures (−40 °C and 85 °C).
- Thermal cycle with a total stay of 12 h, 13 min, and 20 s (44,000 s in total) for each of the two temperatures (−40 °C and 85 °C).

For completeness, the graphs related to thermal shocks and thermal cycles are provided in Figure 2.

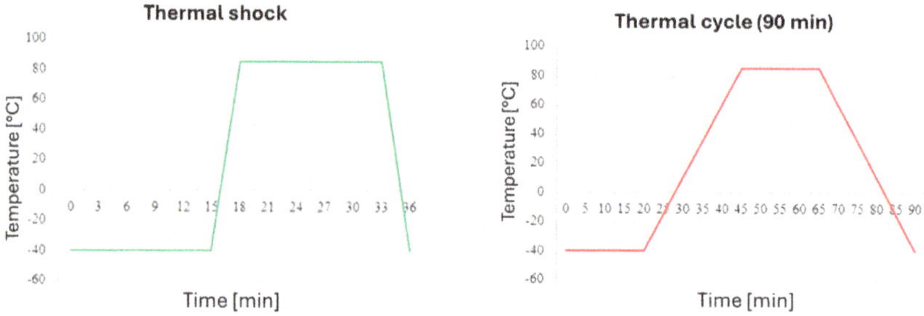

Figure 2. Thermal profile for each cycle (phase 2 on the **left**, phase 3 on the **right**).

Thermal cycling parameters are detailed in terms of duration and temperatures in Table 5.

Table 5. Total exposure time (in seconds) and temperature for each test phase.

Steps	Cycles (n)	Temp (°C)	Minutes per Cycle (min)	Seconds per Cycle (s)	Total Time (s)
Phase 1 (Const Temp)	1	125	720	43,200	43,200
Phase 2 (Low Temp)	50	−40	15	900	45,000
Phase 2 (High Temp)	50	+85	15	900	45,000
Phase 3 (Low Temp)	36.66	−40	20	1200	44,000
Phase 3 (High Temp)	36.66	+85	20	1200	44,000

Referring back to Formula (3), the time parameters (in seconds) and absolute temperature are substituted into the equation to predict the maximum achievable ΔX, with Q set to 50 kJ/mol (this value will be converted to J/mol in the formula). It is reiterated that this apparent activation energy value determines the maximum observable delta; thus, we consider the worst-case scenario, which predicts the maximum growth of the intermetallic thickness. The predicted maximum thickness deltas are summarized in Table 6, detailing the individual thickness variations calculated for each test phase and temperature, then summed in the final row.

Table 6. Intermetallic layer increase (prediction) for each acceptance test phase.

	Temperature (K)	Time (s)	ΔX Max (μm)
Industrial oven	398	43,200	1.254566946
Thermal shock	358 233	45,000 45,000	0.236857009 0.0000288759
Thermal cycle	358 233	44,000 44,000	0.234105235 0.0000285404
			1.725586605
			1.73

In this last table, it is worth noting that exposure to −40 °C does not generate significant variations in the intermetallic thickness, while exposure to the high temperatures of 125 °C and 85 °C largely influences it. Therefore, the expected maximum ΔX is the sum: 1.73 μm.

5. Comparison of the Estimates with the Values of Intermetallic Thickness Observed through Metallographic Sections

At this point, the maximum ΔX just obtained is added to the maximum initial thicknesses X_0 downstream of the previously calculated process (it is recalled that $Xtot = X_0 + \Delta X$). This estimates an upper limit of prediction for the total thickness downstream of the accelerated test to which the board is subjected, to be compared with the thickness data found in the metallographic sections. Table 7 shows the comparison between the estimates and findings.

Table 7. Post-inspection intermetallic layer thickness values, estimates, and real results obtained from metallographic sections.

	Estimate Value	Response Value
	X_{tot} (Maximum Value)	X_{tot} (Maximum Value)
Terminal SMD	3.2 μm	2.5 μm
Pad SMD	3.5 μm	2 μm
Terminal THT	2.7 μm	1.2 μm
Hole THT	2.5 μm	1 μm

It can be observed that the maximum thickness values found in the metallographic sections fall below the upper limits estimated by the mathematical models. It is also interesting to note that the thickness values of intermetallic layers observed for THT components are lower than those observed for SMD components, which is consistent with the estimation. The upper limits observed in the tests are approximately 1 μm lower than those estimated, which is a positive result. However, the literature suggests that an intermetallic thickness of up to 4 μm is still acceptable. In the case at hand, even after the accelerated test conducted following the soldering process, the thickness remains below this 4 μm threshold. To provide an overview, some images of the metallographic sections are reported in Figures 3 and 4, which contain the measurement of the intermetallic thickness downstream of the tests.

The above figure shows in detail the point measurement of the intermetallic thickness. The comb-like structure of the intermetallic layer is clearly evident, as well as how the solder matrix presents additives and dispersed intermetallic phases of Ag_3Sn (more elongated grains) and Cu_6Sn_5.

Figure 3. Metallographic section of a THT component terminal in contact with the solder joint.

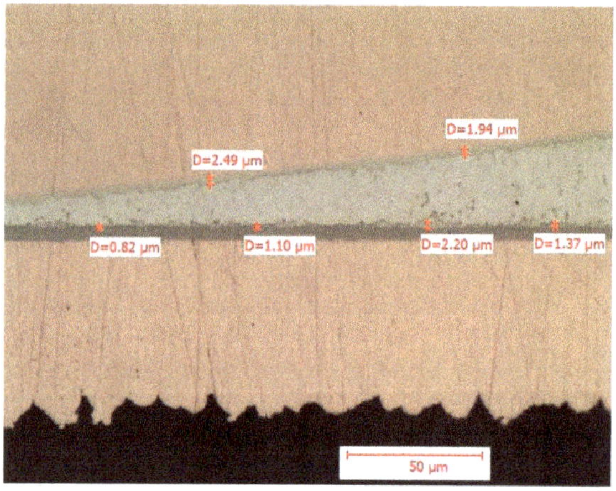

Figure 4. Metallographic section of an SMD component with a J-shaped terminal soldered to a PCB pad.

The above image shows both the pad/solder interface and the terminal/solder interface, detailing the point measurements of the intermetallic thickness.

An additional estimation of the intermetallic thickness delta during operation was made, involving the use of Formula (3) with years and ambient temperature as parameters, adapting the numerical model used to design aging test cycles. The results in the worst-case scenario, where the apparent activation energy Q is assumed to be 50 kJ/mol, are shown in Table 8. This table presents the maximum delta obtainable (expressed in μm) once again.

This growth delta during operation, solely due to ambient temperature, needs to be added to the initial intermetallic thickness downstream of the soldering process. It is worth noting that the boards in use have not undergone any testing following the soldering processes, unlike the samples analyzed earlier. Growth deltas, stemming from operation or ambient temperature alone, are acceptable if the soldering process is of high quality and leaves an initial intermetallic thickness of a few micrometers. Conversely, if the process is not controlled, the growth delta could significantly affect joint integrity. From the results in Table 8, it is anticipated that a board operating at ambient temperatures of 20–30 °C will not experience excessive intermetallic thickness variation, with significant changes becoming apparent after 30 years. However, in the extreme range of 40–60 °C found in desert environments with intense sun exposure, noticeable variation is observed after just one year of operation, while excessively significant changes should arise after ten years or more. These effects theoretically compound with thermal effects related to current passages, although they consist of short, non-continuous electrical pulses that should not significantly alter the joint's lifespan. Further investigation into this aspect could be worthwhile.

Table 8. Prediction of the intermetallic layer thickness growth delta (values expressed in μm) with exposure to various ambient temperatures.

T (°C)	1 Year	5 Years	10 Years	20 Years	30 Years
20	0.17	0.39	0.57	0.81	1.00
30	0.34	0.78	1.11	1.60	1.97
40	0.63	1.46	2.10	3.01	3.72
50	1.15	2.65	3.81	5.46	6.74
60	2.01	4.64	6.66	9.55	11.79
70	3.40	7.86	11.27	16.16	19.96
80	5.59	12.92	18.52	26.56	32.80
90	8.94	20.66	29.62	42.47	52.44
100	13.95	32.21	46.18	66.22	81.77
110	21.25	49.06	70.35	100.89	124.57
120	31.68	73.16	104.91	150.44	185.75
130	46.32	106.95	153.36	219.92	271.54
140	66.47	153.50	220.12	315.64	389.72
150	93.79	216.58	310.57	445.34	549.87
200	294.0	604.5	973.5	1243.0	1395.9

Finally, it is necessary to make a qualitative comparison between the data obtained from lead-free alloys and those obtainable through the use of a generic SnPb alloy, albeit obsolete for current environmental requirements. It is well known that in terms of performance, alloys that involve the use of lead have less pronounced intermetallic thicknesses downstream of the process and during operation compared to those found for lead-free alloys. However, the differences in thickness are minimal, and lead-free alloys have a greater tendency to form intermetallic compounds. This is partly because a SAC-type alloy has a tin (Sn) percentage of approximately 95%, and copper (Cu) is also present in the alloy, in addition to that of the base metal of the pad or hole. Both are the main reactants that will form Cu_6Sn_5 and Cu_3Sn. In contrast, a SnPb alloy has a tin content of 63% and reacts only with copper (Cu) from the base metal. Therefore, it is intuitive to conclude that, quantitatively, the formation of compounds and intermetallic thicknesses will be more pronounced for lead-free alloys, presenting ingredients in greater measure. However, from what we have observed, the differences still appear negligible [18]. Nevertheless, production processes, in accordance with the RoHS directive, have shifted towards the use of lead-free alloys for soldering, and it is appropriate to focus on these, as performed in the present work. Conducting a microscopic analysis on alloys that use lead is useful to predict the behavior of devices already installed or still on the market. In this regard, Hitachi provided us with some metallographic sections of SnPb solder joints present on electronic boards still in use. Three electronic boards were analyzed: a new board, a board that was used in an Italian environment (standard environment), and finally a board that was used in a different climate environment and in the presence of various corrosive agents (hostile environment). For each board, three solder joints of SMD components mounted in different areas were analyzed and, in this case, the copper pads have a nickel surface finish. The average intermetallic thickness values found are reported in Table 9 and Figures 5–7.

Table 9. Average intermetallic layer thicknesses in µm observed for SnPb alloy solders on various components and analyzed boards.

Component	1	2	3
New electronic board	1.04 µm	0.92 µm	0.74 µm
Electronic board in standard environment	1.18 µm	1.43 µm	2.33 µm
Electronic board in hostile environment	1.48 µm	1.32 µm	1.50 µm

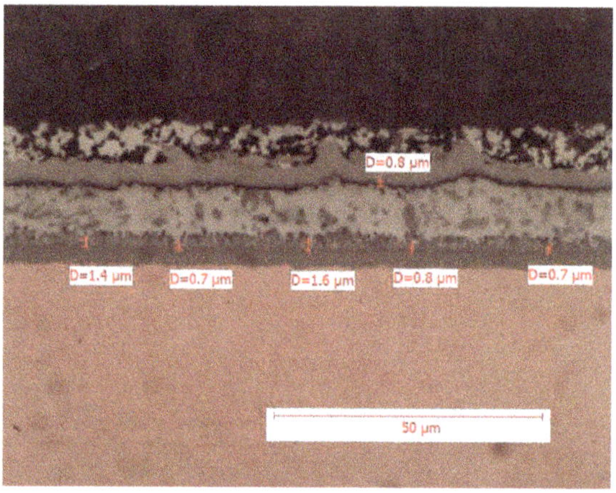

Figure 5. Intermetallic thicknesses found for SnPb alloy soldering on component 1 (new board).

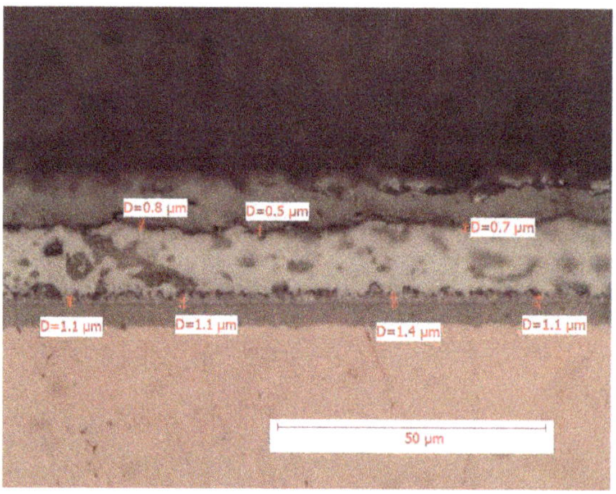

Figure 6. Intermetallic thicknesses found for SnPb alloy soldering on component 1 (standard environment).

It can be observed that, concerning the new board, the intermetallic thickness values found downstream of the soldering process are consistent with those predicted by the estimates previously performed for a lead-free, SAC-type solder alloy. As expected for a SnPb-type alloy, used in this case, the intermetallic thickness is slightly lower compared to what is expected for the lead-free SAC alloy. Regarding the two boards used in standard and hostile environments, assuming an initial intermetallic thickness downstream of the soldering process equal to that found in the new board, a contained thickness variation during operation is evident, in line with what is also expected for a SAC alloy. However, applying the constants of SAC alloys to SnPb alloy solder joints for the prediction model of ΔX (Formula (3)) would lead to an overestimation of the thickness variation compared to the new board.

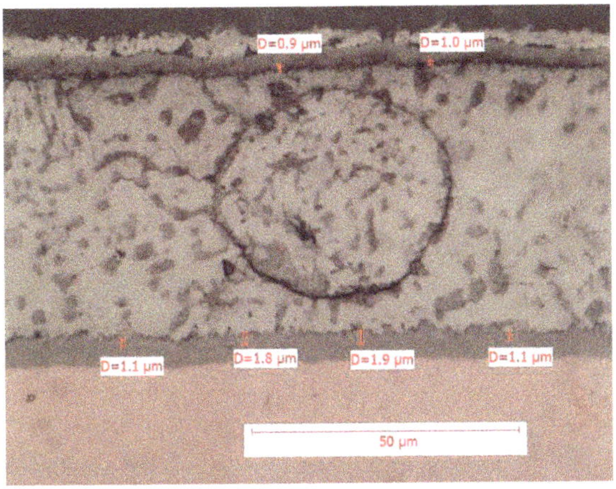

Figure 7. Intermetallic thicknesses found for SnPb alloy soldering on component 1 (hostile environment).

Similarly, the total intermetallic thickness observed in the exercised boards is contained and aligns with expectations, with no significant differences from the experimentally observed intermetallic thickness. It should be noted that although one of the two boards operated in a hostile environment and was exposed to a higher degree to chemical agents such as sulfur, it likely benefited from exposure to lower temperatures (Northern European railway route). Finally, the morphology of the intermetallic compound grains observed for SnPb alloy solder joints consistently appears comb-like, but the intermetallic thickness, particularly on the new board, exhibits more jagged edges and less smooth contours than those observed in the analysis conducted for SAC alloy solder joints.

To complete the discussion on predicting the thickness of the intermetallic layer, one can make a reverse reasoning, setting a limit value for ΔX beyond which the total intermetallic thickness becomes excessive and compromises the mechanical strength and reliability of the solder joint permanently. If, as previous stated, an intermetallic layer thickness of 4 µm is deemed acceptable, assuming an initial thickness X_0 downstream of the soldering process of 1 µm, a limit value of ΔX of 3 µm can be defined. Assuming again the same constants seen for the analysis of SAC alloy solder joints and fixing ΔX, the exposure and operational times at various temperatures that would lead to a 3 µm variation can be calculated from Formula (3), thus defining a useful life of the solder joint and therefore of the board. Estimates of useful life are reported in Table 10.

As the reader can notice, exposure to room temperature ensures an "infinite life". It is instead the temperatures reached by the soldering during operation that, if kept constant, lead to a rapid drift in the growth of the intermetallic thickness. Generally, temperatures up to 40 °C are not concerning. However, even though current flow and board usage may bring soldering temperatures up to 85 °C for very short periods, it must still be considered that at high temperatures the growth of intermetallic thickness is greatly facilitated. Consequently, the solder joints of an electronic board, if subjected to an excessive, repeated, and continuous number of work cycles, in terms of mechanical strength and reliability, could last less than desired and expected.

Table 10. Expected lifetime of electronic boards under constant temperature exposure/use, based on intermetallic layer thickness (ΔX limit of 3 μm).

Temperature (°C)	Expected Lifetime in Years (ΔX = 3 μm)
10	984.1
20	244.0
30	66.3
40	19.6
50	6.2
60	2.1
70	0.8
80	0.3
90	0.1

6. Conclusions

This research begins with a comprehensive overview of the intermetallic compounds present in the soft soldering processes of joints used in the assembly of electronic boards. The models proposed in the literature are used to estimate the intermetallic thickness.

The values obtained from the theoretical models are compared to experimental results of a new electronic board with an SbPb alloy solder and a SAC (lead-free) alloy solder joint on a board subjected to accelerated aging cycles.

Experimental results have shown intermetallic compound (IMC) values always below 3 μm, with a variation from a minimum value of 0.7 μm for a SnPb alloy solder joint on a new board to a maximum value of 2.7 μm observed on a SAC (lead-free) alloy solder joint on a board subjected to accelerated aging cycles. In both types of soldering processes, whether reflow soldering or wave soldering, no criticalities have emerged that lead to uncontrolled growth of IMCs. These phenomena could only materialize in an advanced stage of operation, where the electronic board approaches its end of life or in situations where the board is subjected to particular environmental stresses, without appropriate protection.

The original contribution of this work is the use of mathematical models, employing an inverse approach compared to what is described in the scientific literature.

The mathematical relationships referred to are those typically used to design aging tests for electronic boards. In this work, these relationships have been readapted to assess reliability during the operating period, up to the end of life; for this purpose, the test thermal parameters have been replaced by those imposed by the different installation environments. The use of mathematical models has allowed us to obtain a reliable estimate of the intermetallic layer thicknesses, both with reference to the initial ones, just after the process, and their evolution over time.

Utilizing the models, it has also been found that the intermetallic layer is sensitive to high temperatures, with thickness values that can grow exponentially (in relatively short times compared to a typical lifespan of 10–20 years) if the soldered joints are subjected frequently or constantly to operating temperatures well above 40 °C, with increasingly drastic thickness variations from 50 °C onwards.

From an industrial perspective, studies like this allow for the prediction of the reliability of printed circuit boards already at the design stage, minimizing the need for destructive (and/or aging) tests required to monitor the quality of processes/products and maintain greater control over a phenomenon as microscopic as it is relevant to that of intermetallic compounds. Moreover, the reduction of thermal aging tests and related destructive tests means a reduction in energy consumption and a lower environmental impact, in a context where manufacturing companies are devoting increasing attention with regard to life cycle assessment (LCA) of products and processes.

Future developments of this work will involve the use of models to explore the evolution of the intermetallic layer with a higher level of detail, by densifying the temporal observation points with varying temperatures, or even by artificially increasing the operating temperatures of the boards and their joints during testing through current passages. We aim to expand our study to include a broader range of lead-free solder alloys and conduct a more detailed analysis of discrepancies in subsequent projects. This will allow us to provide a more comprehensive understanding of the model's predictive capabilities across different materials and conditions. A greater amount of data would, for example, enable interpolating the values of k and n at each temperature to predict the growth of the intermetallic layer by simulating the conditions and operating temperatures of the solder joints.

Author Contributions: Conceptualization, P.R., V.R. and M.A.; methodology, T.R. and M.A.; formal analysis, T.R.; writing—review and editing, P.R., V.R. and M.A.; supervision, V.R. and P.R. All authors have read and agreed to the published version of the manuscript.

Funding: This research received no external funding.

Data Availability Statement: Data are unavailable due to industrial privacy.

Conflicts of Interest: Michele Ambrico and Vito Romaniello are employed of Hitachi Rail STS, but any financial or commercial are related to the research of this manuscript; therefore, I think that no conflicts of interest are present.

References

1. Chada, S. Topics in lead-free solders: Restriction of hazardous substances recast (RoHS2). *JOM* **2013**, *65*, 1348–1349. [CrossRef]
2. Menon, S.; George, E.; Osterman, M.; Pecht, M. High lead solder (over85%) solder in the electronics industry: RoHS exemptions and alternatives. *J. Mater. Sci. Mater. Electron.* **2015**, *26*, 4021–4030. [CrossRef]
3. Subramanian, K.S.; Connor, J.W. Lead contamination of drinking water: Metals leaching from soldered pipes may pose health hazard. *J. Environ. Health* **1991**, *54*, 29–32.
4. Deubzer, O.; Baron, Y.; Nissen, N.; Lang, K.D. Status of the RoHS directive and exemptions. In Proceedings of the Electronics Goes Green 2016+ (EGG), Berlin, Germany, 6–9 September 2016; pp. 1–6.
5. Su, S.; Akkara, F.J.; Thaper, R.; Alkhazali, A.; Hamasha, M.; Hamasha, S. A state-of-the- art review of fatigue life prediction models for solder joint. *J. Electron. Packag.* **2019**, *141*, 040802. [CrossRef]
6. Cheng, S.; Huang, C.-M.; Pecht, M. A review of lead-free solders for electronics applications. *Microelectron. Reliab.* **2017**, *75*, 77–95. [CrossRef]
7. Ma, H.; Suhling, J.C. A review of mechanical properties of lead-free solders for electronic packaging. *J. Mater. Sci.* **2009**, *44*, 1141–1158. [CrossRef]
8. Sona, M.; Prabhu, K.N. Review on micro structure evolution in Sn–Ag–Cu solders and its effect on mechanical integrity of solder joints. *J. Mater. Sci. Mater. Electron.* **2013**, *24*, 3149–3169. [CrossRef]
9. Jung, D.-H.; Jung, J.-P. Review of the wettability of solder with a wetting balance test for recent advanced microelectronic packaging. *Crit. Rev. Solid State Mater. Sci.* **2019**, *44*, 324–343. [CrossRef]
10. Kahar, H.; Abd Malek, Z.A.; Idris, S.R.A.; Ishak, M. Influence of second reflow on the intermetallic compound growth with different surface finish. *Key Eng. Mater.* **2016**, *701*, 2–31. [CrossRef]
11. Fan, Y.; Wu, Y.; Dale, T.F.; Lakshminarayana, S.A.P.; Greene, C.V.; Badwe, N.U.; Aspandiar, R.F.; Blendell, J.E.; Subbarayan, G.; Handwerker, C.A. Influence of pad surface finish on the microstructure evolution and intermetallic compound growth in homogeneous sn-bi and sn-bi-ag solder interconnects. *J. Electron. Mater.* **2021**, *50*, 6615–6628. [CrossRef]
12. Dale, T.; Singh, Y.; Bernander, I.; Subbarayan, G.; Handwerker, C.; Su, P.; Glasauer, B. Fatigue life of sn3. 0ag0. 5cu solder alloy under combined cyclic shear and constant tensile/compressive loads. *J. Electron. Packag.* **2020**, *142*, 041001. [CrossRef]
13. Wang, Y.; Yao, Y.; Keer, L. An Analytical Model to Predict Diffusion Induced Intermetallic Compounds Growth in Cu-Sn-Cu Sandwich Structures. *Theor. Appl. Mech. Lett.* **2020**, *10*, 33–37. [CrossRef]
14. Ramli, M.I.I.; Salleh, M.A.A.M.; Abdullah, M.M.A.B.; Zaimi, N.S.M.; Sandu, A.V.; Vizureanu, P.; Rylski, A.; Amli, S.F.M. Formation and Growth of Intermetallic Compounds in Lead-Free Solder Joints: A Review. *Materials* **2022**, *15*, 1451. [CrossRef] [PubMed]
15. Peng, X.W.; Wang, Y.; Wang, W.L.; Ye, Z.; Yang, J.; Huang, J.H. Kinetics of Cu_6Sn_5 and Cu_3Sn intermetallic compounds growth and isothermal solidification during Cu-Sn transient liquid phase sintering process. *J. Alloys Compdounds* **2023**, *949*, 169631. [CrossRef]
16. Ismail, N.; Atiqah, A.; Jalar, A. A systematic literature review: The effects of surface roughness on the wettability and formation of intermetallic compound layers in lead-free solder joints. *J. Manuf. Process* **2022**, *83*, 68–85. [CrossRef]

17. Scarano, V.L. Misure per la Caratterizzazione di Materiali Saldanti senza Piombo: Nuovi Processi di Saldatura per Sonde Biomediche ad Ultrasuoni. Ph.D. Thesis, Università degli Studi di Firenze, Firenze, Italy, 2008.
18. Pan, J.; Chou, T.C.; Bath, J.; Willie, D.; Toleno, B.J. Effects of reflow profile and thermal conditioning on intermetallic compound thickness for SnAgCu soldered joints. *Solder. Surf. Mt. Technol.* **2009**, *21*, 4. [CrossRef]
19. Dariavach, N.; Callahan, P.; Liang, J.; Fournelle, R. Intermetallic Growth Kinetics for Sn-Ag, Sn-Cu, and Sn-Ag-Cu lead free Solders on Cu, Ni, and Fe-42Ni Substrates. *J. Electron. Mater.* **2006**, *35*, 7. [CrossRef]
20. Zhang, L.; Fan, X.Y.; He, C.W.; Guo, Y.H. Intermetallic compound layer growth between SnAgCu solder and Cu substrate in electronic packaging. *J. Mater. Sci. Mater. Electron.* **2013**, *24*, 9. [CrossRef]

designs

MDPI

Article

Design of an Environment for Virtual Training Based on Digital Reconstruction: From Real Vegetation to Its Tactile Simulation

Alessandro Martinelli [1], Davide Fabiocchi [2], Francesca Picchio [1], Hermes Giberti [2] and Marco Carnevale [2,*]

[1] Dipartimento di Ingegneria Civile e Architettura, Università di Pavia and DICAr, 27100 Pavia, Italy; alessandro.martinelli01@unipv.it (A.M.); francesca.picchio@unipv.it (F.P.)

[2] Dipartimento di Ingegneria Industriale e dell'Informazione, Università di Pavia, DIII, 27100 Pavia, Italy; davide.fabiocchi01@universitadipavia.it (D.F.); hermes.giberti@unipv.it (H.G.)

* Correspondence: marco.carnevale@unipv.it

Abstract: The exploitation of immersive simulation platforms to improve traditional training techniques in the agricultural industry sector would enable year-round accessibility, flexibility, safety, and consistent high-quality training for agricultural operators. An innovative workflow in virtual simulations for training and educational purposes includes an immersive environment in which the operator can interact with plants through haptic interfaces, following instructions imparted by a non-playing character (NPC) instructor. This study allows simulating the pruning of a complex case study, a hazelnut tree, reproduced in very high detail to offer agricultural operators a more realistic and immersive training environment than those currently existing. The process of creating a multisensorial environment starts with the integrated survey of the plant with a laser scanner and photogrammetry and then generates a controllable parametric model from roots to leaves with the exact positioning of the original branches. The model is finally inserted into a simulation, where haptic gloves with tactile resistance responsive to model collisions are tested. The results of the experimentation demonstrate the correct execution of this innovative design simulation, in which branches and leaves can be cut using a shear, with immediate sensory feedback. The project therefore aims to finalize this product as a realistic training platform for pruning, but not limited to it, paving the way for high-fidelity simulation for many other types of operations and specializations.

Keywords: virtual training; immersive scenarios; virtual reality; vegetation modeling; photogrammetry; haptics; human–machine interface; 3D vision-based human pose estimation

1. Introduction

For years now, the primary sector has been enjoying the ever-increasing development of technology: robotics [1,2], Internet of Things (IoT) [3,4], and remote sensing [5] are established technologies in the world of agriculture, which is increasingly taking on the meaning of Smart Agriculture. However, the agriculture industry sector is suffering in many countries due to the lack of labor force, due to the aging of the population, the rural–urban migration, and the lack of interest among younger generations about agricultural careers.

The high cost of training skilled operators, the low flexibility of the conditions necessary for training, like the strong dependence on seasonality, the need to train culturally multi-faceted personnel among non-homogeneous operators, and the delayed feedback on results are the reasons why the development of new solutions for training operators in the agricultural industry sector is required. These new solutions can address these issues by

https://doi.org/10.3390/designs9020032

incorporating cutting-edge technologies, such as virtual reality (VR) and haptics, into the primary sector, developing a training platform available throughout the year with high flexibility and no safety issues.

This work takes place in the context of the POC NODES "DEMETRA" multidisciplinary research project for technological innovation in the agro-industry sector. It is aimed at evaluating the reliability of reality-based productions and offers optimal digital tools for training agricultural operators. DEMETRA aims to develop a platform for training agricultural operators in a virtual environment, leveraging human-in-the-loop techniques and offering innovative levels of immersiveness. As proof-of-concepts (PoCs) of the related technologies, the work deals with the design of an environment for virtual training, dedicated to pruning activities. Compared to existing simulators in the agriculture industry, DEMETRA's platform features highly detailed and realistic 3D reconstructions of trees, enhancing the immersiveness of the training simulator and enabling operators to engage in complex agricultural scenarios with true-to-life interactions, significantly improving the verisimilitude of the training experience.

This paper develops from the complex task of creating parametric digital twins of real trees and on the advanced techniques used for the accurate simulation of the agricultural environment for operator training. Additionally, it describes various methods employed to simulate interactions with the trees, as well as the VR and haptic techniques used to enhance immersive experience. Finally, an assessment on the sense of presence (SoP) perceived by the user is presented; the interaction between human and virtual environments plays a pivotal role in enhancing perceived immersion, the learning effect, and the realism during the training activity [6], so a SoP assessment is crucial when developing interactive systems, such as VR-based training platforms. In this study, the level of presence felt within the virtual environment has been measured through the Igroup Presence Questionnaire (IPQ) [7].

The current state of research in the panorama of the digital reconstruction of greenery presents wide margins for investigation and in-depth analysis due to the organic and complex nature of the arboreal elements. The objective of the experiments is to validate the creation of a dynamic VR scenario with the highest possible level of truthfulness to real vegetation, which will then be exploited through multisensory tools (such as haptic gloves and vocal instructions) to allow maximum immersion with minimum response times.

The simulation takes place by cutting the model of a grafted hazelnut tree, reconstructed through an integrated survey between laser scanner and photogrammetry, followed by the parameterization of the individual branches and the insertion of the model into a suitable virtual environment. This project stands out because, unlike the most common laser scanner acquisitions of immobile objects, it addresses the age-old and difficult challenge of surveying during micro-movements caused by air displacement. The high precision of laser tools, integrated with the reconstruction functions of software such as SpeedTree [8], has recently allowed an alternative workflow to convert imperfect photogrammetric models into realistic reconstructions down to the details, whose impact in virtual training can improve understanding of techniques, immersion, involvement, training accuracy, and various other qualities.

This study is therefore a first step in continuous experimentation to test high-fidelity simulations and optimize virtual training processes. This activity therefore presents an interdisciplinary character between graphic topology and simulation operations, exploiting the collaboration between researchers and agronomists in the innovation of specialist work.

2. State of the Art

Radhakrishnan analyzed 78 different studies in the field of industrial training to assess the effectiveness of immersive virtual reality (IVR) training solutions [9–11]. The study shows how most of the IVR training solutions are implemented in the manufacturing and healthcare fields, with almost half of the solutions providing procedural skills to the operators rather than perceptual-motor or decision-making skills. It is noteworthy to underline that none of the analyzed studies were developed in industry, thus highlighting the need to incorporate these technologies into the primary industrial sector. However, the agro-industry is not entirely new to VR applications, and early developments, such as FARMTASIA [12] and AgriVillage [13], have already helped introduce virtual training programs aimed at agricultural operators. Most of the solutions presented in the literature suffer from low realism and immersivity offered using screens and manual controllers instead of head-mounted displays (HMDs) and haptic interfaces. In fact, the use of these technologies could enhance the user experience and the transfer of real-life skills. Buttussi compared learner performance in a task under three conditions: high-fidelity IVR, medium-fidelity IVR, and desktop VR. Higher fidelity was shown to significantly enhance learner engagement and sense of presence compared to medium-fidelity IVR and desktop VR, proving their superior effectiveness for training purposes [14]. This highlights the importance of high-fidelity reconstruction of environments through advanced 3D modeling techniques to maximize the benefits of IVR in training scenarios.

To date, high-quality instrumental digital reconstruction procedures are mainly practiced in professional or R&I sectors, often for large works such as engineering, architecture [15] or archaeology [16]. Attempts to detect the natural landscape [17,18] using laser scanners or ground cameras have always left unsolved problems of detailed vegetation reconstruction [19]. In fact, current scanning tools do not allow acquiring stable data on organic elements subject to wind movement, making this task one of the most sought-after objectives of current research in the field of green representation.

A promising alternative to manual modeling, which would be excessively demanding and imprecise in the segmentation of complex databases, is the segmentation of shapes using AI [20]. In recent years, this methodology has been developed and implemented to support various professions [21], including vegetation [22,23], but to date it is not sufficiently complete or transferable to industrial application.

For these reasons, the most valid alternative for immersive simulation purposes seems to be a multidisciplinary approach that passes through scanning, semi-automatic post-production, and model parameterization [24]. This procedure has already been tested in other areas of botany [25] but apparently never for the purpose of creating virtual training or applications more associated with industry.

After creating a faithful simulation environment, it is necessary to develop an instrument to direct the operator who is using the simulator for training. For this purpose, marker-based motion capture systems are today among the most structured and widespread techniques for reproducing human movement, both in the professional field and in the entertainment industry [26]. In many studies, motion capture technologies have been used to improve social presence and communication effectiveness [27], providing many insights for designing more natural VR interactions [28]. It has been assessed how the absence of avatars reduces engagement and degrades communication quality [29]. For these reasons, the development of a virtual instructor is considered a suitable option also in a simulator for agriculture. The creation of a virtual instructor (i.e., NPC) requires the recording of the spatial movements of an expert botanist and has been achieved using optoelectronic technology. To complement this system, haptic gloves have been shown in the literature to be the most suitable control/navigation system to reproduce such manual tasks based on hand sensoriality [30].

3. Materials and Methods

The use of advanced digital technologies, such as scanning and photogrammetry, can generate vast amounts of three-dimensional data, making it difficult to extract an archetypal representation of vegetation. Transforming the point cloud into simplified geometric models requires advanced knowledge in agronomy and botany fields, as well as targeted methodologies for efficient and meaningful digitization procedures.

In this sense, the use of parametric processes for the predictive development of tree branching enables highly realistic and efficient simulations, offering a significant advantage in green space representation for planning and management. This approach combines integrated surveys, such as photogrammetry and ultra-high-resolution laser scanning, with reality-based parametric remodeling. For the study of the hazelnut tree considered in this work, the workflow was structured to ensure maximum fidelity to reality, allowing for an accurate analysis and verisimilitude to allow botanical operators maximum immersion during cutting training. After preliminary field tests, the survey was conducted in the most suitable season to capture the detailed evolution of the canopy and branches. The first phases were immediate tests and checks in the field to obtain cognitive outputs, on which to structure the actual survey in the right season with respect to the evolution of the tree.

The aim, following the digital reconstruction of an identical hazelnut, is to insert it into a multisensory simulation for training botanical operators in the pruning and cutting phase of the branches. This action, to be taught with care and awareness in the simulation, has been agreed upon and supervised by experts in general arboriculture and tree cultivation: a virtual trainer shows the technique to the user, who must repeat it. During the demonstration, the trainer explains the functioning of the servo-controlled shear and shows how to approach the plant with caution, the shape of the branches to be selected (i.e., the thinner ones that develop with a twisted direction towards the inside and which would hinder the correct expansion of the plant), how to bend down, extend the arm to reach the distant branches, at what height to make the cut (i.e., at the last internodes highlighted by blooms and changes in thickness), how to stand up, and finally how to put the tools back. All these actions, with simulated training in a safe environment, allow the user to make multiple attempts without harming himself or the plants, providing clear benefits for the introductory phases of this type of work.

The complete workflow up to the simulation, schematized in Figure 1, therefore, has a varied interdisciplinary nature:

- Data acquisition campaign with integrated data acquisition survey;
- Scan to model to full texturing process—parameterization of the model with completion of the missing parts;
- Insertion of the parametric model into the graphics engine that will simulate the collisions;
- Spherical panorama shooting for HDRI (high dynamic range image) and environmental composition;
- Creation of the suit and motion capture session;
- NPC modeling and animation;
- Integration of haptic gloves into the virtual environment and development of the related logics;
- Visual Programming Language (VPL) programming of branch cutting interactions;
- VPL programming of synchronization between instruments (i.e., virtual shear and haptic gloves);
- Tests and validation.

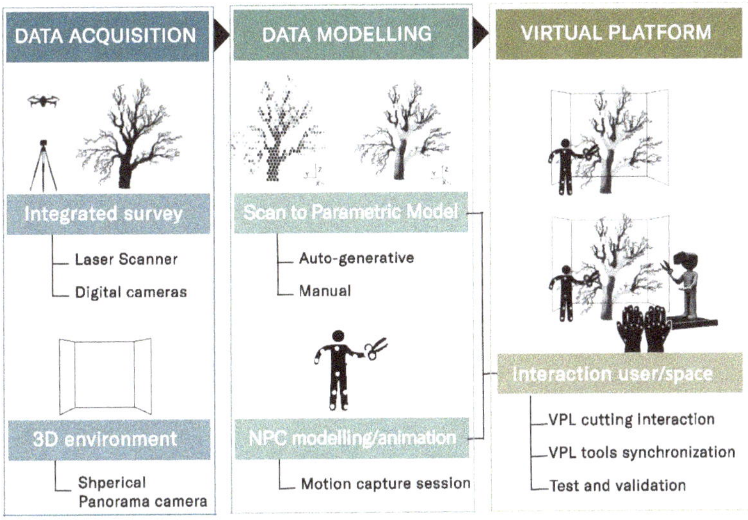

Figure 1. Research diagram of the workflow stages.

The structured method of study thus investigates the potential of scan-to-BIM procedures for the development of informative digital twins describing plants. Specifically, for this purpose, the research explores innovative methodological protocols, designed as PoC that can be replicated in other contexts, aiming at an effective representation of greenery through parameterized prototypes. The study then explores the interaction between 3D models, photogrammetric meshes, and point clouds within game engine platforms, aiming to create a single BIM-based (building information modeling) virtual scenario composed of highly efficient assets in terms of metric accuracy, visualization, and information. To further enhance this immersive experience, the research also presents a VR experiment based on Visual Scripting programming, designed to develop virtual tools for real-time querying of information associated with parametric models. The investigations and experimental applications within this work aim to demonstrate the potential impact of a multidisciplinary workflow designed for the large-scale use of informed and testable models.

4. Development

4.1. Survey Campaigns Experiments

The hazelnut detected is part of a culture of various specimens on the campus of the Università Cattolica del Sacro Cuore of Piacenza, and two acquisition sessions were carried out at different moments of its growth, at the beginning of October and at the end of January. The latter provided the cleanest data thanks to winter fading, and it is the most important part of the project, since pruning operations are carried out in wintertime (i.e., January, February).

The irregular shape of the arboreal architecture, the ascending branching, and the presence of a low and dense canopy led to the early exclusion of a single-sensor acquisition method, favoring a multi-resolution documentation approach suitable for describing the heterogeneous morphology of the vegetation [31]. In particular, the survey was conducted by immediately identifying various environmental risk factors, which were addressed through diversified methodologies and by integrating range-based and image-based acquisitions.

To ensure maximum data integration during post-processing, control points were placed around the survey area on both vertical and horizontal planes. The initial processing results provided a representative database of the vegetation system, characterized by high metric accuracy but also an approximated digital morphological representation. The cause of this outcome lies, on one hand, in an excess of acquired digital data, which generated excessive noise around the canopy, and, on the other hand, in a general discontinuity of the point cloud due to shadowed areas created by the foliage.

Additionally, some issues related to canopy reproduction were identified, caused by the micro-movement of smaller shoots and the geometry of the leaves themselves, which are too thin and therefore complex to document.

With the aim of minimizing errors in the digital duplicate and addressing the identified challenges, an additional acquisition campaign was conducted on completely leafless trees. The second survey was carried out by applying new control points directly on the trees, thus optimizing the processing of photogrammetric datasets and laser scans.

The acquisition from the lidar laser scanner was taken using an RTC 360 model (Figure 2), while a Nikon D 5600 was used for the photographic sets. Branches most subjected to wind, as expected, were not acquired correctly, but are subsequently hypothesized in post-production by the SpeedTree software. In total, six photographic sets of 300 photos each and seven lidar laser stations were carried out in the final survey campaign.

Figure 2. Use of the laser scanner and photographic campaign.

The acquired data, both range-based and image-based, were treated, respectively, with post-production software: Leica Cyclone for laser scanner data (Figure 3) and Agisoft Metashape for 3D photogrammetry (Figure 4). The various laser positions were aligned with manual verification of the homologous points [32]. Their accuracy was ascertained with Cloud Compare software (the alignment error was less than 3 mm for most of the branches (Figure 5), and finally they were used as a basis for the georeferencing of the photogrammetric model. The latter saw the alignment of 200,000 points for each set, and the final dense cloud counted around 30 million for a 5 million polygon mesh model. The dense cloud was first automatically cleaned with a color filter to erase movement noise, after which the imperfections were manually cleaned, and two versions were extracted from the most accurate model among those of the various sets: a complete digital twin (to be used later as a reference) and a clean model of trunks only (for subsequent parameterization).

Figure 3. Laser scanner data cleaning process. The tops of the branches, colored white, are subject to too much noise and therefore appear to be unreliable data.

Figure 4. Evolution of photogrammetric data. Sparse point cloud, dense cloud, textured 3D model, clean and trimmed 3D model for import into SpeedTree.

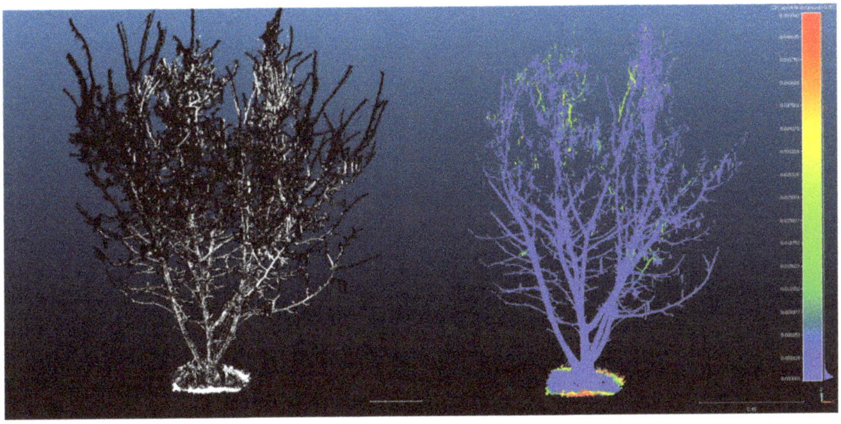

Figure 5. Comparison of the point cloud from the laser scanner with that from photogrammetry, using the Cloud Compare software. As for the deviation, the graph highlights a gap constantly lower than 3 mm on the whole surface, therefore a highly reliable data in terms of geometric precision.

After obtaining the most optimal static model, its transformation into a parametric model (i.e., a model that can be controlled in its various characteristics, such as length of the branches, radius, number of branches, leaves, and many others) is then necessary to allow live branch cutting and interactions based on real physics, but also to solve the problem of the impossibility of scanning the smallest branches, which would therefore leave gaps if the survey model were used.

The parameterization of a discrete model can take place in various ways. In particular, the construction phase of the three-dimensional model required a series of experiments to determine which of the processing options found in the literature [23,33] best suited the case study, allowing for the creation of a model with precise geometric and dimensional characteristics. In particular, different modeling approaches were evaluated and compared, including semi-automatic geometric modeling (reverse modeling and mesh from 3D scanning), manual modeling (low-poly mesh based on point cloud), and generative modeling (semi-automatic, AI, deep learning), in a process that goes from the most automatic (such as retopology or AI shape detection) to more manual ones. In this case study, the latter was preferred to maintain a level of detail and verisimilitude not achievable otherwise. The SpeedTree software boasts a series of functions that can be used to read the shape and color of the bases of the shoots to develop their continuations and optionally indicate the exact position/direction of the individual branches to generate a totally controllable digital twin.

This phase (Figure 6) was therefore divided into assisted reading of the cylindrical sections of each branch, extraction of the textures from the cylindrical mapping and transformation to seamless, clearance (cleaning of the edges of the mesh), extension of the branches, shaping (torsion and orientation of the branches), manual tracing of the positions of the original branches, and placement of the leaves.

Figure 6. Generation process on SpeedTree. Starting from the photogrammetric data, the trunks are further cut to be replaced with parametrized branches, initially straight, to which the texture extracted from the first model is then applied, and the direction of the real branches is given (taken from the most complete cloud of the laser scanner) until the exact original conformation is obtained.

The process took approximately 20 min per branch up to medium-sized ones, followed by multi-step branching of the main segments with appropriate parameters of twisting, internode, welding angle, and forces experienced (such as gravity and wind). The leaves and inflorescences were reproduced with a short photogrammetric campaign on real leaves in a closed environment on approximately 10 varieties to make the foliage heterogeneous (Figure 7). The final model (Figure 8) has approximately 500,000 polygons, of which approximately 200,000 are leaves, and the textures were arranged in a 2048 × 2048-pixel format for each branch, i.e., a format light enough not to burden the subsequent simulation.

Figure 7. Assignment of low-res mesh to the leaves of a hazel tree atlas. Placing the leaves and activating collisions on the final model.

Figure 8. Final version of the two trees detected and reproduced in the parametric software: grafted (on the right) and non-grafted (on the left).

The virtual environment designed for the user to practice gives professional and convincing context to the training while maintaining lightweight computing performance (Figure 9). A spherical panorama of the green environment in the urban area of Pavia was then acquired with an Insta 360, automatically processed, and extracted in HDR, which allowed the simulated space in Unreal Engine to be wrapped with an HDRI theme.

Figure 9. View of the virtual environment in Unreal Engine.

Finally, sounds and furnishings were specifically inserted in line with the natural context of the setting to recreate the illusion of an agricultural operation, and performance improvements such as post-process volumes and super resolution temporal anti-aliasing were applied.

4.2. Haptic Interactions Within the Virtual Environment

The user starts the simulation in the middle of the virtual environment with the possibility of exploring it using the teleport function or by walking inside an empty room, using the built-in tracking of the headset. Once the user reaches the virtual instructor, the teaching demonstration starts, enabling the user to grab the pruning tool once finished. At this point the user can start to cut the branches of the chosen tree by following the suggestion previously given by the instructor. The haptic interface used for the development of the application is the haptic glove "Nova 2′ by SenseGlove. This device allows the reproduction of the movements of the operator's fingers within the virtual environment, to actuate virtual pruning shears, and to provide force feedback and/or vibrotactile feedback to the user. The device works via wireless connection (Bluetooth), and its programming is implemented within Unreal Engine 5.1, using the plugin made available by the manufacturer.

To reproduce the movements of the user's fingers within the virtual environment, it is necessary to extract the flexion data for each finger. The Nova 2 haptic glove is equipped with a system of three cables for each finger wrapped around pulleys inside the device's case. Consequently, the extension of the cables is relative to the flexion of the fingers, which is indicated by a 'normalized flexion' value ranging between 0 and 1.

To reproduce the movement of the fingers, the normalized flexion value was used to compute the rotation value of each finger's joints. As each finger has three joints (two in the case of the thumb), the normalized flexion value alone is not sufficient to define their rotation values, but considering the natural closing gesture of the hand, certain relationships between the joints and the normalized flexion can be imposed, thus defining an unambiguous correspondence.

As previously introduced, the thumb has two joints: the meta-carpophalangeal (MCP) joint and the inter-phalangeal (IP) joint, while the other four fingers have three joints: the meta-carpophalangeal (MCP), the proximal inter-phalangeal (PIP), and the distal inter-phalangeal (DIP); see Figure 10.

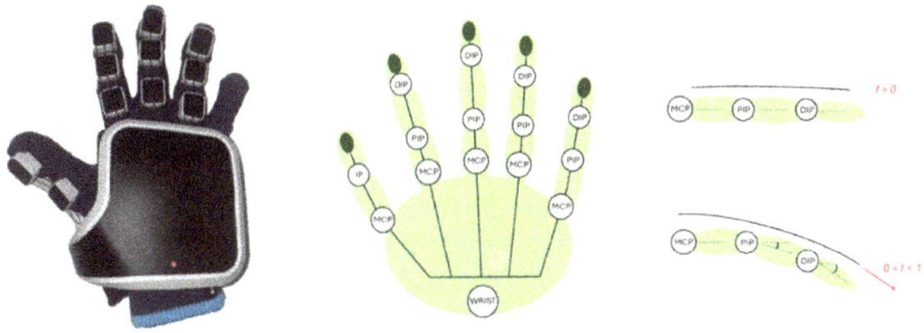

Figure 10. On the left: SenseGlove Nova 2 used in the simulation. On the right: hand joint model used for finger movement reconstruction, where f is the normalized flexion provided by the haptic gloves, and it depends on the finger flexion. The proposed model computes the finger's joint angles starting from the f value.

Given that each joint has a range of motion, it was decided to link normalized flexion (between 0 and 1) to the angle of rotation of the MCP joint of each finger, which typically has a range of motion between 0 and 90°. The following equations from the literature [34] were used to calculate the rotation values of the finger joints to be applied to the avatar's hand in the virtual environment:

$$MCP = f \cdot 90$$

$$PIP = 6 \cdot 10^{-6} \cdot MCP^4 - 0.0015 \cdot MCP^3 + 0.11 \cdot MCP^2 - 0.96 \cdot MCP + 5.6$$

$$DIP = 0.006 \cdot PIP^2 + 0.674 \cdot PIP + 0.104$$

The value indicated with f represents the normalized flexion of each finger, while three values—MCP, PIP, and DIP—describe the rotation angles of the respective joints.

Therefore, the user can interact with the virtual environment using Meta Quest 3 headset controllers and SenseGlove Nova 2 haptic gloves. Hand movement within the virtual environment is enabled by the headset's controllers, while finger movement is managed through the haptic gloves, as previously explained.

The logic managing the interaction between the user's hand and the virtual shear is based on the use of two Boolean variables: isGrabbing and isOverlapped. The variable isOverlapped describes the relative position between the hand and the shear, and it is true if the avatar's hand is inside a collision box built around the shear, false otherwise. The variable isGrabbing describes the state of the fingers of the hand responsible for grabbing the shear. The value of the variable depends on the mean normalized flexion of the three fingers responsible for grabbing (thumb, middle finger, ring finger), so the variable is true if the mean value exceeds a certain threshold, false otherwise.

The shear is initially placed on the avatar's right hip. A third variable, isEquipped, is necessary to describe the shear's equipment status. The variable isEquipped becomes true as soon as the variable isGrabbed switches from false to true, while isOverlapped is true. When the variable isEquipped becomes true, the shear is attached to the avatar's hand. If the user releases the grip, the variable isGrabbed becomes false again as the variable is Equipped and the shear returns to its initial location.

This grabbing logic is accompanied by vibration and force feedback to make the experience more realistic and immersive. Specifically, force feedback on the three fingers responsible for grasping is used to simulate the stiffness of the tool, blocking their movement, while vibrotactile feedback on the fingertips and palm is provided to the user to reproduce the sensation of contact at the beginning of the grasping. The actuation of the shear is regulated by the index finger, which is not subjected to force feedback. The normalized bending value of the index finger is used to control the actuation of the shear trigger once a limit value has been exceeded. Once the trigger is actuated, the shear blade closes, with an imposed law of motion, and, if the action corresponds to the cutting of a branch, vibro-tactile feedback is delivered to the user's palm to reproduce the vibration of the tool caused by the cut.

4.3. Cutting Tree Branches Logic

The virtual model of the leafless tree, once uploaded into Unreal Engine 5.1, was further processed to allow the interaction between the user and the branches and to develop the methodology necessary to enable the cutting of branches using virtual pruning shears.

The virtual tree, selected among the various developed models, was divided into several meshes, one for each branch. Different procedural meshes were generated, equal in number to the branches obtained from the previous subdivision of the model.

A procedural mesh is a three-dimensional data structure representing the shape of an object, with the characteristics of being generated, when necessary, through an algorithm, rather than being manually created and stored on a static file. This type of mesh is particularly useful in the reproduction of objects that need to be created dynamically and modified during a simulation. Using the 'Copy static mesh to procedural' command, the same mesh corresponding to the branches of the tree is associated with each procedural mesh. Subsequently, dependencies between the various branches were set up using parent/child relationships; thus, each branch is defined as a child of the branch to which it is attached. These dependencies are useful during the development of the cutting methodology, as they allow all branches belonging to the cut branch to be instantly identified, enabling the subsequent simulation of the physics and their simultaneous fall to the ground (Figure 11).

Figure 11. Simulation of branch cutting within VR.

Once the cutting command is executed through the haptic gloves, the shortest distance between each branch and a specific point between the scissor blades, defined as 'CuttingPoint', is computed. The branch corresponding to the shortest distance is designated as 'BranchToCut'. An additional check is carried out to ensure that the distance is below a certain threshold to prevent accidental cutting of branches if the scissors are operated away from the tree.

Once the branch to be cut is identified, the Blueprint function 'Slice Procedural Mesh' is used to make the cut. This function requires inputting the point in space to which the cut is to be applied and the plane secant to the mesh that will generate the cut. The CuttingPoint of the scissors is used as input, while the cutting plane is identified by its normal, corresponding to the vector starting from the CuttingPoint and orthogonal to the blades. The normal of the cutting plane is identified in such a way that the new procedural mesh created by the function corresponds to the cut part of the branch, which must then fall.

To allow the cut part of the branch and all branches attached to it to fall to the ground, all branches defined as children of the BranchToCut and above the cutting plane are attached to the new mesh generated by the cut by means of a physical constraint, with which all degrees of freedom are blocked. The physical simulation of the cut branch and all branches attached to it will allow the entire block to fall to the ground, with high levels of realism.

This methodology makes it possible to perform the cut on any branch of the tree, whether it is a single branch or a branch with several children, with realistic feedback related to the generation of the cut surface and the consequent fall of the cut part of the branch. Furthermore, this cutting logic allows branches already subjected to the operation to be cut an infinite number of times, thus placing no limits on any refinements and operations.

4.4. User Avatar and Virtual Instructor

The simulated environment was completed by inserting the user's avatar and a character instructor of pruning techniques (i.e., NPC), together with background and explanation audio assets. To enhance the immersiveness and the sense of presence perceived by the user, it has been decided to represent the whole upper body of the avatar. Since only position data from HMD and controllers were available, the Unreal Engine plugin Mimic Pro [35] has been used to simulate the movement of the upper body. The Mimic Pro plugin uses a control rig to provide stable, realistic full-body estimation that does not impede the player's movement and has minimal impact on performance. The Inverse Kinematic (IK) solver uses the position of the hands, extracted from the HMD's controllers, and the position of the head, extracted from the HMD sensors, to reconstruct the upper body movement in a realistic way. Full-body IK has been applied to a very realistic human body skeletal mesh created through the MetaHuman plugin [36], which allows creating, animating, and using highly realistic digital human characters.

The NPC, on the other hand, required a pre-recorded animation and voice that are activated when the user who is learning the pruning activity approaches. To achieve maximum coherence, expert project partners from the Università Cattolica del Sacro Cuore of Piacenza were recorded and consulted. The recording of the movements to be taught was taken with a motion capture system based on 6 OptiTrack Prime 13 W [37] high-speed cameras (Figure 12) with an infrared detection system, tracking 25 reflective markers on the upper body and 17 on the legs, to calculate the 3D position of the body in real-time.

Figure 12. First tests on instrument calibration, and virtual instructor animation registration.

The motion capture recording allowed for precise documentation of the movements necessary for pruning instruction. These movements, synchronized with an audio recording of guidance and instructions, were performed in a controlled environment without direct reference to a tree, instead mimicking the gestures required for pruning.

The recorded data was saved in Motive software and streamed in real-time to a skeletal mesh within Unreal Engine via the OptiTrack Live-Link plugin. This plugin not only enabled live streaming of the animation but also facilitated the application of the necessary rigging to align the recorded movement data with the corresponding bones in the virtual character's skeleton.

Specifically, the rigging process ensured that the skeletal mesh in Unreal properly interpreted the motion capture data, maintaining realistic joint movement and anatomical fidelity. Once the animation was successfully transferred to Unreal Engine, it was recorded as an fbx animation file. This file was then processed to refine the animation by correcting potential issues, such as unwanted vibrations or unnatural movements, that may have occurred during the motion capture session. After cleaning up the animation, it was reimported into Unreal Engine, where additional modifications were made to align it with one of the hazelnut trees present in the virtual scene. This adjustment ensured that the instructional sequence took place in direct proximity to the tree, enhancing the realism and effectiveness of the training session.

Figure 13 shows a comparison between the pose of the virtual instructor within the virtual environment and the raw skeletons acquired with the OptiTrack motion capture system.

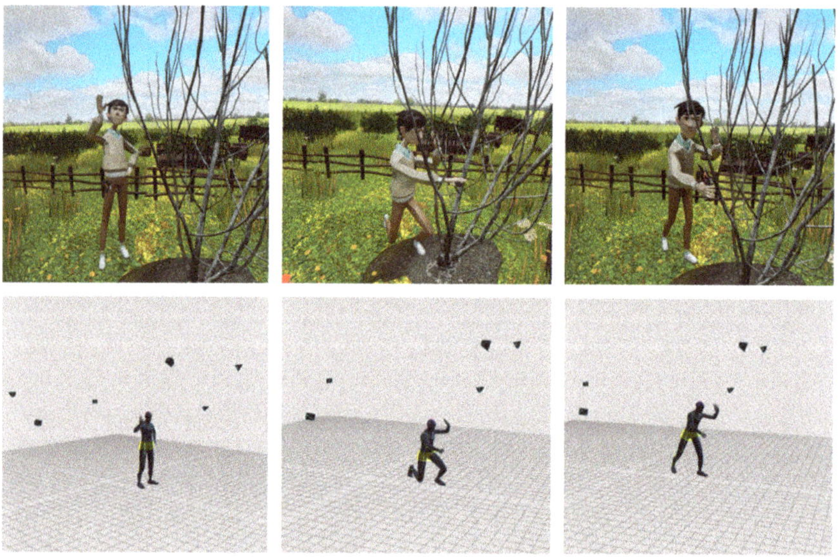

Figure 13. Comparison between the poses of the virtual instructor within the simulation, after the post-processing of the motion capture animation, and the raw registrations of the skeleton within the Motive Software before post-processing.

Furthermore, a pruning shear was added to the NPC's hand, and functionality was implemented to allow the shears to interact with the environment. Specifically, the plant model's behavior during the NPC tutorial is not directly integrated into the recorded NPC animation but is controlled through special branches that have an additional feature compared to the standard ones of the tree. These branches include a collision check that is triggered exclusively by the NPC's shears. When a collision with the shears is detected during the animation, the branches detach from the tree and fall to the ground, simulating a natural cutting behavior.

4.5. User Presence Assessment

To measure the level of presence perceived by the user within the virtual environment, during the simulation, the IPQ [7] was used. The questionnaire comprises 14 items with a 7-point Likert scale and it is divided into three subscales along with an additional general item (G). The subscales address spatial presence (SP)—the sensation of being physically present in the virtual environment; Realism (REAL)—the subjective experience of realism; and Involvement (INV)—the degree of attention dedicated to the virtual environment. A group of 30 volunteers (9 females and 21 males) aged between 22 and 44 years (M = 28.83, SD = 5.00) was recruited within the university community using a convenience sampling approach [38]. These participants, who had little or no prior experience with VR and haptic technologies, were instructed to perform the virtual pruning training session. The training comprised an initial instructional segment led by a virtual instructor, followed by a hands-on phase where users were required to cut tree branches according to the given instructions, using a haptic glove in conjunction with a VR headset and controllers. Immediately after the training session, the participants were asked to complete the IPQ questionnaire. Table 1 shows the questions of the IPQ with the items grouped by category.

Table 1. The IGroup Presence Questionnaire (IPQ).

Variable	Item	Question
G	G	In the computer generated world I had a sense of "being there".
P	SP1	Somehow I felt that the virtual world surrounded me.
	SP2	I felt like I was just perceiving pictures.
	SP3	I did not feel present in the virtual space.
	SP4	I had a sense of acting in the virtual space, rather than operating something from outside.
	SP5	I felt present in the virtual space.
INV	INV1	How aware were you of the real world surrounding while navigating in the virtual world? (i.e., sounds, room temperature, other people, etc.)?
	INV2	I was not aware of my real environment.
	INV3	I still paid attention to the real environment.
	INV4	I was completely captivated by the virtual world.
REAL	REAL1	How much did your experience in the virtual environment seem consistent with your real world experience?
	REAL2	How much did your experience in the virtual environment seem consistent with your real world experience ?
	REAL3	How real did the virtual world seem to you?
	REAL4	The virtual world seemed more realistic than the real world.

5. Results and Discussion

The first testing was carried out in a laboratory environment, with an environment simulated via a 1.32 Gb executable file. The test was performed on a system running Windows 11, equipped with an Intel Core i9-K processor and an NVIDIA RTX 4090 GPU, and no glitches or slowdowns were found upon subjective analysis. The cutting operation was successful according to the expected parameters; the frame rate was set at 90 fps, the resolution (as per the headset settings) was 2064 × 2208, and no bug reports were found from the engine. The simulation, therefore, fully reaches the relevant standards and allows the training of agricultural operators in complete safety and with maximum immersion while providing selected and always updatable theoretical knowledge.

The immersive simulator environment was then proposed to early adopters and at a trade fair for the agro-industry (EIMA 2024), with good user response regarding the immersiveness of the pruning action.

The construction of a three-dimensional model, especially in the context of simulating vegetation, requires careful consideration of the processing methods that will best meet the specific needs of the case study. A series of experiments were conducted to evaluate and compare different modeling approaches. These included semi-automatic geometric modeling (such as reverse modeling and mesh generation from 3D scanning), manual modeling (low-poly mesh based on point clouds), and generative modeling (using AI and deep learning techniques). Among these, manual modeling was preferred in this case study to ensure a high level of detail and realism, which would be difficult to achieve with automated approaches. The selected software (SpeedTree 9.5.2) was used to develop fully controllable digital duplicates of the hazelnut tree. However, it is important to emphasize that the methods used, while ensuring a high level of detail and realism, are rather costly in terms of time, required expertise, or replicability on other plant specimens.

Manual modeling, for example, requires significant effort and technical knowledge, not only for the creation of the three-dimensional model but also for the optimization of textures and geometries.

The accurate reproduction of finger movements and the manipulation of virtual pruning shears via force feedback were achieved by translating finger flexion data into joint rotations within the virtual environment. Overall, this approach demonstrates the importance of blending various modeling and interactive technologies to create immersive, detailed simulations that can be used for practical applications, such as virtual training or environmental simulations. The combination of manual modeling, generative techniques, and haptic feedback creates a highly controllable, engaging experience that stands out for its realism and attention to detail.

5.1. User Virtual Environment Interaction

To transfer real-life skills through VR-based training, it is fundamental to ensure high levels of embodiment and presence to the user, enhancing his experience. To this end, every aspect of this study has been meticulously designed to maximize immersion and realism, ensuring a more effective and impactful training experience using cutting-edge technologies, such as haptic interfaces. One of the key components of this system is the step-by-step virtual instructor, which is directly referenced to the tree, ensuring precise guidance without time constraints, language barriers, or variability in teaching approaches. The integration of haptic gloves further enhances realism by providing a tactile experience that closely mimics real-world pruning interactions, significantly reducing the gap between virtual training and real-life fieldwork, ensuring that the skills acquired in the simulation remain applicable and effective in real-world scenarios.

Additionally, the realistic simulation of branch-cutting mechanics, including the dynamic behavior of falling branches, contributes to an improved sense of immersion, along with the photorealistic representation of branches and their interaction with the environment, and reinforces the efficacy of the experience, further strengthening the user's engagement and learning retention. Preliminary evaluations indicate that this technology has significant potential to improve skill acquisition and training efficiency in agriculture by offering a scalable, highly immersive solution and representing a valuable tool for operator education. Future developments may include the usage of the motion capture system to analyze the operator's movements during training, enabling real-time corrections of improper techniques and postures. Moreover, the wealth of data collected from various sensors could facilitate in-depth ergonomic studies, made possible by markerless, AI-based, human pose tracking systems [39], thus optimizing training strategies to enhance both efficiency and user comfort.

5.2. Sense of Presence Assessment Results

The collected data have been manipulated; specifically, the responses have been converted into the scale 0–6, and reversed items (REAL1, SP2, and INV3) have been appropriately modified to match the other items.

A first simple analysis was conducted on the collected data using descriptive methods and computing the mean, standard deviation, median, and range for each item across all 30 participants. Table 2 shows the results obtained in this analysis.

Table 2. Mean, standard deviation, median, and range of each item across the whole sample.

Variable	Item	Mean	Standard Deviation	Median	Range
G	G	4.73	0.83	5.00	3–6
SP	SP1	4.67	0.88	5.00	3–6
	SP2	4.40	1.57	5.00	0–6
	SP3	4.50	1.48	5.00	1–6
	SP4	4.10	1.45	4.50	1–6
	SP5	4.57	1.17	5.00	1–6
INV	INV1	3.77	1.41	4.00	1–6
	INV2	3.70	1.47	4.00	0–6
	INV3	3.53	1.57	3.50	0–6
	INV4	4.67	0.88	5.00	2–6
REAL	REAL1	3.53	1.41	4.00	0–6
	REAL2	4.03	1.33	4.00	1–6
	REAL3	2.40	1.48	2.00	0–5
	REAL4	1.37	1.22	1.00	0–4

By looking at the table, it is evident that the general item G, measuring the "sense of being there", has a relatively high mean (4.73) and median (5.00), indicating general positive responses, while the range (3–6) indicates how the responses are concentrated in the upper half of the scale. All the items of the SP category have high mean scores (4.10 to 4.67), showing a general positive spatial presence perceived by the user, with a relatively good consistency in the responses (standard deviation between 0.88 and 1.57). The category INV, related to involvement, shows moderate means, with the item INV3, which has the lowest mean (3.53) and the greatest variability in the responses, with a standard deviation equal to 1.57. By looking at the ranges obtained for the different items, it is evident how the respondents had a wider spread of opinions. The Realism category shows the lowest mean, particularly for the item REAL4 (1.37). The low medians (1.00 and 2.00 for REAL3 and REAL4) suggest that many participants provided very low ratings, indicating that this aspect may be perceived less favorably.

Interestingly, the items with the highest variability (INV3 and SP2) correspond to two of the three reversed items in the questionnaire. This could suggest that the reversal process may have contributed to the increased variability in responses. Participants might have found these items more challenging to interpret, leading to a wider distribution of scores.

To have a deeper understanding of participants' experience, the average scores for each category (G, INV, SP, and REAL) per participant have been computed, obtaining four aggregated scores for participants. Figure 14 shows a boxplot visualization of these data to compare central tendency, variability, and potential outliers across the four categories.

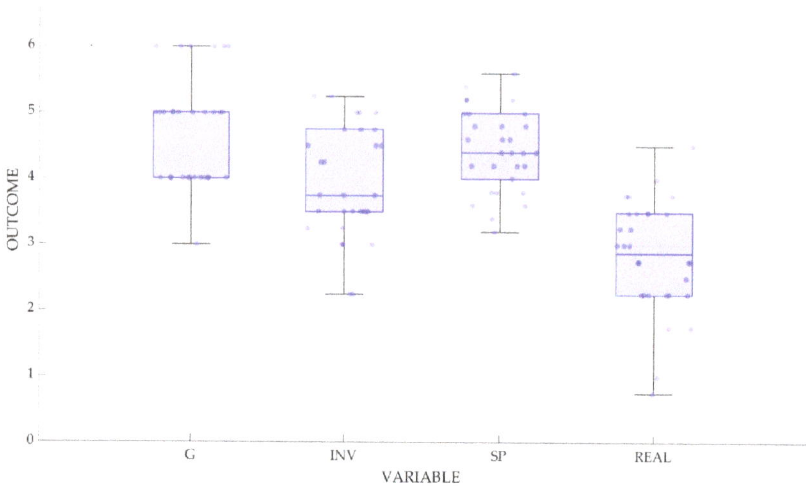

Figure 14. Boxplots for the four categories (G, INV, SP, and REAL), providing an overview of how participants responded to each dimension.

The items related to spatial presence (SP = 4.44) and the sense of "being there" (G = 4.73) showed the highest and most consistent scores, suggesting a clear and convincing perception of the surrounding environment. The perceived involvement was relatively high (INV = 3.92), but it showed a slightly lower median with greater variability, indicating differences in the degree of immersion among participants. Finally, realism received the lowest scores (REAL = 2.83), indicating a weaker perception of realism compared to the other dimensions.

Although effort was made to model the plant faithfully, as evidenced by the low deviance values obtained when compared to real plants, the deliberate decision to introduce purely virtual elements, such as the instructor, and some necessary components for constructing the virtual environment, may have compromised the realism perceived by the users. Nevertheless, the values of spatial presence are high, as are those of involvement, emphasizing how immersive the virtual environment is. Overall, the experience helps the user immerse themselves in the pruning activity, which is also thanks to the high fidelity of the plant reconstruction and the perception of stimuli through the use of haptic interfaces.

5.3. Potential File Exchanges with Predictive Software

During the research, preliminary studies were also carried out on the interaction and bidirectional data exchange methods between the virtual training scenario developed in this work and a predictive model of tree growth, i.e., a digital version of the hazelnut used in botany for the digital simulation of the state of the hazelnut over time [40,41].

However, nowadays, the quality of the two models is very different:

- Predictive growth models are based on a strong abstraction of the tree shapes and summarize the key elements of the hazelnut in simple geometries: cylinders, spheres, and surfaces.
- On the other hand, the survey model is a reconstruction faithful to detail, with the aim of providing maximum immersion to the agricultural operators who will use it.

Both formats can be in obj but have different polygon counting orders: from hundreds for the first to hundreds of thousands for the second. A first exchange was therefore attempted by sending the predictive model into the realistic simulation, then cutting it with haptic gloves and returning it to the growth simulation software. This exchange direction was successful, while the opposite (i.e., sending the detailed model to the growth simulation) was not feasible, since it was hindered by the reduced compatibility of the graphics engines.

We then proceeded with an analysis of the possible resolution strategies, as in the following: Starting from the high-resolution model of the core version, it is assumed that it is possible to apply scripts aimed at reducing the complexity of the model while preserving some of its properties. The process begins with separating all non-continuous geometries, ensuring that distinct parts of the model are treated independently. Once divided, the elements are categorized into three main groups—leaves, branches, and the main trunk—since each requires a different approach to simplification. Naming conventions are applied to help distinguish these components, with leaves, branches, and the trunk labeled accordingly based on their vertex count. To manage the excessive number of leaves (yet not present in the wintertime tree on which pruning is carried out), the script applies a controlled reduction by selecting a specified percentage of leaf objects and eliminating them. This process significantly decreases the number of leaves from approximately 20,000 to around 200, making the model more lightweight and manageable. Similarly, the branches undergo a restructuring process where their overall inclination is preserved, but unnecessary vertices are removed. By isolating the outermost edge loops and eliminating the inner vertices, the branches are simplified into a more streamlined form. By implementing these optimization steps, the script can drastically reduce the total polygon count to one hundredth. This makes the model significantly more efficient for rendering, exporting, or real-time applications while retaining its fundamental visual characteristics.

Although this meets the intended objectives, it remains to be resolved that most growth simulation software treats the leaves as solid objects, which are therefore deleted already at the first export in obj format to the cutting simulation. Furthermore, the main stem, which is the part of the mesh with the highest density of vertices, has loops and mesh directions that are different and irregular, as it is composed of multiple branches. The scripts described cannot completely solve these problems, and it will be the subject of subsequent research to find innovative solutions with the available technologies.

6. Conclusions

In conclusion, the development of a parametricly generated simulation has enabled the creation of an innovative model for highly realistic virtual training in pruning. The resulting model is dynamic and interactive, allowing real-time cuts and modifications within the simulation. Parametric modeling adds significant value by enabling adaptive design solutions. These models facilitate the analysis of tree development over time, improving maintenance strategies, environmental impact assessments, and biodiversity conservation. By integrating parametric tools, researchers and professionals can make data-driven decisions, ensuring resilient and efficiently managed green spaces. This research serves as a starting point for further exploration of topics and methodologies related to the parametric modeling of surveyed data, interactive virtual access to informational metadata, and model-user interoperability.

This model could make a significant contribution to the field of precision agriculture and plant research, as it combines high-resolution data from multiple sources into a cohesive and comprehensive representation. Such an integrated approach enables a variety of applications ranging from agricultural optimization to environmental monitoring. Likewise, the model-based simulation has proven to be sufficiently functional and responsive to successfully test a pruning training prototype with agricultural workers. The ability to simulate pruning in a virtual environment provides a safe and cost-effective method to train individuals in the appropriate techniques without the risk of damaging real trees. This is particularly important in educational and vocational training contexts, where the ability to repeatedly practice and learn from mistakes without real-world consequences can significantly improve learning outcomes. Furthermore, the possibility of using a training platform structured in a virtual environment and therefore available throughout the calendar year would separate the learning path of an agricultural activity from the seasonality connected to it, with obvious benefits in terms of costs and speed of training operations for operators. The workflow presented can be extended to other interdisciplinary research involving the creation of digital twins for monitoring agricultural management and can be extended to agricultural operations on different plant species or other difficult-to-reconstruct specimens belonging to the third sector. Similarly, simulation with such rapid and complete sensory feedback tools can offer potential support to any professional and/or industrial field not yet facilitated by immersive realities.

Author Contributions: Conceptualization, M.C., F.P. and H.G.; methodology, M.C., F.P., H.G., A.M. and D.F.; software, A.M. and D.F.; validation, M.C. and F.P.; formal analysis, A.M and D.F.; investigation, A.M, D.F. and M.C.; resources, M.C., F.P. and H.G.; data curation, A.M. and D.F.; writing—original draft preparation, A.M. and D.F.; writing—review and editing, M.C. and F.P.; visualization, A.M. and D.F.; supervision, M.C. and F.P.; project administration, M.C.; funding acquisition, M.C. All authors have read and agreed to the published version of the manuscript.

Funding: This paper is part of the project NODES which has received funding from the MUR—M4C2 1.5 of PNRR funded by the European Union—NextGenerationEU (Grant agreement no. ECS00000036).

Data Availability Statement: The raw data supporting the conclusions of this article will be made available by the authors on request.

Acknowledgments: We would like to thank Università Cattolica Del Sacro Cuore, Department of Sustainable Plant Production Sciences (DIPRO.VE.S), for having made the fields available on which to carry out the surveys and for the valuable support of Sergio Tombesi, scientific director of "DIGINUT—Development of digital twin tools for decision support in the hazelnut production chain", complementary project to the DEMETRA project. We also extend our gratitude to Pellenc Italia S.R.L. for providing the 3D model of the pruning shears and for allowing us to present the application at EIMA 2024.

Conflicts of Interest: The authors declare no conflicts of interest.

References

1. Bechar, A.; Vigneault, C. Agricultural Robots for Field Operations: Concepts and Components. *Biosyst. Eng.* **2016**, *149*, 94–111. [CrossRef]
2. Rodenburg, J. Robotic Milking: Technology, Farm Design, and Effects on Work Flow. *J. Dairy Sci.* **2017**, *100*, 7729–7738. [CrossRef] [PubMed]
3. Liu, W.; Shao, X.-F.; Wu, C.-H.; Qiao, P. A Systematic Literature Review on Applications of Information and Communication Technologies and Blockchain Technologies for Precision Agriculture Development. *J. Clean. Prod.* **2021**, *298*, 126763. [CrossRef]
4. Akhigbe, B.I.; Munir, K.; Akinadé, O.O.; Àkànbí, L.A.; Oyedele, L.O. IoT Technologies for Livestock Management: A Review of Present Status, Opportunities, and Future Trends. *Big Data Cogn. Comput.* **2021**, *5*, 10. [CrossRef]
5. Weiss, M.; Jacob, F.; Duveiller, G. Remote Sensing for Agricultural Applications: A Meta-Review. *Remote Sens. Environ.* **2020**, *236*, 111402. [CrossRef]
6. Choi, I.; Culbertson, H.; Miller, M.R.; Olwal, A.; Follmer, S. Grabity: A Wearable Haptic Interface for Simulating Weight and Grasping in Virtual Reality. In Proceedings of the 30th Annual ACM Symposium on User Interface Software and Technology, Quebec City, QC, Canada, 22–25 October 2017.
7. Igroup Presence Questionnaire (IPQ) Overview | Igroup.Org—Project Consortium. Available online: https://www.igroup.org/pq/ipq/index.php (accessed on 20 February 2025).
8. SpeedTree—3D Vegetation Modeling and Middleware. Available online: https://store.speedtree.com/ (accessed on 20 February 2025).

9. Radhakrishnan, U.; Konstantinos, K.; Chinello, F. A Systematic Review of Immersive Virtual Reality for Industrial Skills Training. *Behav. Inf. Technol.* **2021**, *40*, 1310–1339. [CrossRef]
10. Thach, N.; Hung, N. Research and Design a Lifeboat Virtual Reality Simulation System for Maritime Safety Training in Vietnam. *J. Mar. Sci. Technol.* **2024**, *32*, 4. [CrossRef]
11. Heibel, B.; Anderson, R.; Swafford, M.; Borges, B. Integrating Virtual Reality Technology into Beginning Welder Training Sequences. *J. Agric. Educ.* **2024**, *65*, 210–225. [CrossRef]
12. Cheung, K.; Jong, M.; Lee, F.; Lee, J.; Luk, E.; Shang, J.; Wong, M. FARMTASIA: An Online Game-Based Learning Environment Based on the VISOLE Pedagogy. *Virtual Real.* **2008**, *12*, 17–25. [CrossRef]
13. Yongyuth, P.; Prada, R.; Nakasone, A.; Prendinger, H. AgriVillage: 3D Multi-Language Internet Game for Fostering Agriculture Environmental Awareness. In Proceedings of the International Conference on Management of Emergent Digital EcoSystems, MEDES'10, Bangkok, Thailand, 26–29 October 2010; pp. 145–152.
14. Buttussi, F.; Chittaro, L. Effects of Different Types of Virtual Reality Display on Presence and Learning in a Safety Training Scenario. *IEEE Trans. Vis. Comput. Graph.* **2017**, *24*, 1063–1076. [CrossRef]
15. Parrinello, S.; Miceli, A.; Galasso, F. From Digital Survey to Serious Game. A Process of Knowledge for the Ark of Mastino II. *Disegnarecon* **2021**, *14*, 17.1–17.22. [CrossRef]
16. Barber, D.; Dallas, R.; Mills, J. Laser Scanning for Architectural Conservation. *J. Archit. Conserv.* **2006**, *12*, 35–52. [CrossRef]
17. Fonstad, M.; Dietrich, J.; Courville, B.; Jensen, J.; Carbonneau, P.E. Topographic Structure from Motion. In Proceedings of the AGU Fall Meeting Abstracts, February 2011; p. 5. Available online: https://onlinelibrary.wiley.com/doi/10.1002/esp.3366 (accessed on 7 February 2025).
18. Bangen, S.; Wheaton, J.; Bouwes, N.; Bouwes, B.; Jordan, C. A Methodological Intercomparison of Topographic Survey Techniques for Characterizing Wadeable Streams and Rivers. *Geomorphology* **2014**, *206*, 343–361. [CrossRef]
19. Berra, E.; Peppa, M.V. Advances and Challenges of UAV SFM MVS Photogrammetry and Remote Sensing: Short Review. *ISPRS-Int. Arch. Photogramm. Remote Sens. Spat. Inf. Sci.* **2020**, *XLII-3/W12-2020*, 267–272. [CrossRef]
20. Muhar, A. Three-Dimensional Modelling and Visualisation of Vegetation for Landscape Simulation. *Landsc. Urban Plan.* **2001**, *54*, 5–17. [CrossRef]
21. Wu, J.; Zhang, C.; Xue, T.; Freeman, B.; Tenenbaum, J. Learning a Probabilistic Latent Space of Object Shapes via 3D Generative-Adversarial Modeling. In *Proceedings of the Advances in Neural Information Processing Systems*; Lee, D., Sugiyama, M., Luxburg, U., Guyon, I., Garnett, R., Eds.; Curran Associates, Inc.: Red Hook, NY, USA, 2016; Volume 29.
22. Digumarti, S.T.; Nieto, J.; Cadena, C.; Siegwart, R.; Beardsley, P. Automatic Segmentation of Tree Structure From Point Cloud Data. *IEEE Robot. Autom. Lett.* **2018**, *3*, 3043–3050. [CrossRef]
23. Liu, Y.; Guo, J.; Benes, B.; Deussen, O.; Zhang, X.; Huang, H. TreePartNet: Neural Decomposition of Point Clouds for 3D Tree Reconstruction. *ACM Trans. Graph.* **2021**, *40*, 1–16. [CrossRef]
24. Oxman, R. Thinking Difference: Theories and Models of Parametric Design Thinking. *Des. Stud.* **2017**, *52*, 4–39. [CrossRef]
25. Favorskaya, M.N.; Jain, L.C. *Handbook on Advances in Remote Sensing and Geographic Information Systems: Paradigms and Applications in Forest Landscape Modeling*; Intelligent Systems Reference Library; Springer: Dordrecht, The Netherlands, 2017; Volume 122, ISBN 9783319523064.
26. Nogueira, P.A. Motion Capture Fundamentals A Critical and Comparative Analysis on Real-World Applications. 2012. Available online: https://www.semanticscholar.org/paper/Motion-Capture-Fundamentals-A-Critical-and-Analysis-Nogueira/051361b591b4f7d981d5a31f67ee353104c86119 (accessed on 20 February 2025).
27. Aseeri, S.; Marin, S.; Landers, R.; Interrante, V.; Rosenberg, E. Embodied Realistic Avatar System with Body Motions and Facial Expressions for Communication in Virtual Reality Applications. In Proceedings of the 2020 IEEE Conference on Virtual Reality and 3D User Interfaces Abstracts and Workshops (VRW), Atlanta, GA, USA, 22–26 March 2020; pp. 581–582.
28. Buck, L.; Chakraborty, S.; Bodenheimer, B. The Impact of Embodiment and Avatar Sizing on Personal Space in Immersive Virtual Environments. *IEEE Trans. Vis. Comput. Graph.* **2022**, *28*, 2102–2113. [CrossRef]
29. Smith, H.J.; Neff, M. Communication Behavior in Embodied Virtual Reality. In *CHI '18, Proceedings of the 2018 CHI Conference on Human Factors in Computing Systems, Montreal, QC, Canada, 21–26 April 2018*; Association for Computing Machinery: New York, NY, USA, 2018; pp. 1–12.
30. Perret, J.; Vander Poorten, E. Touching Virtual Reality: A Review of Haptic Gloves. In Proceedings of the ACTUATOR 2018; 16th International Conference on New Actuators, Bremen, Germany, 25–27 June 2018.
31. Verim, Ö.; Sen, O. Application of Reverse Engineering Method on Agricultural Machinery Parts. *Int. Adv. Res. Eng. J.* **2023**, *7*, 35–40. [CrossRef]
32. La Placa, S.; Doria, E. Digital Documentation and Fast Census for Monitoring the University's Built Heritage. *Int. Arch. Photogramm. Remote Sens. Spat. Inf. Sci.* **2024**, *XLVIII-2/W4-2024*, 271–278. [CrossRef]
33. Okura, F. 3D Modeling and Reconstruction of Plants and Trees: A Cross-Cutting Review across Computer Graphics, Vision, and Plant Phenotyping. *Breed Sci.* **2022**, *72*, 31–47. [CrossRef] [PubMed]

34. Park, Y.; Lee, J.; Bae, J. Development of a Wearable Sensing Glove for Measuring the Motion of Fingers Using Linear Potentiometers and Flexible Wires. *IEEE Trans. Ind. Inform.* **2015**, *11*, 198–206. [CrossRef]
35. Mimic Pro—VR Body IK System for Unreal Engine. Available online: https://jakeplayable.gumroad.com/l/MimicPro (accessed on 20 February 2025).
36. MetaHuman | Realistic Person Creator—Unreal Engine. Available online: https://www.unrealengine.com/en-US/metahuman (accessed on 20 February 2025).
37. OptiTrack—Motion Capture Systems. Available online: https://www.optitrack.com/ (accessed on 20 February 2025).
38. Robinson, O.C. Sampling in Interview-Based Qualitative Research: A Theoretical and Practical Guide. *Qual. Res. Psychol.* **2014**, *11*, 25–41. [CrossRef]
39. Giulietti, N.; Todesca, D.; Carnevale, M.; Giberti, H. A Real-Time Human Pose Measurement System for Human-In-The-Loop Dynamic Simulators. *IEEE Access* **2025**, *13*, 24954–24969. [CrossRef]
40. Lei, X.; Chang, M.; Lu, Y.; Zhao, T. A review on growth modelling and visualization for virtual trees. *Sci. Silvae Sin.* **2006**, *42*, 123–131. [CrossRef]
41. Jinasena, K.; Sonnadara, U. A Dynamic Simulation Model for Tree Development. In Proceedings of the International Conference on Mathematical Modeling, Colombo, Sri Lanka, 10–21 March 2014.

designs

MDPI

Article

Development of Impact Factors Reverse Analysis Method for Software Complexes' Support Automation

Andrii Pukach *, Vasyl Teslyuk, Nataliia Lysa and Liubomyr Sikora

Department of Automated Control Systems, Computer Science and Information Technologies Institute, Lviv Polytechnic National University, 79000 Lviv, Ukraine; vasyl.m.teslyuk@lpnu.ua (V.T.); nataliia.k.lysa@lpnu.ua (N.L.); liubomyr.s.sikora@lpnu.ua (L.S.)

* Correspondence: andriipukach@gmail.com

Abstract: This research represents a corresponding and developed specialized impact factors reverse analysis method for software complexes' support automation; it is intended for the analysis of impact factors affecting the supported software's (or processes of its comprehensive support) subjective perception results, as one of the constituent tasks of the more complex problem of software complexes' support automation. The developed method provides the possibility to restore certain boundaries of impact factors by classifying the multilayer perceptron's hidden layer neurons and calculating the probability coefficients of the belonging of these neurons to the corresponding specific pre-determined impact factors. The problem of determining the influence of impact factors on the subjective perception of the object of support (the supported software or the processes of its comprehensive support) by the relevant subjects (interacting with this object, providing and implementing its support) was resolved through the approach developed and proposed by the authors in the scope of this research. A key feature of the proposed approach is to assign the neurons of the hidden layers (of the multilayer perceptron type of artificial neural networks) functional–semantic meaning(s), which they have been deprived of a priori, performing (before this) by default an exclusively operational (calculation) function mainly for the correctness of the training and functioning of the multilayer perceptron itself. The potential of the developed method allows us to apply it for solving a huge number of applied practical tasks, such as the one provided in the scope of this research, which is as follows: a practical task of the support team members' portrait determination, followed by a further search (detection) of the interchangeable members of this support team to ensure the possibility of quick transfer of the stack of tickets (which are in the middle of the active resolution process) between these members.

Keywords: automation; intellectualization; support; software complex; customer support; subjective perception; impact factors coefficients; artificial intelligence; artificial neural network; multilayer perceptron

1. Introduction

Nowadays, the problem of software complexes' support automation remains one of the urgent scientific and applied problems. One of the main difficulties in solving this problem of software complexes' support automation is the difference in the resulting vision of the supported object (or process) by the various subjects interacting with it, which, directly or indirectly, participate in the processes of operation and the lifecycle of the supported software complex and influence its comprehensive support accordingly. All these differences in the resulting vision of the supported object (or process) are caused by the presence of various subjective impact factors exerting their influence on both the vision of the software's support processes and procedures and the vision of the supported software complex itself. Therefore, analysis of these impact factors (with the possibility of their further representation in any form acceptable for automation) is an extremely important and necessary task in solving the more global problem of software complexes' support automation.

Nowadays, the main (basic or key) directions of automation in the context of a comprehensive support for software complexes (not only a "classic" customer support, but a comprehensive support) are the following:

- Software testing automation;
- Automation of call centers;
- Chatbot automation;
- Automation of service desk requests (both from customers and internally);
- DevOps automation;
- Decision-making automation.

Each of these directions partially (in its own responsibility area) resolves relevant aspects of automation for software complexes' support. The main task of software testing [1] automation is the replacing of routine, often repeated, clearly structured, and algorithmized activities of people (employees of software testing teams and departments) with appropriate means and automation tools in the form of various scripts, drivers, plugins, modules, and other relevant approaches of software testing automation. The main purpose of call center [2] automation is the opportunity to improve and facilitate the work of people—call center operators—as well as to introduce robotic and/or virtual call center operators that are able to process calls from customers and users fully or partially. The main feature of chatbot [3,4] automation is the possibility of the implementation of the primary communication (with clients and users) function using, for example, natural language processing (NLP) [5] or generative pre-trained transformer (GPT) [6] approaches. The main functions of service desk requests' automation are, for example, the registration and processing of external requests (tickets) from customer users of client companies [7,8] and various internal requests (tickets) from users within the development companies [9,10], as well as more complex functionality systems—like the one described in [11], which can automatically process crash GitHub logs, analyze them, and register relevant reports—and other similar complex systems. The main task of DevOps [12–14] automation is the most effective combination of the software complex's development stages and is related to this software complex's stages of IT operations, and the whole of this approach should be manifested in the provision of continuous integration and continuous development (CI/CD) processes [15,16]. The main function of decision-making automation [17–19] is providing an auxiliary function for the staff (from regular employee up to middle and higher levels of management personnel) regarding selection and making certain decisions, depending on the input and output parameters, as well as the relevant criteria.

The developed method fills the gap, consisting in the lack of consideration of the subjective perception of the support object (software product) in the context of the represented existing mechanisms of its comprehensive support automation, thereby ensuring additional intellectualization of this automation.

The rest of this paper is structured as follows:

- A literature review is in Section 2;
- A method development, including its three main components, namely a primary generalized model, an algorithm stage, and an appropriate mathematical model, is given in Section 3;

- The main results, obtained as a result of modeling the developed method on a particular example case, resolved from the very beginning to the very end, with a detailed indication of the intermediate results after each separate step of the developed method's algorithm, including all necessary information about the used means and tools needed for method modeling and approbation, as well as a comparison with the outcomes of the existing similar approaches, are presented and discussed in Section 4;
- The conclusions are in Section 5.

2. Literature Review

The research dynamics of each direction of software complexes' comprehensive support automation is considered in the current chapter of this paper. For example, nowadays software testing automation is considered to be an absolute norm, mainstream, and even a mandatory common trend. However, not so long ago, according to the data presented in [20], only 26% of all test-cases inside the projects among studied companies were covered with the usage of software testing automation approaches. However, only 10 years later, for example, according to the information presented in [21], there is already software testing automation which is based on the technologies and approaches of artificial intelligence [22], which, in fact, is a transition from the stage of automated software testing to the next phase: fully automatic testing. Unfortunately, in the context of software testing automation, the dominant attention is paid primarily to the automation of technical and technological processes (generalizing and depersonalizing them), which only to a certain extent covers some part of the activities of software testing specialists (or so-called "software testers"), while a significant part of the non-technological aspects of software testing (including impact factors that form a "picture" of the personalized perception of the tested software by each of these specialists, identifying, in such a way, the individuality of this specialist) remains out of consideration and/or is not taken into account.

At the same time, automation is also observed in the first line of software complexes' customer support represented by call centers, starting from the call routing automation presented in [23] and continuing with the significant achievements in the field of automatic speech recognition presented in [24] and various speech recognition training techniques like the one presented, for example, in [25], as well as various automated assistants similar to the one presented in [26] or automated translators based on the example given in [27]. The main feature of call centers is the communication factor, which plays a dominant role in the context of intersubjective interaction between the support subjects of any researched software complex. At the same time, the represented research (in the context of call center automation) does not pay attention to the impact factors which lead to the subjective perception of the object of communication interaction (which is, in this case, the supported software complex or the processes related to its support) by these subject communicators.

In addition, remarkable results have also been achieved in the field of chatbot automation, starting from agents with modular architecture, as presented in [28], and continuing with systems for constructing the sentential forms similar to the one presented in [29] and the automated voice control systems similar to the one presented in [30]. The research in [31] lists some of the most popular chatbot platforms, such as Siri, Watson, Google Now, and Cortana, which actually have set the mass market mainstream trend for such types of virtual assistants and chatbots with built-in artificial intelligence technologies inside. The research in [32] lists the main methods of chatbot learning, such as DNN, CNN, RNN, LSTM, decision tree, random forest, SVM, and BiLSTM. At the same time, in practice, a lot of users still experience considerable discomfort when interacting with chatbots and continue to prefer "live" communication (in any of its verbal forms) with relevant customer support specialists (which are living people). One of the reasons for such discomfort is, among other things, the lack of analysis of human factors affecting the interaction between subjects—living people—as well as a lack of consideration of the subjective perception of the support object (the supported software complex or the processes related to its support) by each specific individual user while contacting the support service.

Another area of automation during the development and operations of software complexes is automation in the scope of the registration and initial processing of service desk requests, both "external" (driven from outside by the customers) and "internal" (initiated inside the development company). For example, the research in [33] considers the approach of automated sorting and the assignment of requests related to system defects. In the research in [34], the incident categorization automation process is considered, while in the research in [35,36], system defect categorization automation approaches are presented, which use latent Dirichlet allocation and fuzzy logic, accordingly. The automated assignment of electronic e-mail requests (tickets) sent by customer users and addressed to the appropriate service support team of the developed software complex, using a set of machine learning methods, is considered in the research in [37]. At the same time, as in previously considered cases of software complexes' support automation, within the scope of the represented research on the automation of registration and initial processing of service desk requests, there is no relevant analysis of the factor(s) of the subjective perception (by each of these customers' users) of the object of support, which, in turn, leads to the lack of appropriate corrections necessary to ensure the reliability of information received from these end users.

DevOps automation direction (of comprehensive software support) is also actively developing. One of the most basic and fundamental works in this direction is the research in [38]. A holistic DevOps knowledge management methodology is presented in [39]. According to results presented in the research in [40], 80% of software practitioners (who participated in this research) reported that software building is actually the easiest part of the automation, followed by software packaging and deployment (51.2% and 43.9%, accordingly). The research in [41] additionally emphasizes the importance of DevOps implementation into the development life cycle of each software complex, as well as the fact that DevOps involves not only changes in processes, but also significant changes in the software complexes' development method(s). CI/CD automation (the combined practice of continuous integration and continuous delivery) is presented in [42] as one of the most essential components of the present DevOps methodology. At the same time, in the scope of the above studies in the context of DevOps automation, there is also no analysis of the factors influencing the subjective perception of the object of automation (the supported software product, as well as the relevant processes for its comprehensive support and related operational processes), which (factors), nevertheless, form both the strategic and tactical vision of these operational processes by each of the subjects, which directly support, provide, perform, and implement these processes.

Simultaneously with all the above-mentioned areas of automation, the direction of decision-making automation is actively developing as well. In particular, the research in [43] presents a proactive support decision-making framework called RADAR, which is based on both research fields of automated planning and artificial intelligence. RADAR helps the decision-making person to achieve his/her goals by providing warnings and suggestions about possible plan flaws and resource constraints. This was achieved by generating and analyzing the benchmarks that any successful plan must meet before achieving the goals. Also, the approach proposed in the RADAR framework is fully consistent with the concept of naturalistic decision making. Another example, which is also probably one of the most popular kinds of automated decision-making usage, is the various smart-systems, such as "smart home", which is presented in [44]. However, this direction is also actively developing in the field of software development itself and is considered, for example in the research [45], where the approach of real user feedback-based automatic classification of software products' non-functional requirements is presented, and [46], where the automatic classification of functional and non-functional requirements with the usage of supervised machine learning is presented. The research in [47] represents a comparison of data-driven decision making (DDDM) with automated data-driven model-based decision making (MBDM). In fact, nowadays automated decision-making systems have long left the boundaries of both smart systems and software development and confidently entered the mass market in its various forms. According to the results presented in the research in [48], the European Consumer Organization clearly analyzes the increase in automated decision making, which is based on appropriate algorithms for commercial transactions, and its impact on the functionality of consumer markets and societies. At the same time, the research in [49,50] raises issues caused by the rapidly growing influence of automated decision-making systems and the appropriate challenges and risks caused by these systems. But even despite these risks, interest in these systems continues to grow further; so, they continue to develop and evolve accordingly. In general, decision making is a highly subjectivized process, as each subject (on which the adoption of the relevant decision depends) justifies it in its own personalized way, relying, first of all, on one's subjective vision and perception of the object of support (the supported software complex or the processes related to its support), which (this subjective vision) is shaped by relevant impact factors that, unfortunately, are also not taken into account in the context of the represented studies and research carried out in this direction of decision-making automation in the context of software complexes' support automation.

So, as we can see from the conducted review and analysis, automation has already deeply penetrated inside all key stages of the software complexes' development and operation—starting with their testing and ending with automatic analysis of real feedback left by users (living people) of these software complexes.

Unfortunately, lack of attention to the subjectivization of the support object's (supported software products or processes related to their comprehensive support) perception is a common problem (gap) in all the reviewed and presented research given above, which is aimed at automation of the component processes of the software products' life cycle. In turn, this leads to a depreciation of the impact of the subjectivization of the comprehensive support objects' perception by those subjects (e.g., personnel) who directly implement and provide such comprehensive support fully or particularly.

That is why, within the scope of the current research, a more global approach for the study of software complexes' comprehensive support automation is proposed by the authors: in this research we actually raise up to the next (higher) level, which connects all the described manifestations of automation—from testing to decision making. And at this next higher level, we are mainly interested in the problem of various impact factors having an influence on the representation results of the supported software complexes (or processes) by all the appropriate participants (subjects of interaction) of this support—starting from the testers, continuing with the customer (and/or product) support specialists, and ending with real users at the customer(s) side. Undoubtedly, each of these subjects, directly or indirectly interacting with the supported software complexes, consciously or subconsciously comes across a set of both universal and individual impact factors, which, one way or another, at different degrees of influence, definitely affects their subjective resulting perception of this supported software complex or its support processes. That is why (in context of this research) we are so interested in these impact factors and the possibility of their further research, analysis, differentiation, calculation, and quantitative assessment, as well as the determination of the degrees, levels, and limits of their influence, etc.

Therefore, the main goal of this research is to develop an impact factors reverse analysis method for software complexes' support automation; it is based on a specially developed algorithm, as well as the appropriate mathematics, software, and additional required models and components presented in this research.

3. Method Development

The developed method consists of three main components, namely:

- Primary generalized model;
- Algorithm stages;
- Mathematical model.

The main purpose of the developed and proposed primary generalized model is to provide a simple, understandable, and at the same time universal way of formalizing the representation of the support objects' perception subjectivization.

The main purpose of the developed and proposed algorithm stages is to provide a clear algorithm to enable further automation and software implementation (coding), as well as computer modeling of the developed method.

The main purpose of the developed and proposed mathematical model is to ensure a fundamental mathematical basis for the developed method, which will provide additional opportunities both for mathematical modeling (as an integral part of computer modeling and programming) and for the further use of all existing mathematical tools in order to obtain any proposed, necessary, possible improvements and/or modifications.

Let us start from the consideration that any supported software complex has a number of input characteristics I1. . .In, as well as a number of output characteristics (as a result of its subjective representation) O1. . .Om (where "m" does not necessarily have to be equal to "n"). Input characteristics are the characteristics which are close (as much as possible) to the objective (real or true) ones, while the output characteristics are the results of the subjective perception of the supported software complex (or its support processes) by appropriate subject(s) directly or indirectly interacting with this supported software complex.

In ideal ("vacuum") conditions, those output characteristics are considered as the results of some previously determined and known referential functional dependence from the input characteristics, without any distortions.

However, in real conditions, each input characteristic undergoes certain distortions and "deformations", and, as a result, all of them are transformed into the corresponding resulting (subjective) output characteristics, which form the final perceptional subjective resulting interpretation of the supported software complex (or processes of its support) by the appropriate subject(s) interacting with it. That is why, finally, each subject of interaction with the supported software complex perceives it in its own individualistic (personalized) way/manner.

The main levers of influence (or tools of deformation) in this process of transformation of the input characteristics into output results are the impact factors. There could be any number (and variety) of them, and each of these impact factors affects each input characteristic of the supported software complex (or the processes of its comprehensive support) in its own way.

3.1. *Primary Generalized Model*

The above-described process can be simply interpreted by the developed and proposed (in the scope of this research) primary generalized model of the supported object's subjective perception, a visualized example of which is provided below in Figure 1.

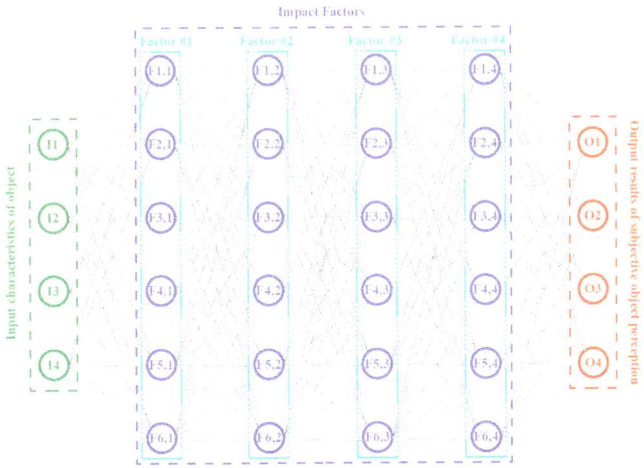

Figure 1. Example of built primary generalized model of the supported object's subjective perception.

Depending on the set goals and objectives, the "supported object" could be represented by the supported software complex itself, as well as the processes of its comprehensive support.

According to the developed and proposed generalized model (of the supported object's subjective perception) each researched supported object should have the following:

- A set of input characteristics;
- A set of sequential impact factors, each of which can influence each of the input characteristics (many-to-many relationship) correspondingly in its own way, gradually (or cascadingly) transforming them into a relevant set of resulting output characteristics;
- A set of resulting output characteristics.

As can be seen in the given graphic representation provided below in Figure 1, the proposed primary generalized model visually reminds the well-known multilayer perceptron (MP), where

- A set of input characteristics is the MP input layer's neurons;
- A set of impact factors is the MP hidden layers' neurons;
- A set of resulting output characteristics is the MP output layer's neurons (with the results of the MP's operation).

However, the multilayer perceptron in its primary basic concept does not assume the presence of any meaningful or cognitive values for hidden layers' neurons. According to the well-known classical concept of a multilayer perceptron, all the hidden layers' neurons (inside any multilayer perceptron) are needed only for purely mathematical calculations and the corresponding training function of the MP. Therefore, it cannot just be assumed that each hidden layer of neurons (in the corresponding constructed multilayer perceptron model) is a separate impact factor, because the structure of a multilayer perceptron itself does not meet such comprehension. Thus, each time when appropriate multilayer perceptron models are implemented for proposed primary generalized models (of the supported object's subjective perception), the boundaries of the impact factors (presented in these primary models) are actually lost (or blurred) in the appropriate MP models.

So, this is actually the problem which should be resolved, and that is actually why we are so interested in the possibility of identifying (or restoring) these certain lost boundaries of the impact factors influencing the supported object(s) while implementing and using the corresponding constructed multilayer perceptron models for them.

Thus, the main goal of this research is the investigation of the possibility of introducing additional meaningful (or cognitive) values (labels) for the neurons of the hidden layers of the corresponding constructed multilayer perceptron models, in accordance with the determined impact factors of the researched supported object(s).

Figure 2 below describes an example of an appropriate multilayer perceptron model corresponding to the primary generalized model of the supported object's subjective perception (which was already provided in Figure 1 above).

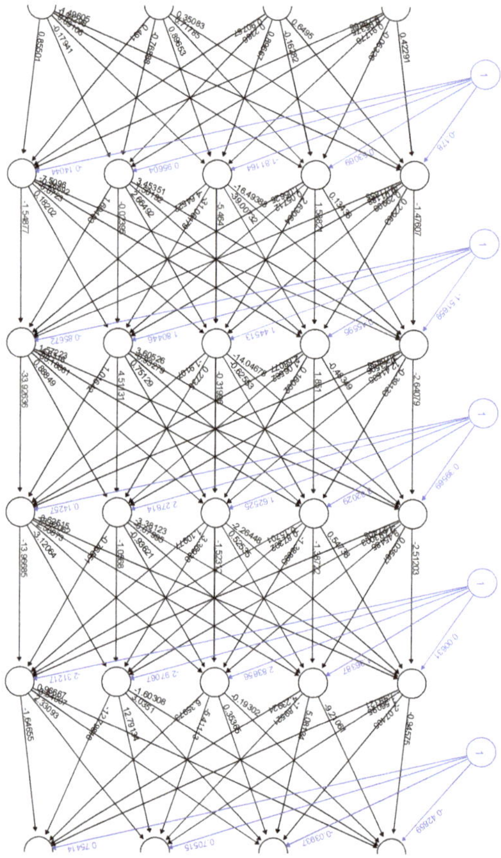

Figure 2. Example of a multilayer perceptron model corresponding to the primary generalized model of the supported object's subjective perception.

As presented in Figure 2, the corresponding multilayer perceptron model is an example of a real model, one of those which have been used for modeling while working on the issue(s) investigated by authors in the scope of this research. Thus, for better clarity and understanding, all further scientific and practical materials, as well as the investigation results, presented in this research are linked to this particular specific research example (in order to ensure the possibility of tracking the entire research chain, from its very beginning to its complete end).

Therefore, the corresponding developed model of the multilayer perceptron, as described in Figure 2, consists of the following:

- Input layer neurons I0–I3;

- Four hidden layers of neurons (five neurons on each hidden layer, + an additional one bias neuron on each hidden layer);
- Output layer neurons O0–O3.

3.2. Algorithm Stages

Let us move on to a consideration of the corresponding developed algorithm (and its stages) of the impact factors reverse analysis method for the software complexes' comprehensive support automation.

The very initial stages of the developed impact factors reverse analysis method are as follows:

Stage 1. Development of a primary generalized model of the supported object's subjective perception (in an arbitrary form but necessarily including strict indication(s) for all input characteristics and output resulting characteristics, as well as all impact factors).

Stage 2. Development of an appropriate multilayer perceptron model fully corresponding to the previously developed primary generalized model of the supported object's subjective perception. At this stage, an extremely important and mandatory requirement is ensuring a unique relationship between each impact factor and the corresponding hidden layer(s) of the multilayer perceptron (in the case that their relationship type is different, it is just "one-to-one").

Stage 3. Training of the developed multilayer perceptron model, performed on a relevant, appropriately prepared, training dataset(s).

For better understanding, Figure 3 below additionally provides appropriate graphical representation of the full process described in all the provided stages of the developed algorithm.

The next stages are the main stages for the analysis of all the defined impact factors present in supported object's subjective perception model, namely:

Stage 4. Preparation or extraction of data for each individual isolated impact factor.

The ideal conditions for this stage (4) are the following:

- The absolute impact of only one (currently considered/investigated) impact factor;
- The zero-level impact of the rest of the impact factors (except the one currently considered/investigated);
- An absolutely clear understanding of how each individual impact factor affects the model in isolation (without the influence of the rest of the impact factors);
- Insight into what will be the input and output data of the model under the isolated influence of only one individual (currently considered/investigated) impact factor.

* Stage 5 (which is an optional stage). Consideration of the need for further additional training (or complete re-training) of separate MP model(s) for individual isolated impact factors. However, this stage (5) is extremely time-consuming, complex, and resource-consuming, and requires a separate consideration (perhaps in separate dedicated research). Therefore, in this research, we use a slightly different approach, which involves using the multilayer perceptron model, already previously obtained in stage 3.

Stage 6. Testing of the trained MP model on each of the datasets obtained in the previous stage (4) for each separate impact factor(s).

Stage 7. Cleaning obtained testing results—as we are only interested in the correct results, while incorrect ones should just be discarded.

Stage 8. In this stage, the corresponding maximal weight reverse chains are built based on each testing result from the previous stage. Each reverse chain is built starting from the correct active MP output layer's neuron, back through all the MP's hidden layers, and finalizing at the corresponding MP input layer's neuron).

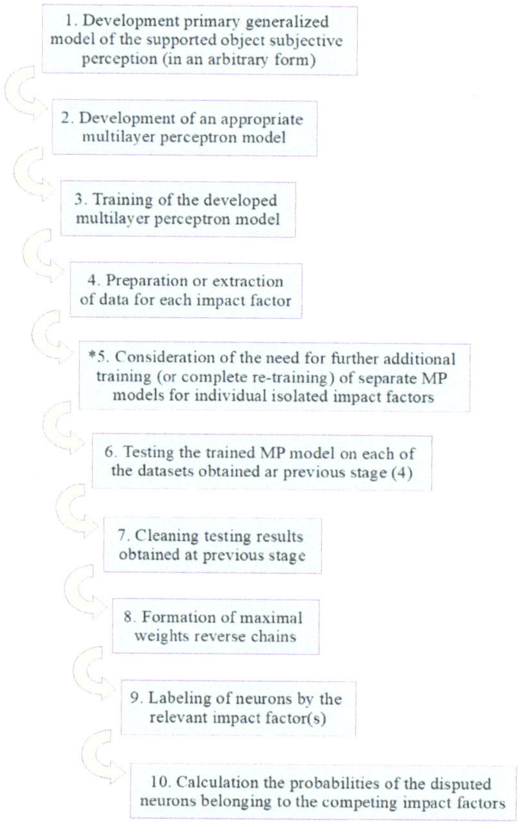

Figure 3. Stages of the algorithm for impact factors reverse analysis method (where *5 is an optional stage).

Stage 9. Labeling of each of the neurons for each of the reverse chains (obtained in previous stage) in accordance with their belonging to the currently considered/investigated specific impact factor.

Stage 10. Calculation of the probabilities for all neurons (obtained in the previous stage) regarding their belonging to those impact factors to which they have been appropriately labeled in the previous stage. * Here, in this stage, it is also important to understand one important thing—the same neuron can simultaneously belong to the reverse chains of several different impact factors (and such a situation is quite normal). Therefore, the main task of this stage is to calculate the probabilities by which each such specific neuron belongs to each of its impact factors. For such neurons, a separate term is proposed and introduced for the developed method in the scope of this research: "disputed neurons". And for such impact factors that include disputed neurons, the developed method also proposes and introduces a corresponding separate term: "competing impact factors".

This stage actually finalizes the description of the developed algorithm.

3.3. Mathematical Model

So, let us move on to the detailed consideration of the developed mathematical component (model) of the impact factors reverse analysis method, according to which the calculation of probabilities (of neurons belonging to the relevant impact factors) is performed in several steps, as provided below. In the first step, the percentage ratio is calculated for the frequency of occurrence of each unique reverse chain (within their impact factors) relative to the total number of all cases (within their impact factors). The appropriate calculations for this first step are performed using the expression given below:

$$\mathrm{FoA_{local}}[i][j] = \frac{\mathrm{count}(\mathrm{URC}[j] \in F[i])}{\mathrm{count}(\mathrm{URC}[\mathrm{all}] \in F[i])}, \tag{1}$$

where *FoA_local*[*i*][*j*]—frequency of appearance of the [*j*]-th unique reverse chain within [*i*]-th impact factor; *count*(*URC*[*j*]∈*F*[*i*])—the number of appearances of the [*j*]-th unique reverse chain (*URC*) within the [*i*]-th impact factor; *count*(*URC*[*all*]∈*F*[*i*])—the total number of appearances of all unique reverse chains existing within the [*i*]-th impact factor.

In the second step, the percentage ratio is calculated for the frequency of occurrence of each unique reverse chain (beyond its impact factors or, in other words, in the general set of cases, a "global dataset") relative to the total number of all cases (within this general set of cases or the same "global dataset"). Appropriate calculations for this second step are performed using the expression given below:

$$\mathrm{FoA_{global}}[i][j] = \mathrm{FoA_{local}}[i][j] \cdot \mathrm{FI}[i], \tag{2}$$

where *FoA_global*[*i*][*j*]—the frequency of appearance of the [*j*]-th unique reverse chain of the [*i*]-th impact factor within the global dataset; *FI*[*i*]—index of the [*i*]-th impact factor (factor index—*FI*).

In turn, the impact factor index is calculated using the following expression provided below:

$$\mathrm{FI}[i] = \frac{\mathrm{count}(F[i] \in [\mathrm{General}])}{\mathrm{count}(F[\mathrm{all}] \in [\mathrm{General}])}, \tag{3}$$

where *count*(*F*[*i*]∈[*General*])—the number of appearances of the [*i*]-th impact factor within the global dataset; *count*(*F*[*all*]∈[*General*])—the total amount of all cases (for all impact factors) of the same global dataset.

In the third step, the values of the probabilities that the disputed neurons belong to the appropriate competing impact factors are calculated by the following expression:

$$\mathrm{PoDNbtCIF}[i][k] = \frac{\sum \mathrm{FoA_{global}}[i][j](\mathrm{Neuron}[k] \in \mathrm{URC}[j])}{\sum \mathrm{FoA_{global}}[i][j](\mathrm{Neuron}[k] \in \mathrm{URC}[\mathrm{all}])}, \tag{4}$$

where *PoDNbtCIF*[*i*][*k*]—probability that the disputed neuron [*k*] belongs to the competing impact factor [*i*]; *FoA_global*[*i*][*j*](*Neuron*[*k*]∈*URC*[*j*])—the frequency of appearance of the [*j*]-th unique reverse chain of the [*i*]-th impact factor within the global dataset (for those unique *URC*[*j*] reverse chains which include the disputed neuron *Neuron*[*k*]); *FoA_global*[*i*][*j*](*Neuron*[*k*]∈*URC*[*all*])—the frequency of appearance of the [*j*]-th unique reverse chain of the [*i*]-th impact factor within the global dataset (for all unique reverse chains of all impact factors, e.g., *URC*[*all*], which includes the disputed neuron *Neuron*[*k*]).

Therefore, in a global sense, the essence of the concept of the developed (provided above) expressions, representing the appropriate mathematical component/model of the impact factors reverse analysis method, is the following:

- First—the probability of the appearance of each unique reverse chain (with its binding to each of the relevant impact factors, of course) in the global dataset should be estimated;
- then—the probability of the appearance of each disputed neuron (in the scope of each unique reverse chain, to which this neuron belongs, and with the binding of each of these chains to the related impact factor) is estimated, and this provides a specific link between this particular disputed neuron and that particular competing impact factor.

Or, in other words (for some additional, or maybe better, understanding), we have a list of neurons (all of them, including disputed ones) and a list of unique reverse chains (which include these disputed neurons), and these unique reverse chains, in turn, are also part of the impact factors (all of them, including competing ones).

So, we could

- Simply count the number of unique reverse chains (which include some currently investigated disputed neurons) of one single specific impact factor (to which these unique reverse chains belong);
- Then divide this number by the total number of all unique reverse chains (which includes this currently investigated disputed neuron), regardless of the binding of these unique reverse chains to a specific impact factor, but instead just considering absolutely all unique reverse chains (which include this currently investigated disputed neuron) for all impact factors.

However, such an approach will be correct only if

- The probability of the appearance of each impact factor in the global dataset is equal for all impact factors;
- The probability of the appearance of each unique reverse chain in the global dataset is equal for all unique reverse chains.

But the real situation in practice is completely different, and there are specific numerical values of both the probabilities of the appearance of each impact factor in the global dataset and the probabilities of the appearance of each unique reverse chain in the same global dataset.

That is why when calculating the ratio

- of the number of unique reverse chains (which include the currently investigated neuron) of one single specific impact factor (to which these unique reverse chains belong)
- to the total number of all unique reverse chains (which include this currently investigated neuron) for all impact factors

we already take into account such pre-calculated numerical values as

- the probability of the appearance of each impact factor in the global dataset,
- as well as the probabilities of the appearance of each unique reverse chain in the same global dataset.

That is actually what all these developed and provided expressions, representing the appropriate mathematical component/model of the developed impact factors reverse analysis method, really mean.

4. Results and Discussion

This section of the paper intends to present the developed method in as much detail as possible through its practical step-by-step use in a particular example study case.

In addition, for better understanding, this section is structured into subsections, as follows:

- Section 4.1. Working with the developed multilayer perceptron model;
- Section 4.2. Formation of maximal weight reverse chains with the following labeling of neurons by the relevant impact factor(s);
- Section 4.3. Calculating the probabilities of the disputed neurons belonging to the competing impact factors;
- Section 4.4. Example of a practical task resolution by the developed method;

- Section 4.5. Comparison with outcomes of existing similar approaches.

Let us move forward with each of the subsections of the current section.

4.1. Working with the Developed Multilayer Perceptron Model

Thus, the appropriate multilayer perceptron model (for the researched experimental model, presented in this research as an example case) was developed and trained in the R 3.6.3 environment using the "neuralnet" package and the LeakyReLu activation function.

The example code of the corresponding developed software (for this part of the investigation) in the "R" programming language is given below in Table 1:

Table 1. The code of the corresponding developed software in the R programming language.

```
support <- read.csv(":/dataset.csv")
str(support)
support_train <- support[1:1000,]
support_test <- support[1001:1100,]
library(neuralnet)
leakyrelu <- function(x) {ifelse(x < 0, 0.01*x, x)}
support_model <- neuralnet(O1 + O2 + O3 + O4 ~ I1 + I2 + I3 + I4,
        data = support_train,
        hidden = c(5,5,5,5),
        #using default algorithm "rprop+"
        act.fct = leakyrelu,
        linear.output = FALSE,
        threshold = 1, stepmax = 1,000,000)
plot(support_model)
support_model$weights
```

As the input data for the training and testing of the developed multilayer perceptron model, data from a CSV-file are used (just as an example presented in this paper, because the developed method itself does not set any limitations regarding data representation formats). Before that, all these data (present in this CSV-file) were obtained as a result of the execution of additional software developed in the Python 3.12 programming language using the Thonny 4.1.4 IDE, which was used as a data generation engine. At the same time, all these data in the CSV-file were additionally depersonalized (in particular, they were deprived of any personalization, sensitivity, or semantic load; instead, only "pure numbers" were left) and normalized (input neuron values: [0.1–0.9]; output neuron values: 0 or 1, where only one of four output neurons can be 1, while the rest must be 0). Depersonalization and normalization of these data were also performed by additional software developed in Python. An example of the data from a CSV-file is shown in Figure 4 below.

	A	B	C	D	E	F	G	H	I
1	I1	I2	I3	I4	Oideal	O1	O2	O3	O4
2	0.3	0.3	0.3	0.2	0.1	1	0	0	0
3	0.4	0.9	0.4	0.6	0.5	0	0	1	0
4	0.8	0.8	0.7	0.2	0.5	0	0	1	0
5	0.4	0.9	0.4	0.4	0.5	0	0	1	0
6	0.8	0.3	0.3	0.4	0.3	0	1	0	0
7	0.6	0.6	0.5	0.8	0.5	0	0	1	0
8	0.5	0.6	0.5	0.3	0.3	0	1	0	0
9	0.6	0.4	0.8	0.5	0.5	0	0	1	0
10	0.8	0.8	0.4	0.3	0.5	0	0	1	0
11	0.4	0.5	0.9	0.4	0.5	0	0	1	0
12	0.2	0.6	0.5	0.8	0.5	0	0	1	0
13	0.6	0.4	0.3	0.7	0.3	0	1	0	0
14	0.9	0.1	0.7	0.2	0.3	0	1	0	0
15	0.6	0.4	0.6	0.7	0.5	0	0	1	0
16	0.7	0.6	0.4	0.3	0.3	0	1	0	0
17	0.6	0.9	0.9	0.8	0.7	0	0	0	1
18	0.2	0.1	0.1	0.3	0.1	1	0	0	0
19	0.2	0.4	0.4	0.9	0.3	0	1	0	0

Figure 4. CSV-file data format, for training and testing of the developed multilayer perceptron.

The accuracy of the developed MP model trained on the testing dataset was achieved and reached up to 96.04%; it could even grow more, but it was specifically decided to not train the model above this level, because for most real cases such accuracy is rarely achieved. So, the main goal was just to make sure that the developed MP model was well trained and could show acceptable, really good, and pretty accurate testing results.

Returning back to our example, in Figure 2 an appropriate trained multilayer perceptron model was obtained corresponding to the primary generalized model of the supported object's subjective perception. That is, in fact, completed in stage 3 of the developed algorithm for the impact factors reverse analysis method.

So, the next stage (4) is actually a preparation (or extraction) of the data for each individual isolated impact factor. In this stage, each of the impact factors is considered in isolation; so, here we must clearly realize and understand what input and output data of the model we will obtain when isolating the influence (performed on our supported software complex or the processes of its support) of each individual impact factor taken from the set of all available impact factors. At the same time, the influence of all remaining impact factors (except the single currently investigated factor) must be reduced to zero (0).

For example, in our specific example case, presented in this research, an appropriate strategy was chosen, according to which each separately taken impact factor (out of four existing in our example model) is responsible for activation of only one specific resulting output neuron (out of four existing in our MP model). (* But this is just an example; in other cases, the strategies of the isolated influence of the impact factors could be completely different; however, we consider such a specific example primarily for better simplicity, clarity, and understanding of the whole of the material represented in this research as an example).

By means of appropriate additional developed Python software, the corresponding CSV-files were prepared with data for testing the isolated influence of each individual impact factor of the developed and considered example model.

The next stage (6) is responsible for the testing of the trained MP model on each of the datasets obtained in the previous stage (4) for each separate impact factor(s). In other words, it means that we should use the datasets from each separate prepared CSV-file (for each separate corresponding individual isolated impact factor) for testing the same trained multilayer perceptron model obtained in stage (3).

In the next stage (7), the results (obtained in the previous stage of testing for each individual isolated impact factor) should be cleared, leaving only the correct results; that is, those where (at the output of our multilayer perceptron model) we obtained the actual correct result which fully corresponded with the expected result present in the appropriate CSV-file.

4.2. Formation of Maximal Weight Reverse Chains with Following Labeling of Neurons by the Relevant Impact Factor(s)

Then, in the next stage (8), we can finally form the corresponding reverse chains of the maximal weights, starting from the correct active output layer's neuron, continuing back through all the hidden layers, and finalizing by the corresponding input layer's neuron. For this purpose, in the scope of the current research, the corresponding additional dedicated software was developed in Python as well. An output results example, obtained by this developed Python software, is presented below in Figure 5.

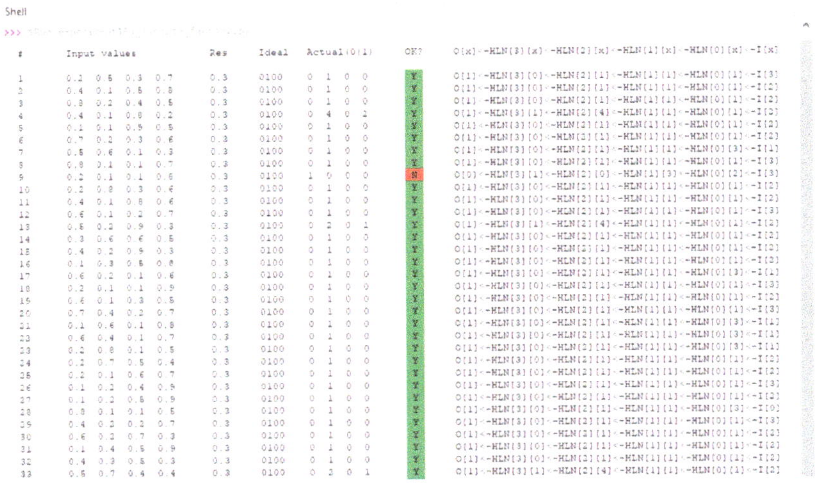

Figure 5. An output results example obtained by the developed software dedicated for implementation of stage 8 of the algorithm for impact factors reverse analysis method.

As the operation result of this dedicated developed Python software, the following results were obtained, as presented in Table 2 below.

Table 2. Obtained reverse chains of neurons for related impact factor(s).

Obtained Reverse Chains of Neurons for Related Impact Factor	Related Impact Factor
O[0] < -HLN[3][1] < -HLN[2][0] < -HLN[1][3] < -HLN[0][2] < -I[2]	Impact factor 1
O[0] < -HLN[3][1] < -HLN[2][0] < -HLN[1][3] < -HLN[0][2] < -I[3]	
O[0] < -HLN[3][1] < -HLN[2][0] < -HLN[1][3] < -HLN[0][2] < -I[1]	
O[0] < -HLN[3][1] < -HLN[2][0] < -HLN[1][3] < -HLN[0][2] < -I[0]	
O[0] < -HLN[3][1] < -HLN[2][1] < -HLN[1][1] < -HLN[0][3] < -I[1]	
O[0] < -HLN[3][1] < -HLN[2][4] < -HLN[1][1] < -HLN[0][1] < -I[2]	
O[1] < -HLN[3][0] < -HLN[2][1] < -HLN[1][1] < -HLN[0][1] < -I[3]	Impact factor 2
O[1] < -HLN[3][1] < -HLN[2][4] < -HLN[1][1] < -HLN[0][3] < -I[1]	
O[1] < -HLN[3][0] < -HLN[2][1] < -HLN[1][1] < -HLN[0][1] < -I[2]	
O[1] < -HLN[3][1] < -HLN[2][4] < -HLN[1][1] < -HLN[0][1] < -I[2]	
O[1] < -HLN[3][0] < -HLN[2][1] < -HLN[1][1] < -HLN[0][3] < -I[1]	
O[1] < -HLN[3][0] < -HLN[2][1] < -HLN[1][1] < -HLN[0][3] < -I[0]	
O[1] < -HLN[3][1] < -HLN[2][4] < -HLN[1][1] < -HLN[0][3] < -I[0]	
O[2] < -HLN[3][3] < -HLN[2][0] < -HLN[1][0] < -HLN[0][3] < -I[1]	Impact factor 3
O[2] < -HLN[3][3] < -HLN[2][2] < -HLN[1][4] < -HLN[0][3] < -I[1]	
O[2] < -HLN[3][2] < -HLN[2][0] < -HLN[1][0] < -HLN[0][3] < -I[1]	
O[2] < -HLN[3][3] < -HLN[2][2] < -HLN[1][4] < -HLN[0][1] < -I[2]	
O[2] < -HLN[3][2] < -HLN[2][0] < -HLN[1][0] < -HLN[0][3] < -I[0]	
O[2] < -HLN[3][3] < -HLN[2][0] < -HLN[1][0] < -HLN[0][3] < -I[0]	
O[2] < -HLN[3][2] < -HLN[2][2] < -HLN[1][4] < -HLN[0][1] < -I[2]	
O[2] < -HLN[3][2] < -HLN[2][2] < -HLN[1][4] < -HLN[0][1] < -I[3]	
O[2] < -HLN[3][2] < -HLN[2][2] < -HLN[1][4] < -HLN[0][3] < -I[1]	
O[3] < -HLN[3][3] < -HLN[2][2] < -HLN[1][4] < -HLN[0][1] < -I[2]	Impact factor 4
O[3] < -HLN[3][3] < -HLN[2][2] < -HLN[1][4] < -HLN[0][3] < -I[1]	
O[3] < -HLN[3][1] < -HLN[2][4] < -HLN[1][4] < -HLN[0][1] < -I[2]	

For better clarity and understanding of the obtained resulting reverse chains, Figure 6 below shows graphic visualizations of a few of them as an example.

Figure 6. Graphic visualization of several examples of reverse chains obtained in stage 8 of the developed impact factors reverse analysis method's algorithm.

Since the input layer's neurons I0–I3, as well as the output layer's neurons O0–O3, are actually of no interest from the point of view of their belonging to the impact factors (according to the strategy of developed method), only the belonging of the hidden layers' neurons (to these impact factors) is of interest. Taking this into consideration, the following unique reverse chains will be obtained (without input and output layers' neurons):

- Obtained reverse chains of neurons for impact factor 1:

  HLN[3][1] < -HLN[2][0] < -HLN[1][3] < -HLN[0][2]
  HLN[3][1] < -HLN[2][1] < -HLN[1][1] < -HLN[0][3]
  HLN[3][1] < -HLN[2][4] < -HLN[1][1] < -HLN[0][1]

- Obtained reverse chains of neurons for impact factor 2:

  HLN[3][0] < -HLN[2][1] < -HLN[1][1] < -HLN[0][1]
  HLN[3][1] < -HLN[2][4] < -HLN[1][1] < -HLN[0][3]
  HLN[3][1] < -HLN[2][4] < -HLN[1][1] < -HLN[0][1]
  HLN[3][0] < -HLN[2][1] < -HLN[1][1] < -HLN[0][3]

- Obtained reverse chains of neurons for impact factor 3:

  HLN[3][3] < -HLN[2][0] < -HLN[1][0] < -HLN[0][3]
  HLN[3][3] < -HLN[2][2] < -HLN[1][4] < -HLN[0][3]
  HLN[3][2] < -HLN[2][0] < -HLN[1][0] < -HLN[0][3]
  HLN[3][3] < -HLN[2][2] < -HLN[1][4] < -HLN[0][1]

HLN[3][2] < -HLN[2][2] < -HLN[1][4] < -HLN[0][1]
HLN[3][2] < -HLN[2][2] < -HLN[1][4] < -HLN[0][3]

- Obtained reverse chains of neurons for impact factor 4:

HLN[3][3] < -HLN[2][2] < -HLN[1][4] < -HLN[0][1]
HLN[3][3] < -HLN[2][2] < -HLN[1][4] < -HLN[0][3]
HLN[3][1] < -HLN[2][4] < -HLN[1][4] < -HLN[0][1]

Thus, as shown in Figure 6 above (with a graphical visualization of reverse chain examples), we can label the neurons included in these chains with markers of the corresponding impact factors based on each of the obtained reverse chains. This actually completes stage 9 of the developed impact factor analysis method.

4.3. Calculating the Probabilities of the Disputed Neurons Belonging to the Competing Impact Factors

So, at this moment, it only remains to perform the final stage 10 with the calculation of the probabilities (for all neurons, obtained in previous stage) of their belonging to those impact factors to which they were appropriately labeled in the previous stage.

As can be seen from the above-presented list of unique reverse chains, a significant number of neurons were labeled with various multiple impact factors. For example:

- Neuron HLN[3][1] simultaneously belongs to impact factors 1, 2 i 4;
- Neuron HLN[3][3] simultaneously belongs to impact factors 3 i 4;
- Neuron HLN[2][0] simultaneously belongs to impact factors 1 i 3;
- Neuron HLN[2][1] simultaneously belongs to impact factors 1 i 2;
- Neuron HLN[2][4] simultaneously belongs to impact factors 1, 2 i 4;
- Neuron HLN[2][2] simultaneously belongs to impact factors 3 i 4;
- Neuron HLN[1][1] simultaneously belongs to impact factors 1 i 2;
- Neuron HLN[1][4] simultaneously belongs to impact factors 3 i 4;
- Neuron HLN[0][3] simultaneously belongs to impact factors 1, 2, 3 i 4;
- Neuron HLN[0][1] simultaneously belongs to impact factors 1, 2, 3 i 4.

This means that there is some number of disputed neurons, for which it is necessary to calculate the probability of their belonging to each of the competitive impact factors.

The main recommendation for minimizing the number of disputed neurons and competitive impact factors is to increase the dimensionality of the hidden layers' neuron matrix and the constant experimental research in this direction (in a similar way to that conducted during the design of any multilayer perceptron models, where the number of hidden layers, as well as the number of neurons on these hidden layers, is often chosen mainly experimentally). However, in this research, we deliberately consider the case with a significant number of disputed neurons (as in our presented experimental case: 50% of all the hidden layer neurons of the MP model are actually disputed, and this is really quite a high rate) in order to better illustrate the process of calculating the probabilities of the belonging of these disputed neurons to appropriate competitive impact factors.

So, for our considered specific experimental case, we obtained the following values, calculated according to the above-mentioned expressions: FI[1] = 0.06476, FI[2] = 0.471928, FI[3] = 0.40652, and FI[4] = 0.056792; such exact values were obtained as an operation result of the appropriate additional specially developed Python software for calculating the influence indexes of each of the four impact factors of the developed model, as presented in this research, obtained on a million-dataset (a dataset of 1 million records).

Table 3 below shows the results of the calculations executed for each reverse chain of each of the impact factors of the presented developed example experimental model considered in this research.

Table 3. Results of calculations for obtained reverse chains of impact factors.

FoA_Global	FI	FoA_Local	Impact Factor #
			Impact factor 1
0.0607	0.06476	0.93737	HLN[3][1] < -HLN[2][0] < -HLN[1][3] < -HLN[0][2]
0.0009464	0.06476	0.014614	HLN[3][1] < -HLN[2][1] < -HLN[1][1] < -HLN[0][3]
0.00310848	0.06476	0.048	HLN[3][1] < -HLN[2][4] < -HLN[1][1] < -HLN[0][1]
			Impact factor 2
0.233935	0.471928	0.4957	HLN[3][0] < -HLN[2][1] < -HLN[1][1] < -HLN[0][1]
0.0707892	0.471928	0.15	HLN[3][1] < -HLN[2][4] < -HLN[1][1] < -HLN[0][3]
0.10194	0.471928	0.216	HLN[3][1] < -HLN[2][4] < -HLN[1][1] < -HLN[0][1]
0.065126	0.471928	0.138	HLN[3][0] < -HLN[2][1] < -HLN[1][1] < -HLN[0][3]
			Impact factor 3
0.026	0.40652	0.064	HLN[3][3] < -HLN[2][0] < -HLN[1][0] < -HLN[0][3]
0.037277884	0.40652	0.0917	HLN[3][3] < -HLN[2][2] < -HLN[1][4] < -HLN[0][3]
0.26505	0.40652	0.652	HLN[3][2] < -HLN[2][0] < -HLN[1][0] < -HLN[0][3]
0.06864	0.40652	0.16885	HLN[3][3] < -HLN[2][2] < -HLN[1][4] < -HLN[0][1]
0.00828488	0.40652	0.02038	HLN[3][2] < -HLN[2][2] < -HLN[1][4] < -HLN[0][1]
0.0011789	0.40652	0.0029	HLN[3][2] < -HLN[2][2] < -HLN[1][4] < -HLN[0][3]
			Impact factor 4
0.04997696	0.056792	0.88	HLN[3][3] < -HLN[2][2] < -HLN[1][4] < -HLN[0][1]
0.00612445	0.056792	0.10784	HLN[3][3] < -HLN[2][2] < -HLN[1][4] < -HLN[0][3]
0.00069598596	0.056792	0.012255	HLN[3][1] < -HLN[2][4] < -HLN[1][4] < -HLN[0][1]

Returning back to our example, the following specific numerical probability values of the belonging of each of the disputed neurons to their respective competing impact factors were obtained; they are given in Table 4 below.

Table 4. The probability result calculations of the belonging of the neurons to the impact factors.

Probability (min = 0; max = 1) of Belonging to the Impact Factor				
4	3	2	1	Disputed neuron
0.0029221	-	0.7252	0.272	HLN[3][1]
0.3	0.7	-	-	HLN[3][3]
-	0.8274	-	0.1726	HLN[2][0]
-	-	0.99685	0.003155	HLN[2][1]
0.0039425	-	0.978449	0.0176	HLN[2][4]
0.327	0.673	-	-	HLN[2][2]
-	-	0.9915	0.00852	HLN[1][1]
0.33	0.67	-	-	HLN[1][4]
0.013	0.6974	0.28766	0.002	HLN[0][3]
0.1086	0.165	0.72	0.0067	HLN[0][1]
4	3	2	1	Not-disputed neuron
0	0	1	0	HLN[3][0]
1	0	0	0	HLN[3][2]
0	1	0	0	HLN[1][0]
0	0	0	1	HLN[1][3]
0	0	0	1	HLN[0][2]

Figure 7 below provides a graphical representation of the obtained probability calculation results of the belonging of the hidden layer neurons to the relevant impact factors of the developed model considered in this research as an example.

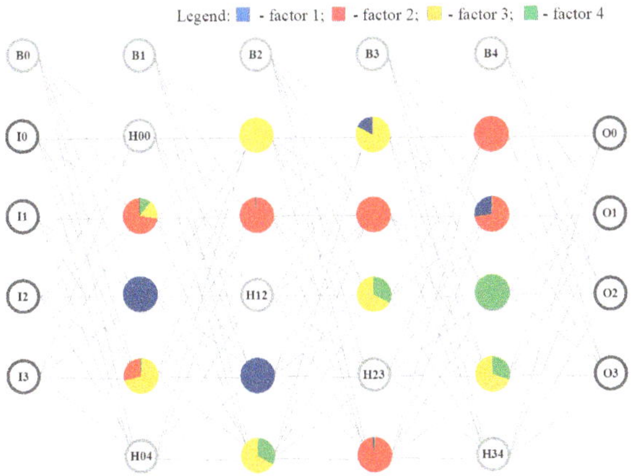

Figure 7. Graphical representation of obtained probability calculation results of the belonging of the hidden layer neurons to the relevant impact factors of the developed model.

Now, having the probabilities of the belonging of the corresponding MP model neurons to the impact factors of the relevant supported object's subjective perception model, we can determine the impact level for each impact factor's influence on the perception results of the investigated support object (supported software complex or the processes of its support) by the relevant subject(s) of interaction (with it), for each individual modeling case. In Figure 8 below, there are also presented visualized examples of the calculated impact factors' fraction(s) as appropriate results of the researched object's perception by the corresponding subject of interaction with it.

Thus, in this way, by means of the developed impact factors reverse analysis method, we managed to form (restore or recreate) certain boundaries of the influence of various impact factors which affect the subjective perception results of the supported software complex or the processes of its comprehensive support, caused by the distortion of its input objective characteristics and their further transformation into the resulting output subjective characteristics of its subjectivized representation.

Also, as was already mentioned earlier in this research, according to the developed method we consider all the input data for modeling in depersonalized (without any functional–semantic meanings) and normalized forms of representation, as, first of all, from a scientific and research point of view, we are mainly interested in the very process of the impact factor analysis itself, in its possibilities and potential. Accordingly, the purpose of the data does not have such an important meaning for us in this case. Moreover, in the absolute majority of cases in real conditions (in real companies and when solving real practical applied problems) all the data usually come for processing in a maximally depersonalized form, caused, first of all, by relevant primarily practical approach (as in most cases, they can represent sensitive personal information or customers' commercial secrets). Of course, the original data for training and testing usually undergo some primary processing, performed either on the client's/customer's side or on the developer's side, but this is conducted only by a limited circle/group of employees provided with special contracted access (to these original data). Later, however, all these original data should mandatorily undergo thorough and complete depersonalization, and only on this basis should the appropriate model's input data be formed. Regarding data normalization, this stage can be carried out on either the client's/customer's side or on the developer's side without any permission restrictions as it is devoid of any sensitivity of incoming information since at this moment it already comes in a depersonalized form.

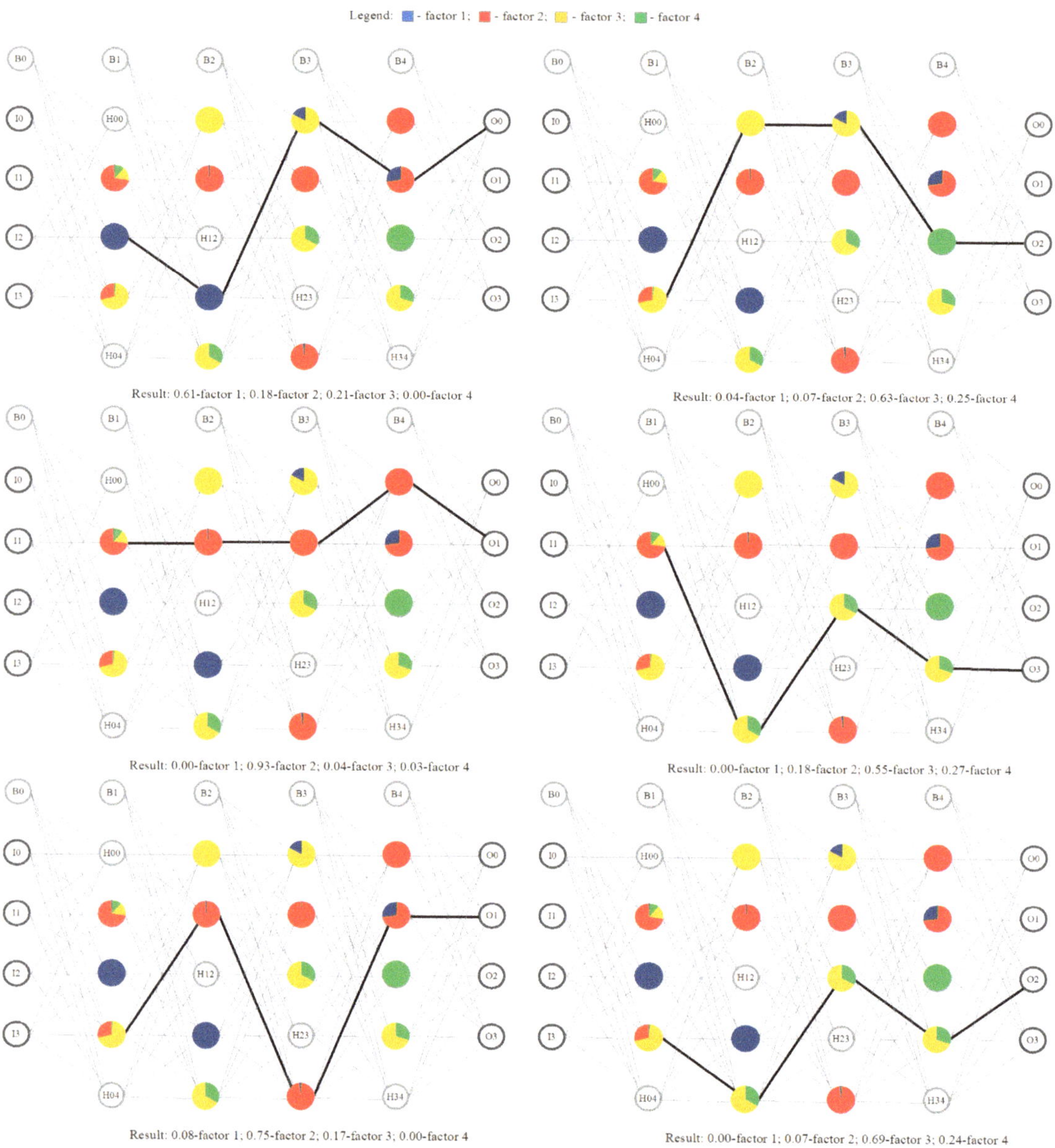

Figure 8. Visualized examples of calculated impact factors' fraction(s), as appropriate results of the researched object's perception by the corresponding subject of interaction with it.

In this way, the boundaries of the impact factors are restored for each individual investigated subject of interaction with the supported software complex. As a result, at the output of the developed method we receive specific numerical values for each investigated subject (which interacts with the supported software complex), and these obtained values actually provide us with a clear specific characteristic for each of these subjects. Later, depending on the applied field(s) of the type of developed method usage, obtained numeric values are used to solve a relevant specific practical applied problem. Examples of such applied areas of the developed method usage include (but are not limited to) such tasks as the following:

- Determination of parametric indicators of subjects interacting with the supported software complex (a kind of specific individual "portrait" of each subject);
- Development of virtual models of the subjects (interacting with the supported software complex and providing its support) based on their parametric indicators;
- Development of virtual assistant(s) with a set of given indicators (with which cooperating with such an assistant would be as comfortable as possible for the subjects);
- Determination of advantages and disadvantages (that is, weaknesses and strengths) of each support team member (which is, at the same time, a subject of interaction with the supported software complex) in order to identify areas, fields, and/or directions of potential self-improvement(s) for considered support team members;
- Simulation (and prediction) of potential conflict situations based on identified weaknesses of researched support team members;
- Selection of the specific support team member(s) most suitable for a specific supported software complex(es);
- Selection of the specific support team member(s) most suitable for the specific client(s)/customer(s) or specific task(s);
- Support team members' categorization according to their parametric indicators;
- Formation of support team(s) based on their parametric indicators and/or parametric indicators of their members;
- Determination of complementary and interchangeable members of support team(s);
- Determination of client(s)/customer(s) employees' (who are the end users of the supported software complex) portraits;
- Development the relevant model(s) of clients and/or clients' users that would allow the provision of further possibilities for automated internal audit of provided support quality for the supported software complex(s);
- And many others.

4.4. Example of a Practical Task Resolution by the Developed Method

In particular, in the scope of this research, as an example, the practical task of the support team members' portrait determination is solved, followed by a further search (detection) of the interchangeable members of this team to ensure the possibility of a quick transfer of a stack of tickets (which are in the middle of the active resolution process) between these members. The main problem during this kind of ticket transferring consists in a colossal loss of the processes' efficiency coefficient, especially during such an unplanned (forced) transfer of tickets between team members. This is mainly caused by the fact that quite often each new recipient of the ticket(s) must actually start solving it from the very beginning, e.g., from scratch (as it is extremely complicated to quickly and carefully figure out and understand what has already been done by the previous "ticket-holder" employee, and what his/her further plans were regarding this ticket). That is why it is so important to choose a new receiver whose point of view on the issue of the considered ticket(s) (and, ideally, view on the whole supported software complex as well) is close (as possible) to the point of view of the previous "ticket-holder".

So, in order to resolve the given practical task, the developed method was applied for all the support team members in order to form their appropriate individual portraits (based on the obtained parametric indicators), which (portraits) are actually the results matrix (for each member) with the calculated probabilities of the hidden layer neurons belonging to the declared impact factors of the developed model. These data are certainly important for further in-depth analysis. However, in our case, they are only intermediate data for obtaining the final result: quantitative characteristics (in percentage) for each of the impact factors. That is, in other words, the summation of the obtained probability values of the belonging of each of the neurons to each of the impact factors is performed. This process can be visualized using the example of Table 4 above, where the values in the columns (representing each of the impact factors) should just be summed up. In this way, the total probability values for the belonging of the neurons to each of the identified impact factors (in our considered research case, the number of such factors = 4—see Table 4) will be obtained. After that, all these total values are converted from absolute to relative in order to provide the possibility of their further comparison with the same values for all the other investigated subjects of interaction with the same supported software complex. Figure 9 below shows an illustrative interpretation of the described process of calculating the researched subject's comparative portrait.

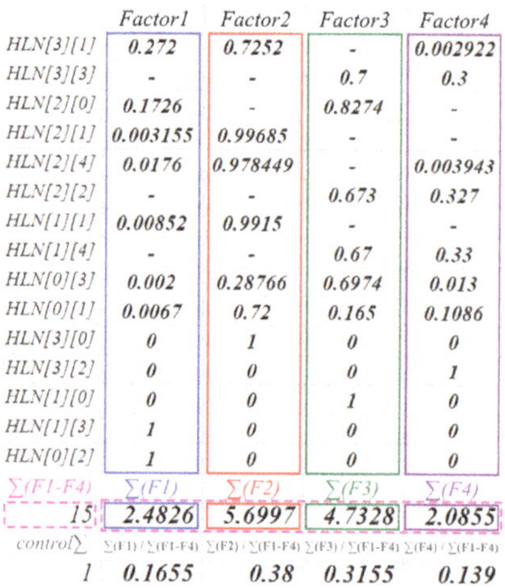

Figure 9. An illustrative interpretation of the described process of calculating the researched subject's comparative portrait.

Following the same scenario, the formation of the comparative portraits of all the other investigated subjects (interacting with the researched supported software complex) is performed as well. Table 5 below represents the data obtained from the calculations of the comparative portraits for all investigated subjects.

Table 5. Obtained comparative portrait calculation data for all investigated subjects.

Factor 4	Factor 3	Factor 2	Factor 1	
0.139	0.3155	0.38	0.1655	Subject 1
0.1062	0.562	0.2069	0.1249	Subject 2
0.1224	0.2277	0.4954	0.1545	Subject 3
0.1354	0.4261	0.1752	0.2633	Subject 4
0.1682	0.2976	0.1981	0.3361	Subject 5
0.2174	0.3256	0.2649	0.1921	Subject 6
0.2311	0.2972	0.2891	0.1826	Subject 7
0.2059	0.4188	0.2684	0.1069	Subject 8
0.2549	0.1672	0.3582	0.2197	Subject 9
0.2739	0.1978	0.403	0.1253	Subject 10
0.1964	0.1774	0.2095	0.4167	Subject 11
0.3529	0.1415	0.3718	0.1338	Subject 12
0.2294	0.3152	0.3641	0.0913	Subject 13
0.0912	0.367	0.4912	0.0506	Subject 14
0.2145	0.1892	0.2912	0.3051	Subject 15

Also, in Figure 10 below, the obtained calculation data (for all investigated subjects) for the comparative portraits are presented in the form of a corresponding general histogram for better visual perception.

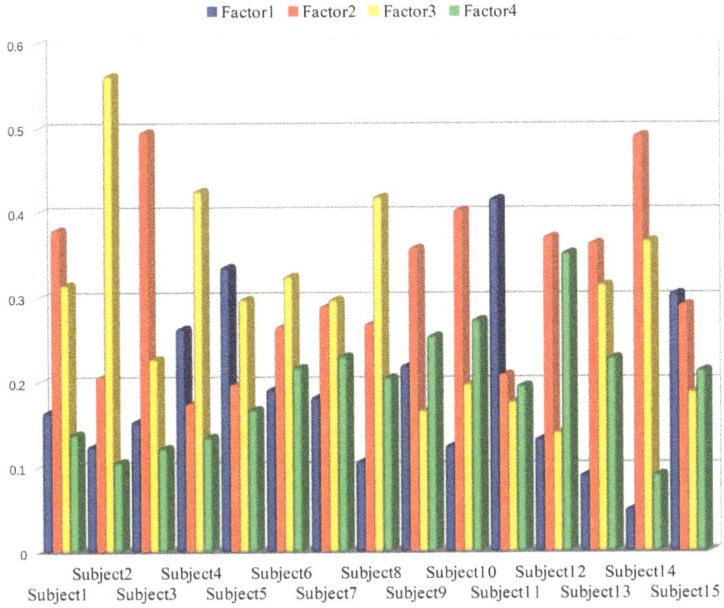

Figure 10. General histogram, representing investigated subjects' comparative portraits.

Table 6 below also presents the final results of the obtained subjects' comparative portraits, as well as the found (detected) most optimal interchangeable subjects.

Table 6. Table of interchangeable subjects' definition.

S15	**S14**	**S13**	**S12**	**S11**	**S10**	**S9**	**S8**	**S7**	**S6**	**S5**	**S4**	**S3**	**S2**	**S1**	
0.43	0.325	**0.18**	0.428	0.617	0.316	0.34	0.34	0.218	0.23	0.4	0.417	0.231	0.493	-	S1
0.746	0.569	0.561	0.841	0.769	0.728	0.79	**0.32**	0.53	0.473	0.546	0.335	0.669	-	0.493	S2
0.485	0.279	0.389	0.461	0.672	0.303	0.395	0.549	0.413	0.461	0.595	0.64	-	0.669	**0.23**	S3
0.474	0.632	0.566	0.828	0.497	0.733	0.605	0.327	0.419	0.343	**0.26**	-	0.64	0.335	0.417	S4
0.279	0.725	0.49	0.717	**0.24**	0.621	0.494	0.458	0.308	0.288	-	0.257	0.595	0.546	0.4	S5
0.279	0.535	0.222	0.485	0.449	0.389	0.317	0.193	**0.08**	-	0.288	0.343	0.461	0.473	0.23	S6
0.249	0.544	0.186	0.409	0.468	0.313	0.26	0.243	-	**0.08**	0.308	0.419	0.413	0.53	0.218	S7
0.459	0.446	0.238	0.555	0.62	0.442	0.503	-	0.243	**0.19**	0.458	0.327	0.549	0.322	0.34	S8
0.215	0.666	0.308	0.223	0.414	**0.19**	-	0.503	0.26	0.317	0.494	0.605	0.395	0.79	0.34	S9
0.36	0.515	0.235	**0.18**	0.583	-	0.189	0.442	0.313	0.389	0.621	0.733	0.303	0.728	0.316	S10
0.22	0.943	0.651	0.638	-	0.583	0.414	0.62	0.468	0.449	0.24	0.497	0.672	0.769	0.617	S11
0.438	0.69	0.347	-	0.638	**0.18**	0.223	0.555	0.409	0.485	0.717	0.828	0.461	0.841	0.428	S12
0.428	0.358	-	0.347	0.651	0.235	0.308	0.238	0.186	0.222	0.49	0.566	0.389	0.561	**0.18**	S13
0.756	-	0.358	0.69	0.943	0.515	0.665	0.446	0.544	0.535	0.725	0.632	**0.28**	0.569	0.325	S14
-	0.756	0.428	0.438	0.223	0.36	**0.21**	0.459	0.249	0.279	0.279	0.474	0.485	0.746	0.43	S15

According to the obtained calculation results, the most optimal interchangeable subjects are as follows: for subject S1—subject S13; for subject S2—subject S8; for subject S3—subject S1; and so on, according to Table 6 above.

It should also be noted that solving the given task automatically solves a number of other additional tasks, such as

- Finding the most optimal pair of team members whose portraits differ minimally compared to others (in the considered case presented above, it is the pair of subjects S7–S6 with the minimum comparison value 0.08 in Table 6);
- Finding the least optimal pair of team members whose portraits have the maximum difference compared to others (in the considered case presented above, it is the pair of subjects S2–S8 with the maximum comparison value 0.32 in Table 6);
- Finding the most optimal candidate for replacement of a team member who is already replacing another previous team member;
- Finding a team member with the minimum value of the total portraits' comparison difference with the rest of the team members (in the considered case presented above, it is subject S7) as the most universal potential "acceptor" of the considered team's tasks' stack;
- Finding a team member with the maximal value of the total portraits' comparison difference with the rest of the team members (in the considered case presented above, it is subject S2) as the least universal potential "acceptor" of the considered team's tasks' stack;
- Providing possibilities for various division(s) of the researched team(s) into appropriate groups and/or subgroups in accordance with obtained portraits of their members;
- Determination of a generalized comprehensive portrait of whole researched team based on the individual portraits of its members, with the possibility of further correction(s);
- Determination of the deviation of team members' portraits from the generalized comprehensive portrait of this whole team (e.g., the whole team's portrait);
- And many others.

Thus, the specific practical task of the determination of the most optimal support team members' replacements (to ensure the possibility of quick transfer of the stack of tasks, which are in the middle of the active resolution phase, between the most interchangeable team members) is solved with the help (and by means) of the developed and presented method.

So, as can be seen from the performed research, the potential of the developed impact factors reverse analysis method (for software complexes' support automation and intellectualization) is extremely significant both in solving the scientific and applied problem of software complexes' support automation and for solving a huge number of relevant related applied practical tasks. Among such applied tasks, there is also the one solved and provided in this research as a demonstrative example—a practical task of the support team members' portrait determination, followed by a further search (detection) of the interchangeable members of this support team to ensure the possibility of quick transfer of the stack of tickets (which are in the middle of the active resolution process) between these members.

4.5. Comparison with Outcome of Existing Similar Approaches

Additionally, comparing the developed and proposed approach with an existing one [51] (where the relevant approach was represented, dedicated for measuring perceptions of various impact factors that affect software development team performance) confirms a number of advantages, including the following:

- The presence of built-in artificial intelligence which is based on multilayer perceptron neural networks;
- The provision of hidden layer neurons (of encapsulated models of a multilayer perceptron) with functional and semantic load/meaning in the context of researched impact factors;
- The availability of research opportunities not only for studying subjective perception of impact factors affecting a supported object (e.g., developed software), but also for studying subjective perception (caused by these impact factors) of the supported object itself.

In the future, additional studies and research will be conducted, related to the application of the developed method for solving other scientific and applied practical problems and tasks in the direction of software complexes' comprehensive support automation and intellectualization.

In addition, one of the potential directions of further investigation is researching potential ways of improvement(s) based on calculating the number of hidden layer neurons (e.g., calculating some "optimal hidden layers' configuration" of an encapsulated multilayer perceptron model in the scope of the developed method) in order to improve the results obtained by the developed method, because at this moment, in general, the quality of the obtained results depends on the number/amount of "disputed neurons" and "competing impact factors". So, the "ideal conditions" for the method involve reducing the total amount of such cases to zero (so that one separate hidden layer neuron would belong to only one exact separate impact factor). But in situation(s) where there is a large (like, for example, $\geq$10) number of "competing impact factors" for only one specific separate "disputed neuron", it might occur that some "belonging probability value(s)" would appear so "small" (like, for example, $\leq$0.001) that they could be treated as a "classical statistical error", but later they could affect the quality of the results obtained by the method. So, theoretically, when the number of neurons in hidden layers is greater, it could increase our chances that less impact factors will "compete" for a single neuron. But the "tricky" thing here is that we never know how many "disputed neurons" and "competing impact factors" we will obtain before finalizing and processing the existing related datasets and analyzing all the obtained reverse chains. So, the situation described here is similar to the "classic dilemma" with a multilayer perceptron: the researcher (e.g., neural network designer) never knows (in advance) exactly how many hidden layer neurons will be needed until a full cycle of training and testing is finalized, and only after that (based on acquired practical heuristic experience) can the researcher finally find the best configuration. That is why calculating an optimal hidden layer configuration of an encapsulated multilayer perceptron model in the scope of the developed method is one of an actual scientific direction for further investigations in the scope of the declared researched area, as presented in this paper.

5. Conclusions

The developed impact factors reverse analysis method for software complexes' support automation is presented in this research. It provides possibilities for analysis of the influence performed on the supported software complexes (as well as on the processes of their comprehensive support) by the relevant pre-determined impact factors. Also, the developed method allows the restoration of the influence boundaries of the impact factors in the corresponding models of software complexes' support, which is developed and presented in this research as well. In addition, the developed method is suitable for solving a batch of applied practical problems (tasks) described in this research. An appropriate primary generalized model of the supported object's subjective perception and the algorithm for the impact factors reverse analysis method, as well as the mathematical component of the impact factors reverse analysis method, have been developed and presented in this research as well. Each of the stages of the developed algorithm is considered in detail in the scope of this research, as well as all the questions, specifics, and aspects related to the format, structure, and processing of the data needed for the developed method's correct and effective functional operation.

According to the developed method, the analysis of impact factors is performed on the basis of multilayer perceptron artificial neural network(s), as well as the constructed reverse chains with maximal weights between

- Output subjective characteristics of the perception's results of the supported object or process (represented by neurons of MP's output layer);
- All intermediate layers of input characteristics' transformation into the relevant output characteristics (but mandatory in reverse order: from the final such layer, back to the initial one) represented by neurons of MP's hidden layers;
- Corresponding input characteristics of the researched supported object or process (e.g., software complex or processes of its comprehensive support) represented by the neurons of MP's input layer.

Subsequently, by means of the developed method, appropriate quantitative and structural classification of the multilayer perceptron(s) hidden layer neurons was obtained based on relation(s) with their belonging to the corresponding pre-determined impact factors. In turn, this provides possibilities for carrying out an automated analysis and quantitative assessment of the level of influence, performed by various impact factors, on the results of the supported object's (the supported software complex, as well as the processes of its comprehensive support) perception by appropriate subjects, which directly or indirectly interact with it.

Thus, the results obtained by resolving the task of impact factor analyses, presented in the scope of this research, are extremely important for further research and solving a much more complex and difficult scientific and applied problem of software complexes' support automation. In addition, the practical task of the support team members' portrait determination, followed by a further search (detection) of the interchangeable members of this support team to ensure the possibility of quick transfer of the stack of tickets (which are in the middle of the active resolution process) between these members, was solved and represented in this research as a practical application example of the developed method.

Therefore, the developed method also solves the problem of restoring the boundaries of the impact factors inside the relevant models of software complexes' support, after these boundaries have been blurred or lost as a result of the introduction and the application of a multilayer perceptron (for ensuring additional self-learning functionality for these models) into these models. Among other things, this significantly contributes to the resolution of the software complexes' support automation problem.

Therefore, in the future, the authors plan to continue the studies and research in the same direction of the software complexes' support automation by further application and improvements of the developed method. In addition, the developed method appeared to be extraordinarily universal; so, it can be additionally used not only in the field(s) of software products' support (and information technologies in general), but also in quite different fields, such as, in particular, psychology, sociology and others, as well as the relevant related interdisciplinary areas like IT + psychology, IT + sociology, IT + human resources, IT management, human–machine interface, human-centered design, human factors evaluation, etc. This can be seen by the obtained method approbation's results in the scope of the resolved applied practical task considered and presented in the scope of this research.

One of potential common solutions for the problem of the aforementioned areas is the adaptation of the developed method to each of these areas by researching, substantiating, selecting, and declaring the appropriate sets of impact factors, since the developed method is quite universal and flexible (which is its additional benefit/advantage); so, depending on the selected set of impact factors, it provides possibilities for researching the influence of these factors (on the researched object) without any strict limitations on the selected area of the research itself.

Author Contributions: Conceptualization, A.P. and V.T.; methodology, A.P.; software, A.P.; validation, A.P. and V.T.; formal analysis, A.P., V.T., N.L. and L.S.; investigation, A.P.; resources, A.P. and N.L.; data curation, A.P.; writing—original draft preparation, A.P.; writing—review and editing, A.P., V.T., N.L. and L.S.; visualization, A.P.; supervision, V.T., N.L. and L.S.; project administration, A.P. All authors have read and agreed to the published version of the manuscript.

Funding: This research received no external funding.

Data Availability Statement: Data are contained within the article.

Acknowledgments: The authors would like to express appreciation for the editors and the anonymous reviewers for their insightful suggestions to improve the quality of this paper.

Conflicts of Interest: The authors declare no conflicts of interest.

References

1. Parsa, S. *Software Testing Automation: Testability Evaluation, Refactoring, Test Data Generation and Fault Localization*; Springer: Cham, Switzerland, 2023; 580p. [CrossRef]
2. Gîrbea, D.; Ciotîrnae, P. Efficient Response Solution for Integrated Command and Control Center Using Automatic Interactive Voice Response System. *J. Mil. Technol.* **2021**, *4*, 35–38. [CrossRef]
3. Yathiraju, N. Investigating the use of an Artificial Intelligence Model in an ERP Cloud-Based System. *Int. J. Electr. Electron. Comput.* **2022**, *7*, 26. [CrossRef]
4. Samoili, S.; López Cobo, M.; Delipetrev, B.; Martínez-Plumed, F.; Gómez, E.; De Prato, G. *AI Watch. Defining Artificial Intelligence 2.0. Towards an Operational Definition and Taxonomy for the AI Landscape*; Publications Office of the European Union: Luxembourg, 2021; 125p. [CrossRef]
5. Khurana, D.; Koli, A.; Khatter, K.; Singh, S. Natural language processing: State of the art, current trends and challenges. *Multimed. Tools Appl.* **2022**, *82*, 3713–3744. [CrossRef] [PubMed]
6. Pawar, S.; Borse, P.; Shah, A.; Chaudhari, K.; Joshi, A. A survey paper on all stream ai chatbot using GPT & NLP. *Int. Res. J. Mod. Eng. Technol. Sci.* **2023**, *5*, 665–668. [CrossRef]
7. Ritonga, Y.S.; Fauzan, M.N.; Habibi, R. Helpdesk Ticketing System. *Int. J. Innov. Res. Technol.* **2023**, *9*, 557–560. [CrossRef]
8. Al-Hawari, F.; Barham, H. A machine learning based help desk system for IT service management. *J. King Saud Univ. -Comput. Inf. Sci.* **2021**, *33*, 702–718. [CrossRef]
9. Karmokar, A.; Kolambe, S.; Mundergi, S.; Kulkarni, M. Bug tracking and reporting system (BTS). *Int. Res. J. Mod. Eng. Technol. Sci.* **2022**, *4*, 447–453. Available online: https://www.irjmets.com/uploadedfiles/paper/issue_4_april_2022/20845/final/fin_irjmets1649767417.pdf (accessed on 24 November 2024).
10. Batra, P.; Jatain, A.; Chaudhary, S. Study of various Bug Tracking Tools. *LC Int. J. STEM* **2020**, *1*, 55–60. [CrossRef]
11. Picus, O.; Serban, C. Connecting Issue Tracking Systems and Continuous Integration/Continuous Delivery Platforms for Improving Log Analysis: A Tool Support. In Proceedings of the 19th International Conference on Evaluation of Novel Approaches to Software Engineering (ENASE), Angers, France, 28–29 April 2024; pp. 379–386. [CrossRef]
12. Onyarin, J. A Complete Guide to DevOps Best Practices. *Int. J. Comput. Sci. Inf. Secur.* **2022**, *20*, 1–6. [CrossRef]
13. Hamza, U.; Abdullah, N.L.; Syed-Mohamad, S.M. DevOps Adoption Guidelines, Challenges and Benefits: A Systematic Literature Review. *J. Adv. Res. Appl. Sci. Eng. Technol.* **2024**, *61*, 115–136. [CrossRef]
14. Tatineni, S. A Comprehensive Overview of DevOps and Its Operational Strategies. *Int. J. Inf. Technol. Manag. Inf. Syst.* **2021**, *12*, 15–32. Available online: https://www.researchgate.net/profile/Sumanth-Tatineni/publication/377434438_A_Comprehensive_Overview_of_DevOps_and_Its_Operational_Strategies/links/65a6839ecc780a4b19bee35a/A-Comprehensive-Overview-of-DevOps-and-Its-Operational-Strategies.pdf (accessed on 24 November 2024).
15. Mahida, A. A Review on Continuous Integration and Continuous Deployment (CI/CD) for Machine Learning. *Int. J. Sci. Res.* **2021**, *10*, 1967–1970. [CrossRef]

16. Bagai, R.; Masrani, A.; Ranjan, P.; Najana, M. Implementing Continuous Integration and Deployment (CI/CD) for Machine Learning Models on AWS. *Int. J. Glob. Innov. Solut.* **2024**, 20. [CrossRef]
17. Alturki, N.; Alharthi, R.; Umer, M.; Saidani, O.; Alshardan, A.; Alhebshi, R.M.; Alsubai, S.; Bashir, A.K. Efficient and secure IoT based smart home automation using multi-model learning and blockchain technology. *Comput. Model. Eng. Sci.* **2024**, *139*, 3387–3415. [CrossRef]
18. Teslyuk, V.; Batyuk, A.; Voityshyn, V. Preliminary Estimation for Software Development Projects Empowered with a Method of Recommending Optimal Duration and Team Composition. *Appl. Syst. Innov.* **2024**, *7*, 34. [CrossRef]
19. Teslyuk, V.; Batyuk, A.; Voityshyn, V. Method of Software Development Project Duration Estimation for Scrum Teams with Differentiated Specializations. *Systems* **2022**, *10*, 123. [CrossRef]
20. Kasurinen, J.; Taipale, O.; Smolander, K. Software test automation in practice: Empirical observations. *Adv. Softw. Eng.* **2010**, *2010*, 620836. [CrossRef]
21. Trudova, A.; Dolezel, M.; Buchalcevova, A. Artificial intelligence in software test automation: A systematic literature review. In Proceedings of the 15th International Conference on Evaluation of Novel Approaches to Software Engineering, Prague, Czech Republic, 5–6 May 2020; pp. 181–192. [CrossRef]
22. Khankhoje, R. AI-Based Test Automation for Intelligent Chatbot Systems. *Int. J. Sci. Res.* **2022**, *12*, 1302–1309. [CrossRef]
23. Jan, E.-E.; Kuo, H.-K.; Stewart, O.; Lubensky, D. A framework for rapid development of conversational natural language call routing systems for call centers. In Proceedings of the Interspeech 2009, Brighton, UK, 6–10 September 2009; pp. 292–295. [CrossRef]
24. Yu, D.; Deng, L. *Automatic Speech Recognition: A Deep Learning Approach*; Signals and Communication Technology; Springer: London, UK, 2015; 321p. [CrossRef]
25. Gadekar, P.; Kaldane, M.H.; Pawar, D.; Jadhav, O.; Patil, A. Analysis of speech recognition techniques. *Int. J. Adv. Res. Ideas Innov. Technol.* **2019**, *5*, 1129–1132. Available online: https://www.ijariit.com/manuscripts/v5i2/V5I2-1584.pdf (accessed on 24 November 2024).
26. Vasilateanu, A.; Ene, R. Call-center virtual assistant using Natural Language Processing and speech recognition. *J. ICT Des. Eng. Technol. Sci.* **2018**, *2*, 40–46. [CrossRef]
27. Vashisht, V.; Pandey, A.K.; Yadav, S.P. Speech recognition using machine learning. *IEIE Trans. Smart Process. Comput.* **2021**, *10*, 233–239. [CrossRef]
28. Pilato, G.; Augello, A.; Gaglio, S. A Modular Architecture for Adaptive Chatbots. In Proceedings of the Fifth IEEE International Conference on Semantic Computing (ICSC), Palo Alto, CA, USA, 18–21 September 2011; pp. 177–180. [CrossRef]
29. Setiaji, B.; Wibowo, F.W. Chatbot using a knowledge in database: Human-to-machine conversation modeling. In Proceedings of the 7th International Conference on Intelligent Systems, Modeling and Simulation (ISMS), Bangkok, Thailand, 25–27 January 2016; pp. 72–77. [CrossRef]
30. Shah, R.; Lahoti, S.; Lavanya, K. An intelligent chatbot using natural language processing. *Int. J. Eng. Res.* **2017**, *6*, 281–286. [CrossRef]
31. Mariciuc, D.F. A bibliometric review of chatbots in the context of customer support. *J. Public Adm. Financ. Law* **2022**, *26*, 206–217. [CrossRef]
32. Alazzam, B.A.; Alkhatib, M.; Shaalan, K. Artificial Intelligence Chatbots: A Survey of Classical versus Deep Machine Learning Techniques. *Inf. Sci. Lett.* **2023**, *12*, 1217–1233. [CrossRef]
33. Shokripour, R.; Kasirun, Z.; Zamani, S.; Anvik, J. Automatic bug assignment using information extraction methods. In Proceedings of the International Conference on Advanced Computer Science Applications and Technologies (ACSAT), Kuala Lumpur, Malaysia, 26–28 November 2012; pp. 144–149. [CrossRef]
34. Silva, S.; Pereira, R.; Ribeiro, R. Machine learning in incident categorization automation. In Proceedings of the 13th Iberian Conference on Information Systems and Technologies (CISTI), Caceres, Spain, 13–16 June 2018; pp. 1–6. [CrossRef]
35. Somasundaram, K.; Murphy, G.C. Automatic categorization of bug reports using latent Dirichlet allocation. In Proceedings of the 5th India Software Engineering Conference, Kanpur, India, 22–25 February 2012; pp. 125–130. [CrossRef]
36. Chawla, I.; Singh, S.K. An automated approach for bug categorization using fuzzy logic. In Proceedings of the 8th India Software Engineering Conference, Bangalore, India, 18–20 February 2015; pp. 90–99. [CrossRef]
37. Mandal, A.; Malhotra, N.; Agarwal, S.; Ray, A.; Sridhara, G. Cognitive system to achieve human-level accuracy in automated assignment of helpdesk email tickets. In *Lecture Notes in Computer Science (Including Subseries Lecture Notes in Artificial Intelligence and Lecture Notes in Bioinformatics), LNCS, Proceedings of the Service-Oriented Computing, ICSOC 2018, Hangzhou, China, 12–15 November 2018*; Springer: Cham, Switzerland, 2018; Volume 11236, pp. 332–341. [CrossRef]
38. Humble, J.; Farley, D. *Continuous Delivery: Reliable Software Releases Through Build, Test, and Deployment Automation*; Addison Wesley: Boston, MA, USA, 2010; 512p. Available online: https://dl.acm.org/doi/book/10.5555/1869904 (accessed on 24 November 2024).

39. Wettinger, J.; Andrikopoulos, V.; Leymann, F. Automated capturing and systematic usage of DevOps knowledge for cloud applications. In Proceedings of the 2015 IEEE International Conference on Cloud Engineering, Tempe, AZ, USA, 9–13 March 2015; pp. 60–65. [CrossRef]
40. Salameh, H. The impact of DevOps automation, controls, and visibility practices on software continuous deployment and delivery. In Proceedings of the 2nd International Conference on Research in Management and Economics, Rome, Italy, 12–14 September 2019; pp. 22–46. Available online: https://www.dpublication.com/wp-content/uploads/2019/09/IME-F793.pdf (accessed on 24 November 2024).
41. Sushma, T.N.; Kumar, A.R. Automation of Software Development using DevOps and its Benefits. *Int. J. Eng. Res. Technol.* **2020**, *9*, 430–432. [CrossRef]
42. Mohanty, L.; Tajammul, M. DevOps CI Automation Continuous Integration. *Int. Res. J. Eng. Technol.* **2022**, *9*, 1221–1226. Available online: https://www.irjet.net/archives/V9/i3/IRJET-V9I3225.pdf (accessed on 24 November 2024).
43. Vadlamudi, S.G.; Chakraborti, T.; Zhang, Y.; Kambhampati, S. Proactive Decision Support using Automated Planning. *arXiv* **2016**, arXiv:1606.07841. [CrossRef]
44. Jaihar, J.; Lingayat, N.; Vijaybhai, P.S.; Venkatesh, G.; Upla, K. Smart home automation using machine learning algorithms. In Proceedings of the 2020 IEEE International Conference for Emerging Technology (INCET), Belgaum, India, 5–7 June 2020; pp. 1–4. [CrossRef]
45. Lu, M.; Liang, P. Automatic classification of non-functional requirements from augmented app user reviews. In Proceedings of the 21st International Conference on Evaluation and Assessment in Software Engineering, Karlskrona, Sweden, 15–16 June 2017; pp. 344–353. [CrossRef]
46. Kurtanovic, Z.; Maalej, W. Automatically classifying functional and non-functional requirements using supervised machine learning. In Proceedings of the IEEE 25th International Requirements Engineering Conference (RE), Lisbon, Portugal, 4–8 September 2017; pp. 490–495. [CrossRef]
47. Landau, I.; Landau, V. From data driven decision making (DDDM) to automated datadriven model based decision making (MBDM). *HAL Open Sci.* **2017**, 8. Available online: https://hal.archives-ouvertes.fr/hal-01527766 (accessed on 24 November 2024).
48. BEUC. Automated Decision Making And Artificial Intelligence—A Consumer Perspective. Report, June 2018. Available online: https://www.beuc.eu/publications/beuc-x-2018-058_automated_decision_making_and_artificial_intelligence.pdf (accessed on 24 September 2024).
49. Larus, J.; Hankin, C.; Carson, S.G.; Christen, M.; Crafa, S.; Grau, O.; Werthner, H. *When Computers Decide: European Recommendations on Machine-Learned Automated Decision Making*; Association for Computing Machinery: New York, NY, USA, 2018; 20p. [CrossRef]
50. Araujo, T.; Helberger, N.; Kruikemeier, S.; de Vreese, C.H. In AI we trust? Perceptions about automated decision-making by artificial intelligence. *AI Soc.* **2020**, *35*, 611–623. [CrossRef]
51. Machuca-Villegas, L.; Gasca-Hurtado, G.P.; Puente, S.M.; Tamayo, L.M.R. An Instrument for Measuring Perception about Social and Human Factors that Influence Software Development Productivity. *J. Univers. Comput. Sci.* **2021**, *27*, 111–134. [CrossRef]

MDPI AG
Grosspeteranlage 5
4052 Basel
Switzerland
Tel.: +41 61 683 77 34

Designs Editorial Office
E-mail: designs@mdpi.com
www.mdpi.com/journal/designs

www.ingramcontent.com/pod-product-compliance
Lightning Source LLC
Chambersburg PA
CBHW040750250926
51850CB00007BB/1457
9783725850631